PRESSWORKING: STAMPINGS AND DIES

Karl A. Keyes
Editor

Published by:
**Society of Manufacturing Engineers
Marketing Services Department
One SME Drive
P.O. Box 930
Dearborn, Michigan 48128**

PRESSWORKING: STAMPINGS AND DIES

**SME wishes to express its acknowledgement and appreciation to the following
publications for supplying the various articles reprinted within this book.**

Best's Safety Directory
A.M. Best Company
Safety Publications Division
Ambest Road
Oldwick, NJ 08858

Machine and Tool Blue Book
Hitchcock Publishing Company
Hitchcock Building
Wheaton, IL 60187

Modern Machine Shop
600 Main Street
Cincinnati, OH 45202

Manufacturing Engineering
Society of Manufacturing Engineers
One SME Drive
P.O. Box 930
Dearborn, MI 48128

Mechanical Engineering
The American Society of Mechanical
Engineers
United Engineering Center
345 East 47th Street
New York, NY 10017

Machinery and Production Engineering
Machpress Limited
1 Copers Cope Road
Beckenham
Kent BR3 1NB
England

Tooling & Production
Huebner Publications, Inc.
5821 Harper Road
Solon, OH 44139

**Grateful acknowledgement is also
extended to:**

Jack Bradley
Manager, Product Certification
T&B/Thomas & Betts
Division of Thomas & Betts Corporation
920 Route 202
Raritan, NJ 08869

CBI Publishing Company Inc.
51 Sleeper Street
Boston, MA 02210

J.V. Manufacturing Company Inc.
Burtner Road
Natrona Heights, PA 15065

Production Engineering Company
Box 215
Elmhurst, IL 60126

PREFACE

Before the development of sheet metal, industry used presses of one design or another to trim, pierce, form and coin parts to manufacture hard goods, weapons, utensils. . .even money. The introduction of flattened or rolled metals in sheet or strip form—combined with the need to economically produce nearly identical parts—pushed pressworking to the manufacturing forefront. After many years of application, pressworking is still considered one of the most dependable manufacturing processes for accuracy, productivity and economy, even when compared with more modern, newly developed methods. Pressworking's greatest limitation is the need to produce enough parts to justify the initial tooling cost.

A power press fitted with a metal stamping die is capable of transforming sheet material into strong, light-weight, inexpensive, intricately shaped precision parts that are so nearly identical that each one possesses the interchangeable characteristics necessary to meet the demand for easy-to-replace items. Pressworking is astonishing in that a die designed to produce one part per press stroke may produce 1000, 6000, 72,000 or more parts. The actual obtainable production depends on the size, shape and weight of the piece part; type and hardness of the material; speed, style and condition of the press; construction of the die; and most importantly, the method of feeding the stock into the die combined with part ejection and scrap removal.

Press-worked parts can take on the appearance and accuracy once associated only with precision machined parts. Fine-blanked parts are normally more dimensionally accurate and flatter than conventional stamping, especially in thicker materials. Gears, cams, side frames, control devices can be stamped without costly secondary operations; only the die roll feature, inherent on all stampings, exposes the use of the fineblanking method.

Today, there is a tremendous demand on the automobile industry to produce fuel-efficient cars. This demand has led to the introduction of high-strength, low-alloy steels, various types of aluminums, as well as non-metallics and other less forgiving materials than the old work-horse, low carbon steel.

No longer can die designers or builders rely only on past experiences and "black-art" cut and try practices. Dies produced by such methods usually can be made to produce an acceptable stamping, but only after costly labor-intensive changes and fixups.

Today, designers or builders must go beyond the point of being satisfied that their tools can produce acceptable parts.

They must understand the needs of new products and be prepared to council with product engineering so the advantages and limitations of pressworking can be considered prior to the part being released for production. The stamping size, shape, tolerance and material must be evaluated to determine how complicated the part will be to produce. This evaluation is essential in selecting the proper die construction and requirements for the stock, press, feed, part ejection and scrap handling. By knowing these elements, accurate predictions can be made about first die cost as well as the production costs related to both labor and material.

Responsibility also must be accepted to improve the conditions which directly affect the overall efficiency. The aim must be to reduce production material and labor costs by carefully controlling material utilization; to improve press room methods; to consider die maintenance needs in the initial designing stages; to incorporate safety standards (which in many cases can actually increase production or reduce operating costs) as well as to make acceptable parts on the first run through systems and modern manufacturing techniques.

This book was organized with these thoughts in mind. The need is growing for pressworking to become a more exacting, predictable science. This volume should aid the designer and the die builder in striving to meet that need.

Chapter One of this volume, "The Stamping," discusses the functional design and the modern sheet steels used in stamping. The book's second chapter reviews "Die Estimating," providing examples as well as concepts for the reader's consideration.

The book's third chapter, "Die Design," presents a series of articles containing facts every die designer should know. There is also a special paper discussing productivity and die design. "Die Building" is the subject of Chapter Four; articles discuss steel-rule dies and the use of traveling wire EDM to cut costs.

Chapter Five, "Maintenance and Repair," outlines some outstanding tips on maintaining carbide dies and dies used in the ironing of cans.

Chapter Six, "Presses," contains data which can improve the capabilities of the punch press and discusses energy in press selection.

The final chapter, "Safety," presents a feature from *Best's Safety Directory—1977* entitled "Safety Considerations in Die Design."

Technical contributions such as those included here will help improve the state-of-the-art. Our hope is for this to be the first in a series of volumes, utilizing a similar format, designed to cover all the elements of pressworking from improved stamping design and materials through tooling, equipment and safety needs to tooling and production economics. You can aid us in our endeavor by passing along technical articles and papers you have found helpful and interesting in your day-to-day association with pressworking. These should be sent to SME for possible publication.

I would like to thank all of the contributors and their publishers for allowing us to reprint their material.

They include: *Best's Safety Directory, Machine and Tool Blue Book, Machinery and Production Engineering, Manufacturing Engineering, Mechanical Engineering, Modern Machine Shop, Tooling & Production.* In addition, my thanks is extended to Jack Bradley of T & B Thomas & Betts, the CBI Publishing Company, the J.V. Manufacturing Company and the Production Engineering Company for supplying material in this volume.

A special thanks to Bob King and Judy Stranahan, of SME Marketing Services, for their cooperation and assistance.

Karl A. Keyes
Editor

SME

The informative volumes of the Manufacturing Update Series are part of the Society of Manufacturing Engineers' effort to keep its Members better informed on the latest trends and developments in engineering.

With 50,000 members, SME provides a common ground for engineers and managers to share ideas, information and accomplishments.

An overwhelming mass of available information requires engineers to be concerned about keeping up-to-date, in other words, continuing education. SME Members can take advantage of numerous opportunities, in addition to the books of the Manufacturing Update Series, to fulfill their continuing educational goals. These opportunities include:

- Chapter programs through the over 200 chapters which provide SME Members with a foundation for involvement and participation.

- Educational programs including seminars, clinics, programmed learning courses and videotapes.

- Conferences and expositions which enable engineers to see, compare, and consider the newest manufacturing equipment and technology.

- Publications including Manufacturing Engineering, the SME Newsletter, Technical Digest and a wide variety of books including the Tool and Manufacturing Engineers Handbook.

- SME's Manufacturing Engineering Certification Institute formally recognizes manufacturing engineers and technologists for their technical expertise and knowledge acquired through years of experience.

In addition, the Society works continuously with the American National Standards Institute, the International Standards Organization and other organizations to establish the highest possible standards in the field.

SME Members have discovered that their membership broadens their knowledge throughout their career.

In a very real sense, it makes SME the leader in disseminating and publishing technical information for the manufacturing engineer.

TABLE OF CONTENTS

CHAPTERS

THE STAMPING

COST ESTIMATING

DIE DESIGN

DIE BUILDING

MAINTENANCE AND REPAIR

PRESSES

SAFETY

INDEX

CHAPTER 1

THE STAMPING

Design of Flat Stampings

By Federico Strasser

Stamped Links

In many cases the holes in a metal stamping are the most interesting and most important features. Their quantity, size, location, tolerances, surface finish and other characteristics are the determining factors in the workpiece design. The surrounding metal connects the holes, locates them and, of course, gives the strength necessary to the part. This means that in these cases the outer shape or contour of the part has no importance at all; the designer has ample liberty of action in this respect. Links are typical of the above description and are used when two identical holes must be located at a given distance.

When estimating tooling costs, stamped links are an important consideration. The designer of mechanical parts is not supposed to design the press-tools which will be used for the production of the stampings he designs; however, he should be able to visualize the dies needed for every production method so he can estimate the cost of producing his stampings. In its turn, the method of producing a part is governed to a great extent by its design as well as by the other characteristics, such as tolerances, quantities, etc.

The straight, square piece illustrated in Fig. 1-1 is the least expensive shape to design and produce. It takes less material to produce straight ends, and they require simpler, less expensive dies. Additionally, straight end designs have cost advantages for the following reasons: slight inaccuracies in the location of the punched holes in the components are practically unavoidable due to the commercial width tolerances of the bar stock and the

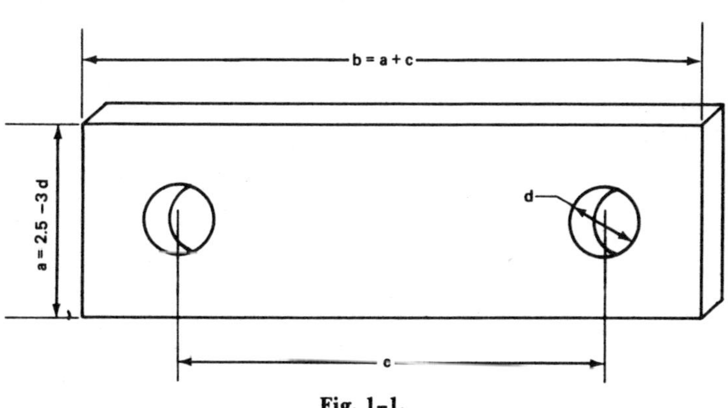

Fig. 1-1.

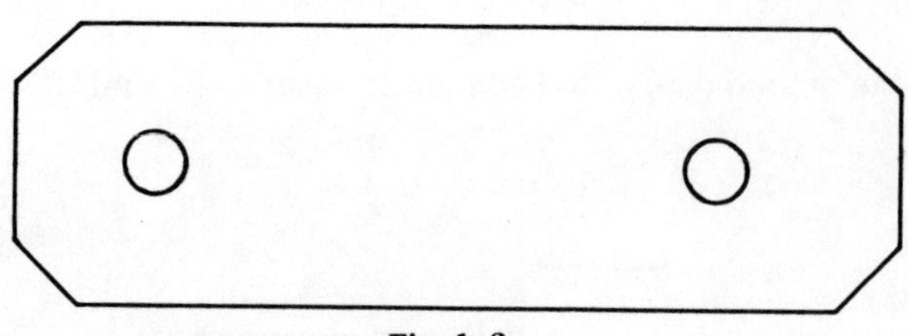

Fig. 1-2.

tolerances required by shearing the strips from the sheets.
Such an eccentric position of
the holes remains unperceived
in the case of straight ends;
in the case of half-round or
other types of ends, it is
quite noticeable.

A very simple esthetic
improvement of the link shape
consists of chamfering the corners (see Fig. 1-2). This
shape requires some additional
notching operations incorporated into the corresponding dies.

With a somewhat more elaborate and more expensive tool,
the corners are rounded instead
of chamfered. In these cases,
never try to blend the rounding
with the sides. Such a design
as illustrated in Fig. 1-3 is
wrong. It requires expensive
tooling, costly operation and
very close tolerances in the
stock width. The corner radii
should always be larger than
the distance from the ends
(Fig. 1-4 is right), or a proper clearance of 10 to 15 deg
should be provided as indicated
in Fig. 1-5.

In some cases it is not
practical to use the above described production methods.
The chief conditions which may
require that the strips have
the transversal cross-sectional
area of the links (thickness x
width, instead of thickness x
length) are these:

   1) Parts are too long.
      The strips become too

heavy and cumbersome to
handle, especially
with heavy gage stock.
   2) The longitudinal edges
      must be rounded or oth-
      erwise mill finished.
   3) One or both ends must
      be specially shaped.
   4) The same tool (univer-
      sal type tooling) must
      be used for different,
      but similar parts.

A "parting" tool (see
Fig. 1-6) is the simplest type
of tool. At each press-stroke,
this tool shapes one end of each
of two subsequent parts and simultaneously punches the holes
in each end so that at each
press-stroke one complete part
is produced.

Referring to the end shape
of the links produced with tools
of the kind shown in Fig. 1-6,
the designer should keep in consideration the same principles
as explained above. First preference should be given to
straight ends, unless for some
important functional reason this
is not possible. Then use chamfered ends or rounded corners.

Many designers prefer fully founded link ends, for example, when the parts are mounted
side by side, such as in electrical bus-bars, and small
length inaccuracies are not so
evident. This design is not
practical because it requires a
die which is more expensive to
build. The parts are unnecessarily weaker, and, besides, the

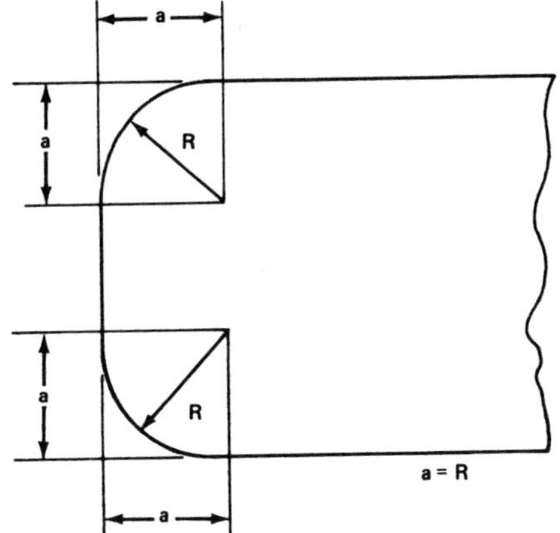

Fig. 1–3.

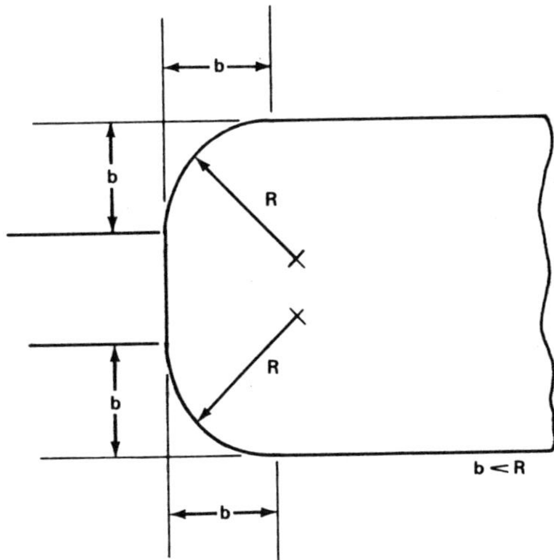

Fig. 1–4.

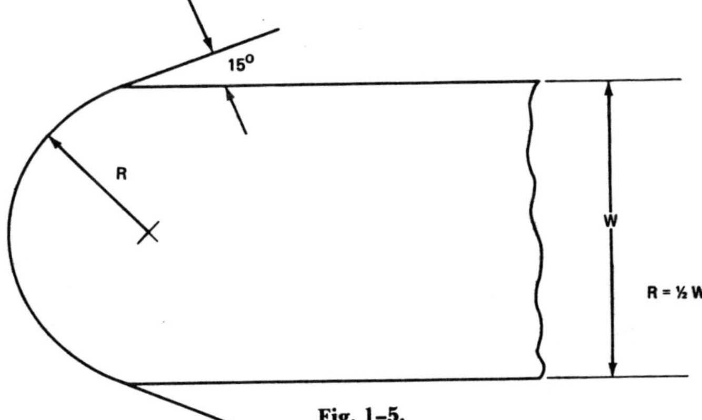

Fig. 1–5.

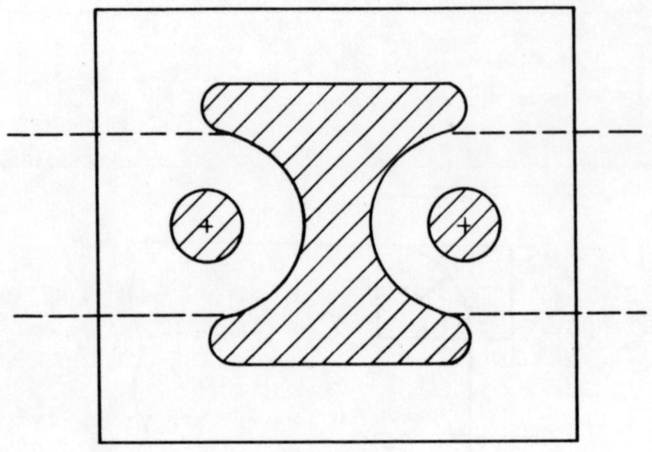

Fig. 1-6.

smallest unavoidable inaccuracy or irregularity in the parting operation has an unfavorable consequence on the esthetic appearance of the workpieces. Therefore, this design should be avoided. If for some special, important reasons the ends must be rounded, then take into account the following basic rule:

Do not blend the rounding with the side contour lines

(Fig. 1-7 is not economical). As in case of Fig. 1-3, this means unnecessarily high tooling and operational costs. Easy solutions are shown in Figs. 1-5 and 1-8. In Fig. 1-8, the rounding radius is larger than half of the part's width. In Fig. 1-5, the radius is equal to half of the part's

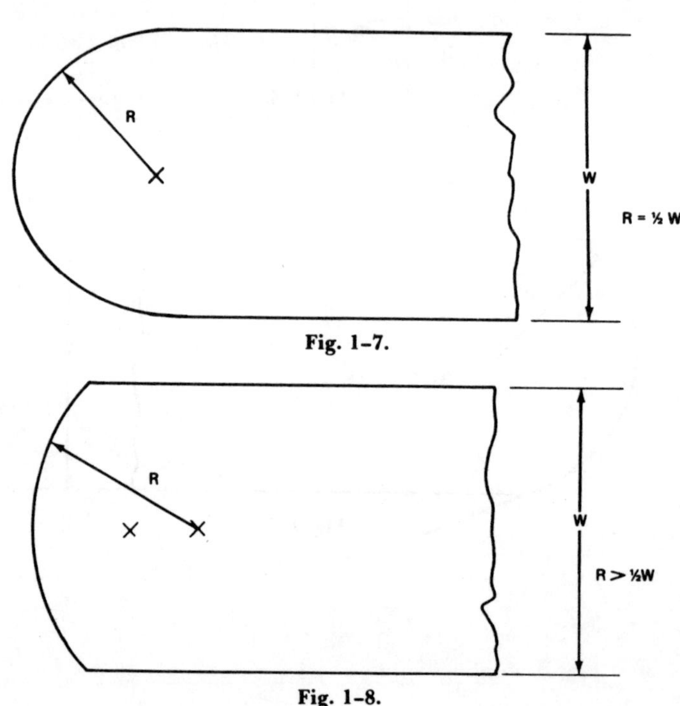

Fig. 1-7.

Fig. 1-8.

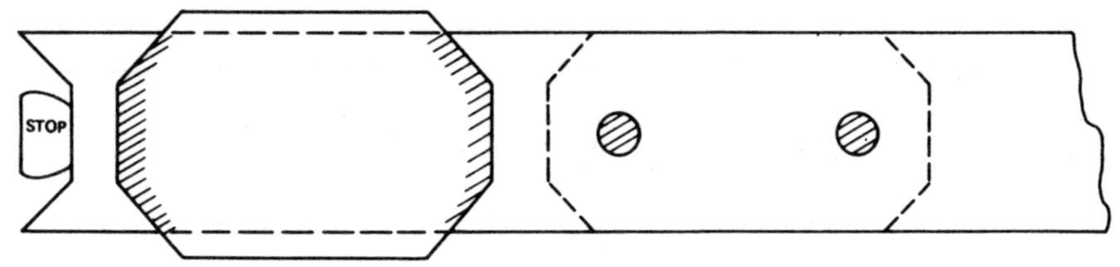

Fig. 1–9.

width, but the end shape is an incomplete semicircle with an inclination of about 10 to 15 deg on each side.

The production methods described up to this point are employed for parts with ample dimensional tolerances in the blank contour and size. This is the usual case. However, sometimes in addition to other specifications, such as hole size and center distances, the part length must also be held within certain tolerance limits. In such cases, the so-called "strip-blanking" method is used. This consists of working with strips of the right dimensions and employing a tool which cuts the two ends of a given part simultaneously. Fig. 1-9 illustrates this principle showing a progressive tool for transversal strips and chamfered ends.

Another case, analogous to link design, is portrayed in Figs. 1-10 and 1-11. Three holes of a given size had to be located in a stamping of a given pattern. This first, elementary solution consists of taking a rectangular piece of sheet metal and punching holes in it in the right quantity, size, shape and location (see Fig. 1-10). This is the way to make a few dozens or hundreds of pieces. For mass production, the stamping's outer contour may be changed in shape. In this case, the outer shape is immaterial making for a very economical "scrapless" strip-layout (see Fig. 1-11) that realizes a substantial stock savings of 40 percent.

Shaped Flat Stampings

Unfortunately, only in the minority of cases is the outer shape of stampings totally immaterial. In the majority of cases, the outer contour becomes partly or totally functional,

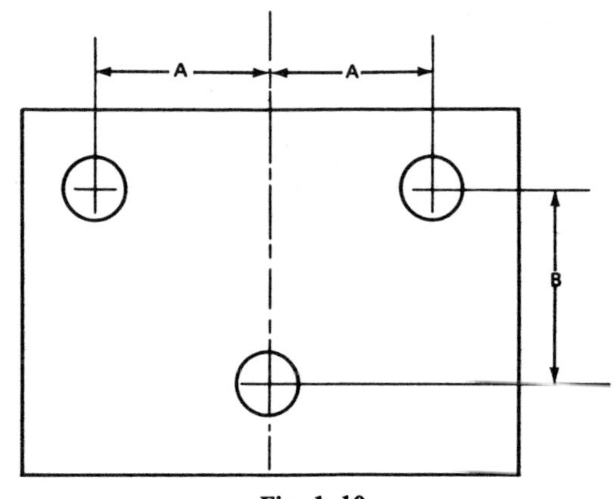

Fig. 1–10.

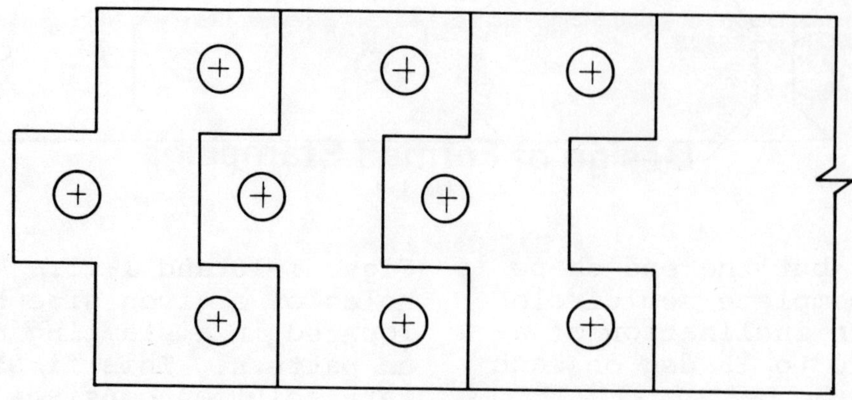

Fig. 1–11.

and thus it must comply with certain requirements or specifications.

Such requirements may be either negative or positive. The first one means clearance for moving or stationary components; the second one means matching, intermeshing or in some way contacting portions of other components.

If the general configuration, accuracy and other characteristics permit, preference should be given to cutting off or parting design because of the several economical and technical advantages offered by such a design (Fig. 1-12).

Blanked Flat Stampings

When the design involves a more complicated outer contour, closer dimensional tolerances, heavier gage stock and other higher performance requirements, it is necessary to fully blank out the stamping from sheet, strip or coil material.

In the design of the outer shape of fully blanked stampings, the designer must consider all the basic rules and recommendations demanded by sound, practical and economic production techniques. The main points to keep in mind in this respect are: simple shapes; preference for straight contour lines; avoidance of sharp corners and details which require weak and/or delicate tool members.

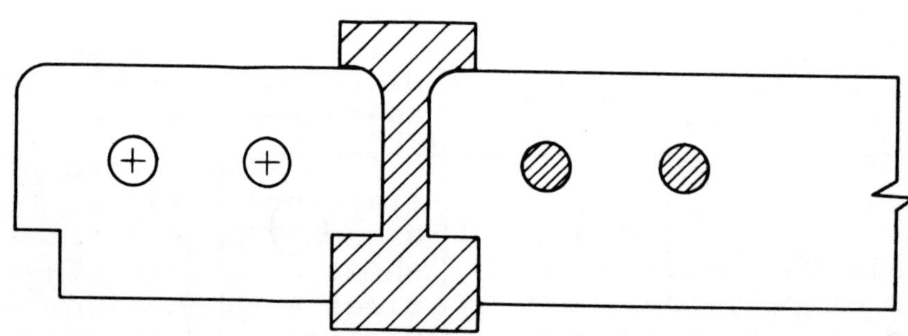

Fig. 1–12.

Design of Formed Stampings

By Federico Strasser

Bent Parts

In the evolution of metal stamping technique, the next step consists in transforming a two-dimensional piece in which the thickness of sheet metal is negligible in comparison to width and length into a three-dimensional object. This may be done by several methods. The easiest and most common way is to bend up one or more sides of the flat blank.

Bent or formed details on a component may be necessary in a stamping for reinforcement for safety to eliminate injury from sharp edges, for ease of assembly or for esthetics.

In the design of formed parts, several basic and special details must be taken into consideration. The chief ones are: sheet lamination direction, burr side, rounding radii, flat blank contour, springback phenomenon, leg height and nesting of stampings.

Formed Boxes

When all the sides of a rectangular or polygonal part are bent up, a box is formed. For light duty, the junction gaps are left open. For heavy duty and/or those cases when the box must be tight, the corners are welded, brazed or soldered.

Drawn Shells and Boxes

Seamless hollow vessels may be formed directly from sheet metal by the process of drawing. Both the cross-sectional and the longitudinal section may be chosen from several shapes. Among possible combinations, the cylindrical, flangeless cup is the most common and most frequent (and the easiest to produce). When designing drawn shells, especially geometrically regular shapes, the designer must strive for a combination of low height, large diameter and heavy gage stock, the most favorable conditions for achieving the easiest operation.

Stacked Stampings

The last step in the evolution of metal stamping techniques is the production of simulated solid three-dimensional objects by means of the so-called stacking method.

A stacked stamping consists essentially of a rigid assembly of several individual sheet metal stampings, sometimes with the addition of some machined or otherwise produced parts. The laminations may be identical or different in shape and/or size. By intelligent combination of these elements, many workpieces which normally are produced by machining, casting, forging or other methods may be made with sheet metal stampings. Thus, the low cost of pressworking the metals is substituted for more costly manufacturing methods.

Stacked stamping is not new. For many years it has been employed in electrical machines, transformers, padlocks and other fields. Lately, however, designers have widely adopted stacked stamping techniques.

Design of Drawn Parts

By Federico Strasser

In designing drawn workpieces, the following information may help to avoid production and tooling difficulties, to reduce manufacturing costs and to solve technical or economical problems.

If lubricated blanks have the tendency to stick together, it is unavoidable that sometimes two blanks will be placed simultaneously in the tool. As a result, accidents, injuries to tools and machines, rejected components, etc., may occur if such blanks are used in subsequent drawing operations. In order to prevent this trouble, a small nonconcentric protuberance or dimple may be stamped in the blanks when possible. The dimple may be stamped with a blanking punch that has a small tip, similar to a reduced pilot pin (see Fig. 6-48).

The open ends of shells are sometimes curled to both stiffen the part and to prevent injuries. Remember that outside curling is simpler, easier and less expensive than inside curling.

High stresses exist in the top or open end of drawn shells, especially if made from comparatively thin stock. If the stresses are released with deep slots (see Fig. 6-49), or with windows located too near the top (see Fig. 6-50), the shells will distort. Slots should be very short (see Fig. 6-51) and windows located well away from the top edge (see Fig. 6-52).

Instead of a bulging operation, a composite construction built up from two comparatively easy drawn parts (see Fig. 6-53) is preferable.

In the edge of the rim of the drawn cup illustrated in Fig. 6-54, a through opening was needed for clearing a wire of a given diameter. In the original design, slots were milled at opposite sides of a fully drawn shell. Since the shape of the slot is unimportant, the cup design was changed so that the slot milling operation was eliminated. First, a round blank with two cut-away segments was considered (see Fig. 6-55). The blank was produced with a round blanking die from a strip narrower than the blank diameter. This blank was

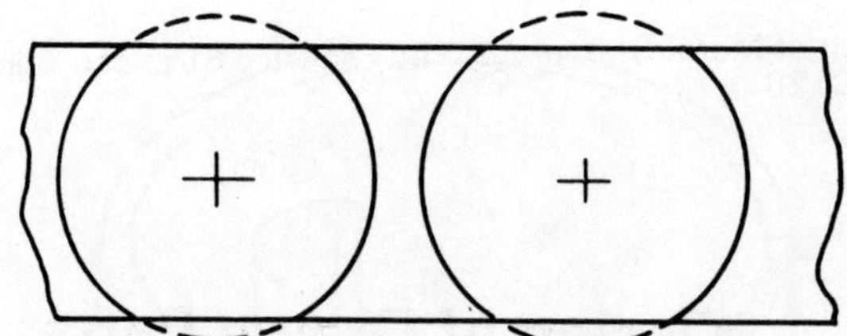

Fig. 6–55. *(Courtesy, Duplicon Company)*

supposed to give a drawn shell according to Fig. 6-56. In this way, the overall cost would have been reduced considerably by decreasing every cost detail: tooling, labor and stock. However, tests showed that the parts came out with stretched ears as illustrated in Fig. 6-57. So a blank had to be developed that was fully blanked from the strip (see Fig. 6-58). Even in this case, an overall savings resulted because supplementary machining was avoided.

The next case involving boxes and similar products has a universal value. At the bottom edge some holes must be designed for the passage of the electric wires. It is customary for designers to specify well-shaped apertures according to Fig. 6-59.

This design requires the holes to be punched after the drawing of the box. Instead of such a design, the following procedure is suggested. At the right locations punch simple round holes in the flat blank before the drawing operation. During the drawing process, these holes will be slightly deformed (see Fig. 6-60), but this will not interfere with introducing the leads into the box. Savings are realized by eliminating multiple operations.

Fig. 6-61 represents a case where the drawing operation can be eliminated. According to the original specification, this workpiece was to be drawn and afterward slit with a thin circular saw. Instead of such a complicated and expensive procedure, the four notches are first made in the flat blank (see Fig. 6-62). Instead of drawing a cylinder, now four segments are bent upward in arcs, and the small collar or boss is formed downward with the same tool in the same

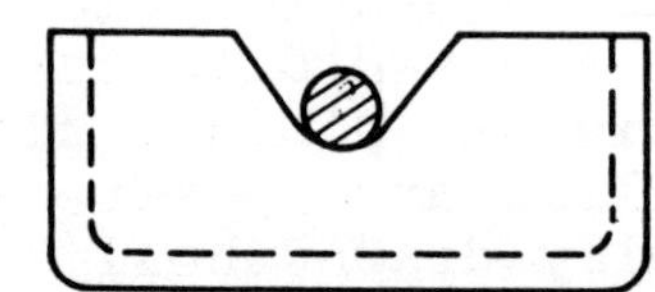

Fig. 6–56. *(Courtesy, Duplicon Company)*

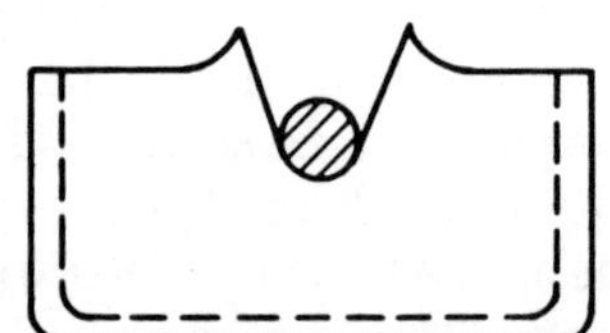

Fig. 6–57. *(Courtesy, Duplicon Company)*

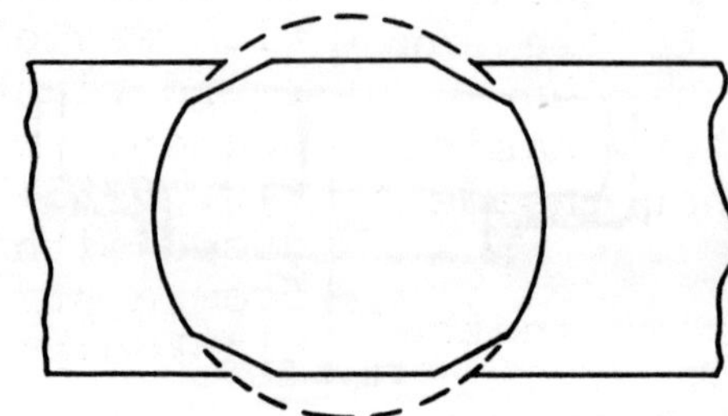

Fig. 6–58. *(Courtesy, Duplicon Company)*

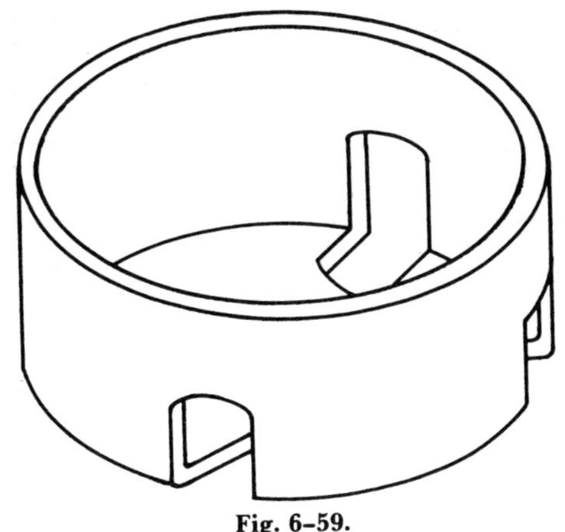

Fig. 6-59.

operation. Thus, a costly multiple machining operation of slitting is totally eliminated, and the chief tooling became simpler and less expensive.

The workpiece shown in Fig. 6-63 is similar. With the original design, a cylindrical shell was drawn, and then the side walls were notched four times. The delicate tooling and the troublesome operation were replaced by the following procedure: a lower, cylindrical shell with a flange was drawn.

The flange was then trimmed with an almost normal multiple-notching tool (see Fig. 6-64). Finally, the remaining ears were turned upward. The production techniques were later modified even further. A specially shaped flat blank (see Fig. Fig. 6-65) was shallow-drawn, and the workpiece was ready with this one operation. Thus, two operations were eliminated with corresponding reductions in tooling and labor cost. In addition, less stock was needed.

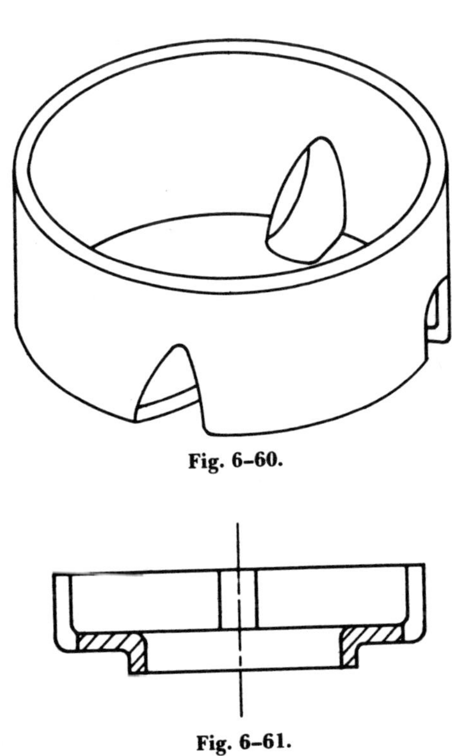

Fig. 6-60.

Fig. 6-61.

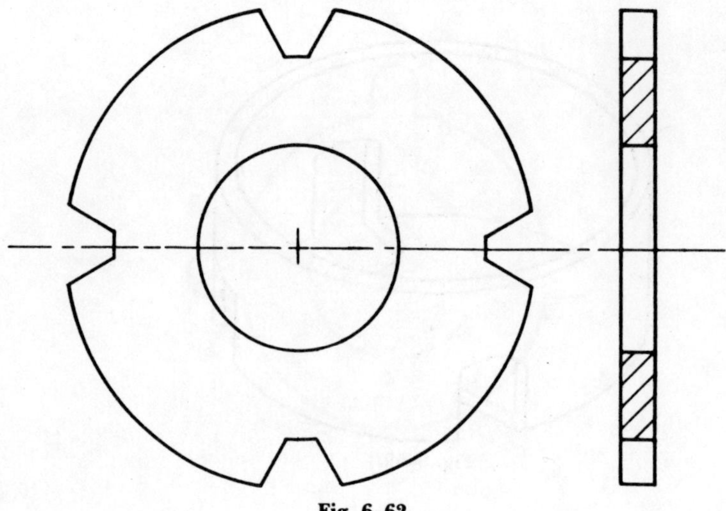

Fig. 6–62.

To produce the part shown in Fig. 6-66, flat strips of sheet metal are simply formed and assembled by spot welding, and then a central hole is punched in the bottom. This part replaces a drawn cup in which four sectors had to be punched out in the bottom. Considerable savings of material, labor and tooling were realized with the fabricated design.

Aldo L. Coen introduced a totally new concept in the manufacturing of light duty perforated basket-like drawn parts. A typical application of this principle is speaker housings for loudspeakers. The new idea consists of using a blank of the same size as the finished housing, punching strategically located kidney-shaped slots in it, and then forming the part into the required three-dimensional speaker housing shape. In the last operation, the slots are enlarged, assuming a diamond-shaped lace pattern which is functionally necessary for unhindered air flow in both directions.

The advantages of the expanded metal design over the conventional method of drawn sheet metal parts may be summed up as follows:

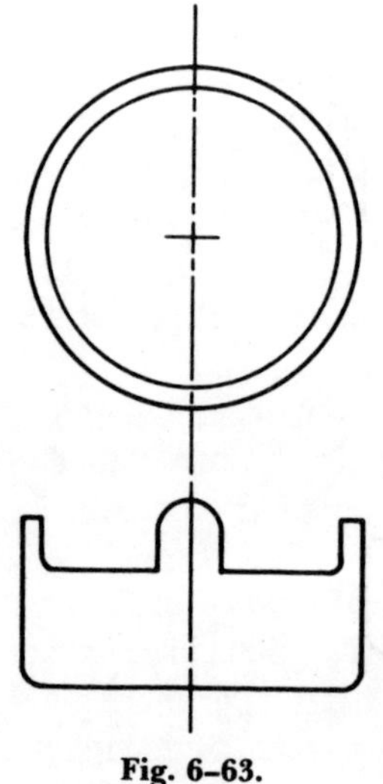

Fig. 6–63.

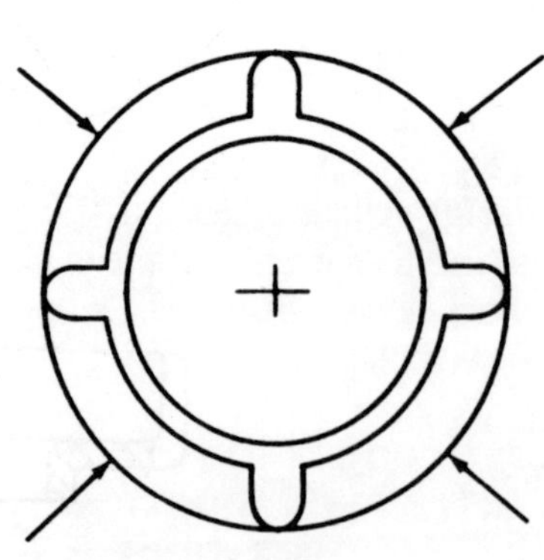

Fig. 6–64.

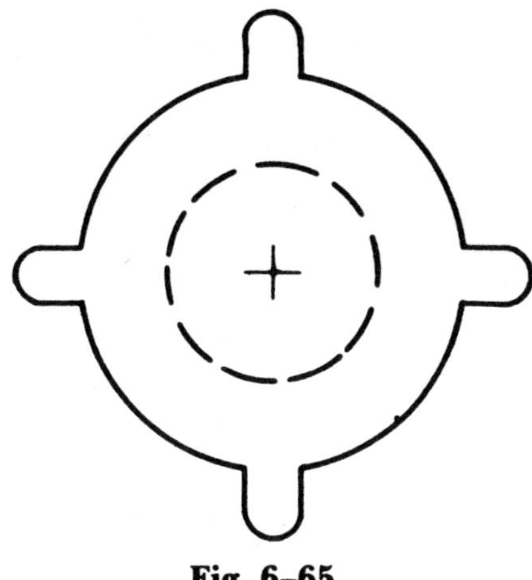

Fig. 6–65.

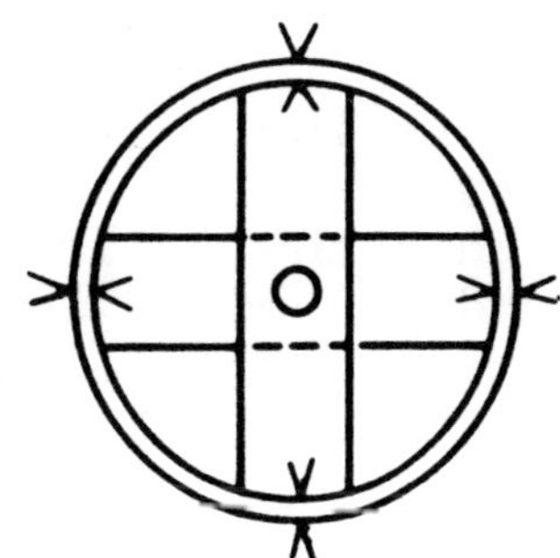

Fig. 6–66.

1) Reduced raw material costs: (a) less stock is used, (b) commercial quality sheet metal can be employed instead of special drawing quality steel
2) Increased production rates: (a) fewer operations, (b) simpler operations
3) Avoidance of internal stresses which are set up in drawn parts by severe plastic flow of metal during the drawing process
4) Flexibility of design in that a forming tool is very easily adapted to different forming depths.

Tolerances in Metal Stampings

By Federico Strasser

TOLERANCES IN METAL STAMPING

The approach to tolerance specifications for stamping should not be the same as the approach for machined parts. Machined parts allow closer tolerances than stampings. Correct dimensioning and determination of tolerances for stampings should be realistic and conform with the true functional requirements.

The best basic procedure consists in taking advantage of empirical, practical data taken from actual products which are functionally similar to the new product in question. If there is no previous experience to work with, then the following simple procedure should be taken as a guide for the determination of tolerances. Minimal clearance between operating components is the space that barely permits assembling of the components and their correct operation under service conditions. Maximum clearance is the distance that still allows proper functioning of the assembly. In this way, the two tolerance limits are established with a good approximation.

Practical Data

Some data on tolerances in metal stampings are presented here as a guide. These data have been collected in actual practice in the field. They are recommendations; deviations from them are quite frequent. They refer to short run jobs in which tools of the lowest cost are anticipated. Closer tolerances are possible in all cases; however, they would require more sophisticated tooling and manufacturing processes which, of course, increase production costs.

Flat Stampings

In every stamping, the die-cut surfaces are never completely straight or perpendicular to the general surface of the stamping. It is composed of a land which is smooth and parallel, and a break which is uneven, rough and tapered (see Fig. 2-2). The land is usually smaller than the break in a hole or on a blank. In both cases always measure or check the land portion.

Tolerances for punched holes and outer contours for blanks vary with stock quality, thickness and hardness with the size and shape of the part, with the condition, design and accuracy of the tool, with the number of stations in progressive dies and with other circumstances.

Fig. 5-23 presents the tolerances suggested by the Small Lot Stamping Institute. The values reported in Fig. 5-23 refer to light gage stock (up to .031 in.). In case of medium gage stock (up to .062 in.), tolerances must be increased by about 25 to 50 percent; for heavy gage stock (over .062 in.) tolerances increase by 100 to 200 percent.

Tolerances for center distances between holes located in

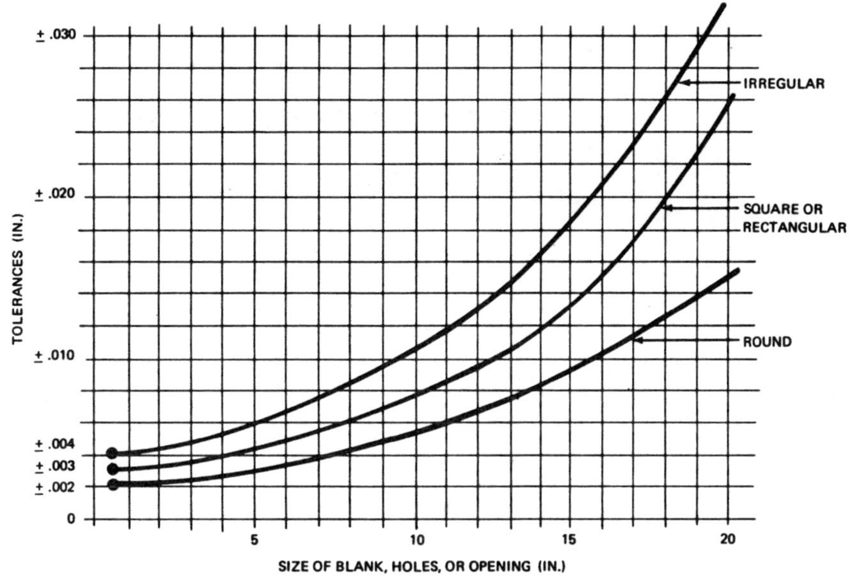

Fig. 5–23.

the same plane depend chiefly upon the corresponding production methods:

   1) If punched simultaneously with the same die: +.003

   2) If punched with separate, single operation dies: +.005

   3) If punched in progressive dies, made in separate stations, according to quantity of stations between operations and type of progression gage: +.005 to +.015

Tolerances for hole location from the edge of a stamping or from a bend are as follows:

   1) Light and medium stock (up to .062 in.): +.008 to +.010

   2) Heavy stock (.062 to .125 in.): +.015 to +.020

   3) Very heavy stock (over .125 in.): +.030 to +.035

Holes in bent legs must always be located from the inside. In this way the variations of sheet metal thickness do not influence the location tolerances.

If closer tolerances are required than is stated in the above paragraph, the holes must be punched after forming. This means lower production rates, higher tooling costs, and, therefore, higher prices per workpiece.

In case of incomplete blanks, the tolerances must be increased because of the inherent lower accuracy of production techniques:

   Parting: Short lengths (up to 6 in.): +.005 to +.008 Long stampings: +.010 to +.015

   Cutting off: Light gage – short lengths: +.015 long parts: +.030 Heavy gage – short lengths: +.025 long parts: +.040

Width tolerances for both

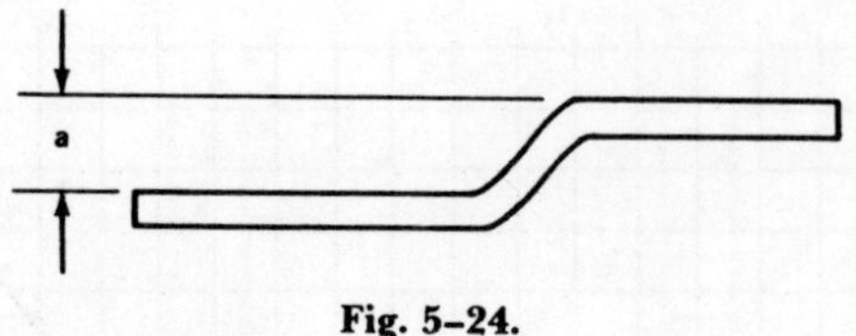

Fig. 5-24.

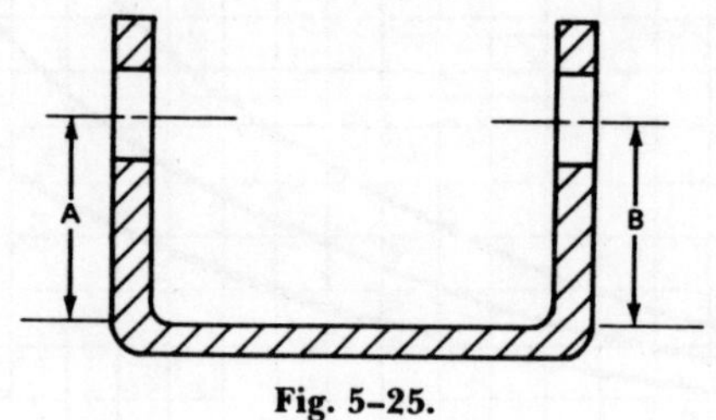

Fig. 5-25.

parting and cutting off correspond to commercial tolerances for slitting strips and coils. Concentricity is the relationship of one dimension to another. Tolerances on concentricity should be specified only if necessary for correct functioning. Commercial concentricity tolerances are: .010 to .020 in. TIR.

The chief governing factors for flatness are the material temper and the tool design. Under ordinary circumstances, a flatness of .005 to .010 in/in TIR can be maintained. If closer tolerances are required, some corrective operation such as coining or spanking, planishing, straightening or grinding is needed. Such corrective operations increase costs. Designers should avoid the frequent mistake of confusing flatness with parallelism.

The designer should specify burr height limits only if the function of the parts demands it. Removal of burrs or removal of sharp edges should be avoided because of the addi-

tional expense.

Formed Stampings

The average dimensional tolerances in workpieces form in <u>press-brakes</u> are:
1) For regular, small parts: $\pm$.030
2) For large and complicated shapes: $\pm$.060 or more

The average dimensional tolerances in workpieces form in <u>dies</u> are:
1) Small workpieces, or portions of work: $\pm$.010
2) Off-sets (see Fig. 5-24): $\pm$.010
3) Channel forming (see Fig. 5-25): tolerances on dimensions A and B can be held within .010
4) Always specify angle tolerances in degrees, not in straight dimensions - usual tolerance limits are $\pm$1 deg, except when one leg is shorter than 1 in.; then the tolerances are $\pm$2 deg
5) Always specify internal

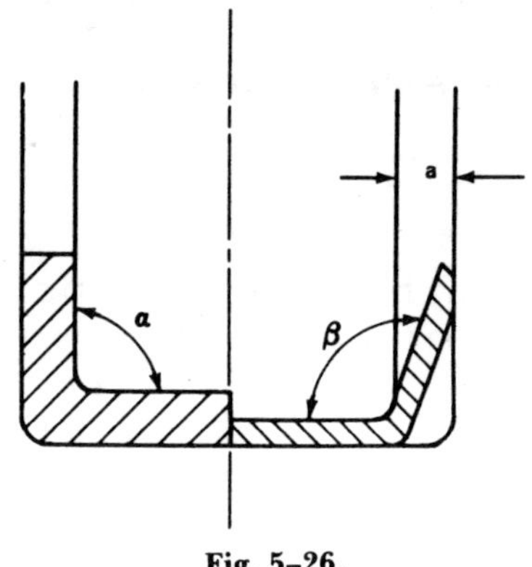

Fig. 5-26.

bending radii, never
outer radii - toleran-
ces on radii should be
liberal: +.010 for
R ≤ .060 and +.020 for R ≥ .060.

Drawn Shells

For drawn shells no stan-
dard, customary tolerance values exist. Every plant has its
own standards. In establishing
such tolerances for drawn
shells, take into account the
following points:
1) Wall thickness of
   drawn shells deviates
   from bottom thickness;
   some portions are thinner, some portions are
   thicker
2) Do not specify both
   O.D. and I.D. with
   close tolerances - only
   one of them can be reasonably held.

Stock Thickness Tolerances

The standard thickness
tolerances given by sheet metal
manufacturers are from about 10
to 20 percent. These allowed
thickness variations prohibit
close tolerance limits on form
and shapes. Fig. 5-26 illustrates how the thickness variations influence the accuracy of
formed stamping shapes.

In channel forming (see
Fig. 5-26), the clearance between the punch and the female
die cavity (α) must be accurately estimated for the highest thickness value. Consequently, the bending angle will
be different according to whether the stock is on the high or
low limit of the commerical
tolerance. The designer should
keep these facts in mind before
setting up tolerance limits for
formed parts.

Standardization of Stamping Design

By Federico Strasser

Only a few details can be standardized in metal stamping designs. These are:
1) Quality of material
2) Thickness of material
3) Strip width
4) Hole sizes

Quality of Material

The designer can make substantial savings through wise selection of material. The specifications of the manufacturers for sheet metal lists a large number of types with the differences in characteristics and performance to be expected. Many times, a compromise choice will enable one type to be used in several production runs. For instance, steel sheets, which are most often used, are classified according to manufacturing processes (cold or hot rolled), analysis, surface finish, temper, drawability, etc. Specific requirements which must be satisfied may demand special material for deep drawing, extrusion, electrical laminations or stainless characteristics; however, practical experience shows that a few steel brands are sufficient for satisfactory performance for the majority of cases. The same is true also for alloy steels, nonferrous metals and their alloys.

If only a few material types can fulfill the needs of a particular company, several advantages are realized:
1) Greater quantities of the same kind of material are purchased each time; this means lower prices because of higher quantity discounts.
2) The overall quantity of material held in storage becomes smaller even if the quantity per brand or size is greater; since the number of types is drastically decreased, the total quantity of stock is lower.
3) Consequently, the corresponding financial investments for material will remain reduced.
4) The probability for supply troubles and mix-ups will be less; easier, clearer control of stock usually results in prompt replacement and timely procurement of new material.

Thickness of Material

It is very convenient to limit the material thicknesses to a few values and gages and to choose from them one for any new component or product. In this way, as with stock quality standardization discussed in the previous paragraph, it is possible to take advantage of warehouse inventories and realize other advantages.

The following sheet thicknesses in inches are recommended for standardization:

.015	.018	.024	.030
1/32	1/16	3/32	1/8

and

.036	.048	.060	.075
5/32	1/4	5/16	3/8

Strip Width

It is good practice to

standardize strip widths with
1/16 in. increments for small
sizes (up to about 1 in.), and
with increments of 1/8 in. for
medium and larger sizes.

Holes

For round holes, the shape
chosen in the majority of cases,
the corresponding ANSI Standard
(ANS Y 14.10-1959) recommends
that the basic diameters should
be governed by standard drill
sizes whenever possible. Unfortunately, this seems rather
difficult to do because the
commercially available standard round punches do not follow this recommendation. The
punches in question are either
in increments of 1/64 in. or of
.010 in. In the first case,
with only one exception (¼ in.),
every standard drill size, both
numbers and letters, are odd
fractions. In the second case,
round decimal fraction sizes,
the situation is somewhat better; about half of the number
sizes coincide, but even with
these it is not possible to do
as the ANSI Standard suggests.
Therefore, it is recommended

that only one supplier of commercial punches be dealt with.
His standard should be made
yours; in actuality, this is
what usually happens in large
companies. Such commercial
punches are available not only
in round shapes, but also in
other shapes, including oblong,
square, rectangular and triangu-
lar.

For larger punches, which
usually must be home-made, take
into account that the outer
"bark" must be removed from the
steel bar. If possible, select
such "net" sizes which are
about 1/8 in. smaller in diameter than the commercial tool
steel bar sizes. The punch diameter should be 1 3/8 in. for
a tool steel bar of 1 1/2 in.

For multiple punching when
the component is punched with
more than one die, all the holes
punched in a given stamping or
at least all the holes in a giv-
en die should be, if possible,
identical in shape and size.
If this is not possible, the
number of different hole sizes
and/or shapes should be reduced
as much as possible.

Functional Design of Metal Stampings discusses design solutions which can help save on die construction, achieve faster stamping
rates, keep scrap losses to a minimum and reduce rework. For more information, contact the Society of Manufacturing Engineers,
One SME Drive, P.O. Box 930, Dearborn, Michigan 48128

Modern Sheet Steels — What They Are and How They're Worked

Reprinted from Manufacturing Engineering, October 1977

Sheet steels are changing, largely as a result of government demands for lighter, safer autos. The beneficial fallout from this is that nonauto users now have better workpiece materials. This article, prepared by the American Iron and Steel Institute, provides detailed information on the steels

HIGHER STRENGTH LEVELS, reduced weight and far longer life . . . These are the criteria by which modern sheet steels are selected. This is particularly true in the automotive and aerospace industries. It's also true of companies that produce small consumer goods. The reason, quite simply, is that the supply of materials is not unlimited and their maximum utilization is mandatory.

This being the case, it follows that steels having higher strength, and better formability and weldability are needed. Steel producers are advancing to meet these needs. Formable and weldable high strength steels in the 50 to 80 ksi (345 to 632 MPa) yield strength are now available — and are now being specified in product design.

New information on these steels indicates their superiority in crashworthiness, as shown by their strain-rate sensitivity. Further use of thinner sections is increasingly possible as new corrosion protection systems continue to become commercially available.

THE FAMILY OF SHEET STEELS

Low Carbon Sheet Steel. Regular low-carbon sheet steel, a type commonly used in the automotive industry, has a typical yield strength in the 25-35 ksi range (172 to 241 MPa). These materials are easily formed and welded with high volume mass production techniques. Additionally, their optimum combination of strength, modulus and fabricability means that the design of even low cost components can meet all of the various performance requirements.

Structurally, low-carbon sheet steel is excellent. It's expected to continue to be the dominant material within the automotive material mix. Typical low carbon sheet steels and their properties are summarized in TABLE 1.

► AVAILABILITY. Low carbon sheet steels are available in cold rolled products down to 0.0142" (0.36 mm) and in hot rolled products up to 0.500" (12.7 mm) and heavier. Specific availability of hot rolled and cold rolled products as well as maximum widths at various thicknesses vary. In selecting them, it's best to consult the producers.

► SAE 1006 AND 1008. Both are soft steels with high ductility. Both are easily formed and shaped. These grades are commonly selected when maximum formability is required and strength is secondary. They are usually produced as rimmed steels or fully aluminum-killed products when optimum formability is required.

► SAE 1010 AND 1012. These steels are slightly stronger than SAE 1006 and

1. LOW CARBON STEELS

Designation	Chemistry C	Mn	Yield Point Typical	Tensile Strength Typical	% Elongation in 2" Typical
SAE 1006	0.08 Max.	0.45 Max.	25-32 ksi	40-53 ksi	35-45%
SAE 1008	0.10 Max.	0.50 Max.			
SAE 1010	0.08-0.13	0.30-0.60	30-35 ksi	45-55 ksi	30-40%
SAE 1012	0.10-0.15	0.30-0.60			

2. QUALITIES AND STRENGTH

Minimum Yield (ksi)	Structural Quality	Semikilled, Capped, Rimmed	High Strength Low Alloy Fully Killed	Killed, Inclusion Shape Control
35	X		X	X
40	X		X	X
45	X	X	X	X
50	X	X	X	X
60		X	X	X
70		X	X	X
80			X	X

Yield Point Minimum (ksi)	Tensile Strength Typical (ksi)	% Elongation in 2" Minimum
35	47/52	28
40	52/57	27
45	57/62	25
50	62/67	22

1008. They're also less ductile and less formable. They're often selected for applications in which forming requirements are not excessively severe and part strength is of some concern.

High Strength Sheet Steels. Low carbon sheet steels provide an effective balance of strength and modulus for many components. In optimum design, they offer significant cost-effective weight reduction.

Many components can be designed more effectively by using higher strength steels at reduced thicknesses, however. A full family of high strength steels is available in strength levels from 35 to 80 ksi (241 to 632 MPa). These steels offer many of the same advantages as the low carbon steels. As such, they are completely compatible with existing manufacturing equipment. They can be formed, joined, and painted at high production rates.

The wide range of qualities of high strength steels permits optimization of selection in terms of cost, formability and weldability. At the same time, these steels meet part-performance requirements of strength, fatigue and toughness. TABLE 2 outlines the various qualities and strength levels to which high strength steels are currently produced. In this table, formability increases from left to right.

Generally, the higher the strength level, the more restricted the limits of sheet product dimensions when compared to low carbon sheet products. The 35 ksi steels are available in cold rolled products down to approximately 0.020" (0.51 mm) and in hot rolled products similar to low carbon steels up to 0.500" (13 mm) and heavier. The 50 ksi (345 MPa) steels are available in cold rolled products down to approximately 0.025" (0.64 mm) and 80 ksi (632 MPa) steels down to approximately 0.035" (0.89 mm). Both of these strength levels are available in plate products up to 0.500" and heavier. A more detailed description of the various qualities and their properties is provided in TABLE 3.

Structural quality steels are produced by three different chemistry approaches: carbon-manganese, nitrogenized and phosphorized. All of these steels are normally produced by semi-killed, capped, or rimmed deoxidation practices. When compared with the high strength low alloy steels they may have less homogeneity, formability, weldability and toughness.

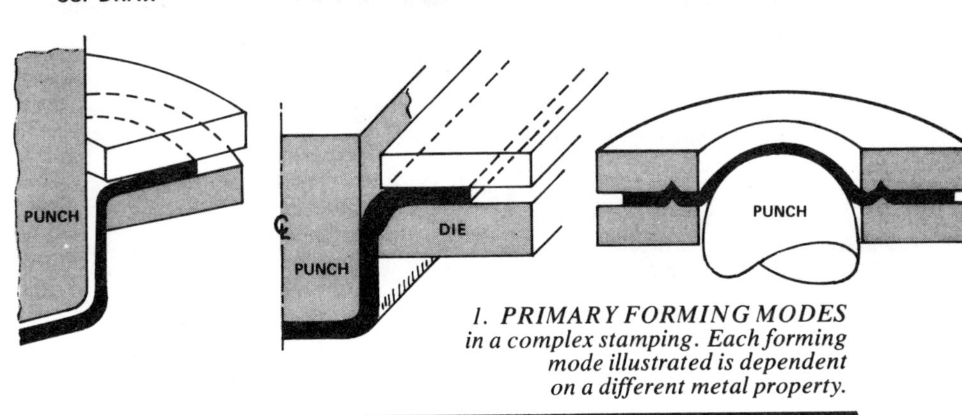

1. PRIMARY FORMING MODES in a complex stamping. Each forming mode illustrated is dependent on a different metal property.

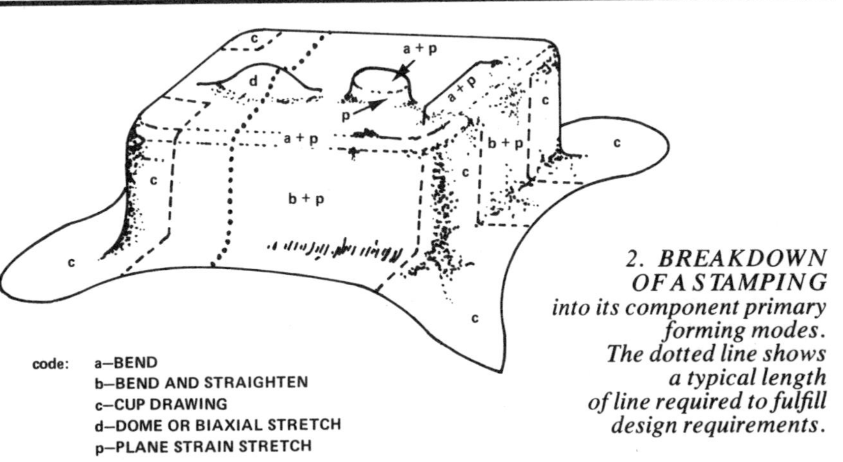

2. BREAKDOWN OF A STAMPING into its component primary forming modes. The dotted line shows a typical length of line required to fulfill design requirements.

Carbon Manganese Steel. Although carbon and manganese are added to increase the strength of steel, they impair its ductility and weldability. Depending on strength level and thickness, maximum carbon levels between 0.15 and 0.25% with manganese levels up to 0.90% are normally encountered. Resistance spot welding at high production rates is not practical when carbon levels are greater than 0.15%.

Nitrogenized Steel. In addition to carbon and manganese, nitrogen is added to these steels to increase strength and hardness. Nitrogen levels in the 0.010 to 0.018% range are typical. The addition of nitrogen enables the producer to use slightly lower carbon and/or manganese levels than in the carbon-manganese grades, thus improving formability.

Nitrogenized steels are also characterized by accelerated strain aging properties. This characteristic enables the steel to attain yield strength increases of as much as 25% in the finished part over the as-received condition. This strength increase depends on the degree of part deformation during straining and the amount of aging at ambient or elevated temperatures. The increase in yield strength is accompanied by a loss of ductility and toughness.

Phosphorized Steel. Like nitrogenized steels, the phosphorized steels are characterized by the addition of a strengthening element — phosphorous. The carbon and/or manganese content of the metal may be reduced slightly from the carbon-manganese grades. Phosphorous levels between 0.03 and 0.15% are typical, depending upon the strength level and thickness. Formability, weldability, and toughness are comparable to carbon-manganese and nitrogenized steels.

High Strength, Low Alloy Steel. The HSLA steels are strengthened by the addition of microalloying elements such as columbium, vanadium, titanium and zirconium, or by low levels of alloying elements such as silicon, chromium, molybdenum, copper and nickel. The use of these elements enables producers to significantly reduce the carbon and/or manganese levels to improve formability, toughness and weldability when compared to structural quality steels.

The major differences among these steels are the deoxidation practices and the spread between yield point and tensile strength.

The major element affecting tensile strength is carbon. Steels with a 20 ksi tensile strength to yield point spread have higher carbon levels. It's generally accepted that they have improved fatigue characteristics. Steels with a 10 ksi tensile strength to yield point spread have lower carbon levels and generally offer better formability and optimum resistance spot weldability.

Deoxidation practices can significantly affect the quality of steel. Semi-killed steels, like capped and rimmed steels, are less homogeneous than killed steels. As a result, they have less formability and toughness. Killed steels are more homogeneous with improved toughness and formability.

Sulfide inclusion control can be obtained in killed steels through the addition of small amounts of zirconium, titanium, or rare earth elements. This results in a steel with optimum formability in both the longitudinal and transverse directions. The more sophisticated the deoxidation practices, the greater the cost of producing the steel.

Ultrahigh Strength Steel. These are the steels to consider when part strength is critical. They're characterized by good weldability and with formability that — while limited — is adequate for roll forming or press-brake operations.

At the lower yield range, specially processed low carbon steels can be produced in a cold rolled condition to minimum yield points of 85 ksi (586 MPa).

Titanium, vanadium, or columbium bearing low carbon steels can be produced in a cold rolled, controlled annealed condition at yield point minimums of 100 ksi (689 MPa), 120 ksi (827 MPa), and 140 ksi (965 MPa). Low carbon martensitic steels are available in strengths up to 200 ksi (1379 MPa) minimum yield strength.

Overview. The preceding are the steels now coming to the fore in numerous industrial applications. Since range of their capabilities has widened considerably in recent years, these are the steels that will enable you to produce better, lighter and far stronger products. Next — How are they worked?

WORKING THE MODERN SHEET METALS

Sheet metal formability has undergone a transition from an art to a science. Formability — within each unique forming mode — can be related to a specific metal formability parameter. These parameters may or may not decrease as the yield strength of a high strength steel increases. The important point — often overlooked — is that these parameters change gradually and predictably as the yield strength of the steel increases. No discontinuous drop in formability is experienced. In fact, certain formability modes are insensitive to yield strength. Therefore, knowing the change in formability parameters expected, compensation can be made in part design, tool design, lubricant selection and press parameters.

A complex forming operation is usually composed of several primary forming modes, each of which is dependent on a different mechanical property. Therefore, the suitability of a sheet steel for an operation has to be decided on the basis of its formability in each of these several modes.

One of the major advantages of sheet steel in terms of design and engineering of automotive parts is the range of types available to meet varied forming requirements. Added to this is the considerable experience of production engineers and the ability of steel suppliers to produce sheet steel tailored to the requirements of a given part.

The Forming Modes. The three most common primary forming modes are cup drawing, bend and straighten, and stretch forming, as shown in *Figure* 1. Blanking, punching, flanging, and trimming are considered secondary forming operations.

Cup or Radial Drawing. In cup drawing, a circular blank usually is drawn into a circular die by a flat bottom, cylindrical punch (*Figure* 1). As the flange is pulled toward the die opening, the decrease in blank circumference causes a circumferential compression of the metal. Unless controlled by blankholder pressure, this circumferential compression can easily generate radial buckles in the flange.

Bend and Straighten. The bend and straighten mode is often confused with cup drawing. In both cases, metal is pulled from a flange, bent over a die radius, and then restraightened. However, in the bend and straighten mode (*Figure* 1), the die line is straight, the flange length does not change, and no circumferential compression or buckles are generated.

During deformation, the outer fiber (convex side of the bend) is first elongated as it bends over the die radius. It is then compressed as the sheet is straightened. The inner fiber (concave side) undergoes the reverse sequence of compression followed by tension. Thus, no radial elongation or sheet thinning is observed in a pure bend and straighten operation.

Stretch Forming. In stretch form-

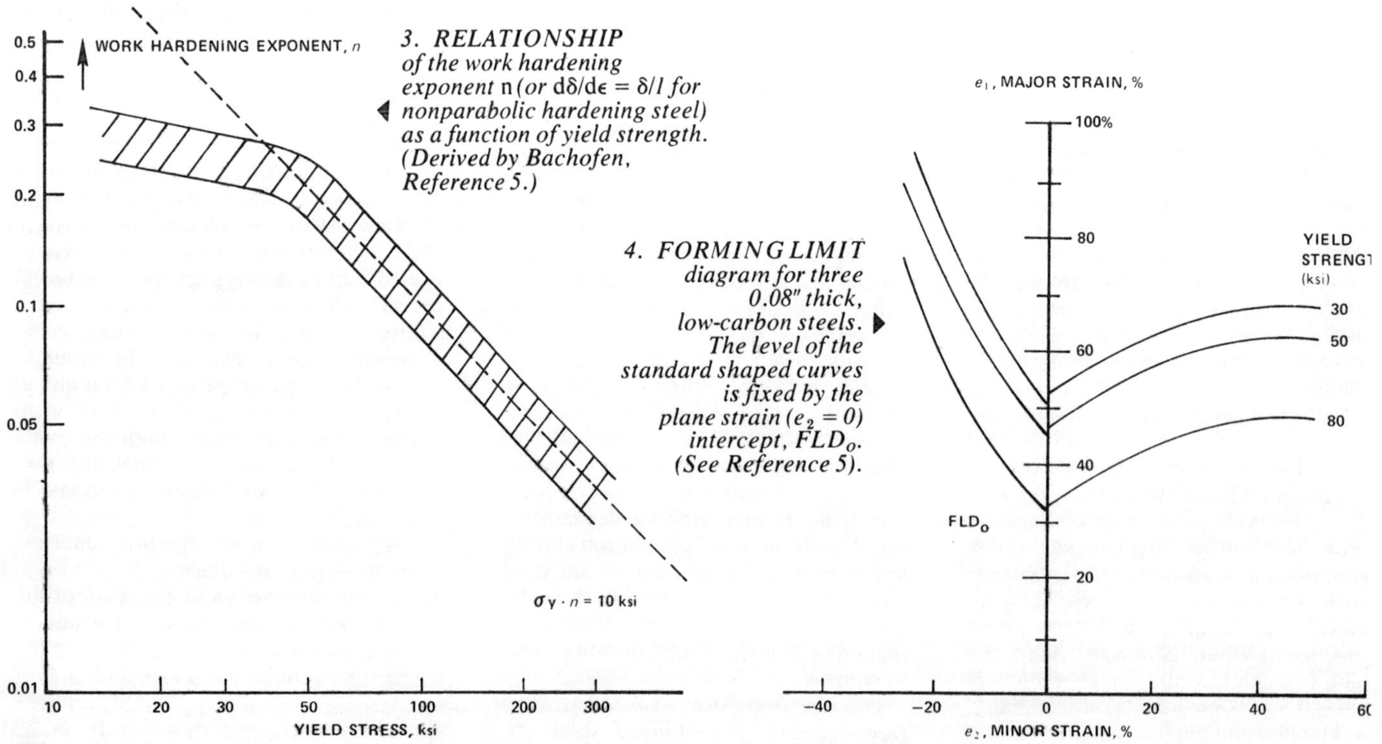

3. RELATIONSHIP of the work hardening exponent n (or $d\delta/d\epsilon = \delta/1$ for nonparabolic hardening steel) as a function of yield strength. (Derived by Bachofen, Reference 5.)

4. FORMING LIMIT diagram for three 0.08" thick, low-carbon steels. The level of the standard shaped curves is fixed by the plane strain ($e_2 = 0$) intercept, FLD_o. (See Reference 5).

ing (*Figure* 1), a blank is clamped at the die ring by hold down pressure or lock beads. A domed punch is pushed into the blank, causing tensile elongation of the metal in all directions of the dome. The thickness of the sheet must therefore decrease. This deformation is called biaxial stretch forming.

If the punch is long compared to its width (for a rectangular die opening), then tensile elongation of the clamped blank occurs only in one direction — across the small punch radius. This tensile elongation is offset by a reduction in sheet thickness. This very common type of deformation is called plane strain stretching.

An important problem is in elongation or stretching of a formed edge. Any damage in the blanking or shearing operation, as evidenced by a burr or rounding of the edge reduces the formability of that edge.

COMPLEX STAMPINGS. The complex stamping shown in *Figure* 2 helps to place the individual forming modes in better perspective. The flange can be segmented into four corners connected by four straight segments. The four corners are created by cup drawing. In fact, the four corners each represent one-quarter of a cylindrical cup. The straight line segments joining the corners are bend-and-straighten forming modes. However, if hold-down pressure or draw beads create a high radial tensile stress over the die radius, a plane strain stretch component will be added to the bend-and-straighten component. The bottom radii are a combination of bending (one-half of the bend and straighten operation) plus plane strain stretching. The dome on the bottom of the pan is formed by biaxial stretching, while the embossment and character line are formed by plane strain stretching and bending.

Forming Limits. In cup drawing, the punch button is pushed against the cup bottom to pull the flange into the cup wall. The cup wall must be able to carry the load required to deform the flange and overcome friction. If the cup wall can carry a larger force without necking down, a larger blank can then be drawn into a deeper cup. One method of characterizing this resistance of the cup wall to necking down is by the normal anisotropy of the metal, or the $\bar{r}$.* The higher the $\bar{r}$, the greater the deep drawability of the metal.

Typical $\bar{r}$ values are as follows
Hot-rolled 1008 . . . . .0.8-1.0
Cold-rolled 1008 rimmed . . .1.0-1.4
Cold-rolled 1008 AK 1.4-1.8
Hot-rolled HSLA 0.8-1.0
Cold-rolled HSLA 1.0-1.4
Based on the $\bar{r}$ values, hot-rolled HSLA steels, whether 50 or 80 ksi (345 or 632 MPa) yield strength, can be drawn to a cup depth equivalent to hot-rolled 1008 steels. Cold-rolled HSLA steels will have cup drawability equiva-

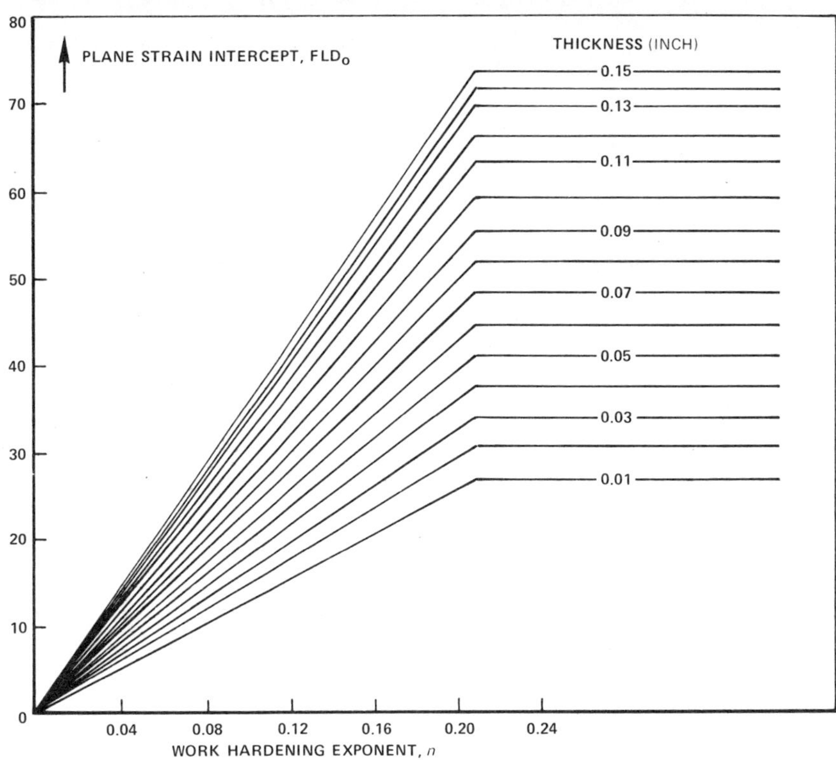

5. INTERRELATIONSHIP along work hardening exponent n, sheet thickness, and the plane strain intercept of the Forming Limit Diagram (FLD₀) for low-carbon sheet steel. (See Reference 5.)

lent to cold-rolled 1008 rimmed steels. A more direct measure of cup drawability is the Limiting Drawing Ratio (LDR) which equals D_b/D_p' where D_b is the maximum blank diameter which can be successfully drawn with a punch of diameter D_p'. A comparison between an 80 ksi (632 MPa) yield strength, cold-rolled HSLA steel with a low $\bar{r}$ of 1.0, and a 27 ksi (186 MPa)

The definition of $\bar{r}$ is the average r value within the sheet where r is the ratio of the change in width to the change in thickness. Thus

$$\bar{r} = \frac{r_0 + 2r_{45} + r_{90}}{4}$$

where the subscripts denote the angle of the applied stress.

yield, cold-rolled AK steel with a high $\bar{r}$ of 1.8 shows only a 25% reduction in the LDR, but an increase in yield strength of 300%. Dramatic increases in yield strength can be achieved with only small reductions in cup drawability.

BEND AND STRAIGHTEN. The limiting factor in the bend-and-straighten mode is the ability of the inner fiber of the metal to withstand the tensile strain in straightening after being subjected to cold work in compression during bending. In the case of bending only, the outer fiber element must withstand the required tensile strain. In both cases, the ability of the metal to withstand the bend and straighten deformation mode

can be correlated to the total elongation of the metal as measured by a tensile test. The higher the total elongation, the sharper the bend radius which can be formed. Typical measurements of total elongation in a 2″ (51 mm) gage length are:

▶ 30 KSI YIELD. Longitudinal elongation — 48%. Transverse elongation — 46%.

▶ 50 KSI YIELD. Longitudinal elongation — 35%. Transverse elongation — 30%.

▶ 80 KSI YIELD. Longitudinal elongation — 20%. Transverse elongation — 20%.

These numbers are important in three ways. First, the gain in yield strength is accompanied by a loss in total elongation. Thus, an increase in yield strength requires an increase in bend radius for equal thickness. As a result, formability in this mode is inversely proportional to yield strength.

Second, the longitudinal total elongation of the 50 ksi (345 MPa) yield strength steel is greater than the transverse total elongation. Thus, the preferred bend-and-straighten axis is across the rolling direction fiber. Blank orientation can significantly improve formability in this mode.

Third, inclusion shape control of the 80 ksi (632 MPa) yield strength steel is important in elevating the level of the transverse total elongation to that of the longitudinal direction. Inclusion shape control can improve the transverse bend-and-straighten capacity of all HSLA steels when required.

STRETCH FORMING. Stretch forming capacity of a metal is related to its ability to delay or resist the onset of a tensile instability or necking. One measure of this resistance to necking is the work hardening exponent or n value (from $\delta = K\epsilon^n$). The higher the n value, the larger the uniform elongation, the greater the resistance to necking.

This relationship between yield strength of steel and the n value is shown in *Figure* 3. The n value influences stretchability in two ways. First, a higher n value improves ability of the metal to resist localization of strain in the presence of a stress gradient. This generates a more uniform distribution of strain and permits more effective utilization of available metal.

Second, the Forming Limit Diagram (FLD) is dependent on the n value. The FLD (*Figure* 4) specifies the maximum strain sheet metal can withstand without necking for a wide combination of strain states. The level of this standard shaped curve for low carbon steel is fixed by the intercept of the $e_2 = 0$ axis; this point is labeled the FLD_0. The dependence of the FLD_0 on the n value for different thicknesses is shown in *Figure* 5.

Note in *Figure* 3 that the n values of a 30 ksi (207 MPa) and a 50 ksi (345 MPa) yield strength steel are approximately equal (within the scatter band). In addition, the FLD's for these two steels are quite close (*Figure* 4). Forming experience has confirmed that stretchability of 30 and 50 ksi (207 and 345 MPa) yield strength steels is approximately equal. A reduction in stretchability is observed for an 80 ksi (632 MPa) yield strength steel because of the lower n value.

Influencing Factors. The previous sections compared the formability of high strength steels with ordinary 1008 steels, assuming the influence of design, tooling, lubricant, and press adjustments does not change. However, all variables in the forming system are closely interrelated, and a change in one variable (steel properties, for example) requires modifications in the other variables. A number of these interactions can be illustrated with the aid of the schematic in *Figure* 2.

The part design requires that a specific length of line be generated, whether it originates by stretch or by bend and straighten. The stretchability of an 80 ksi (632 MPa) yield strength steel is reduced compared to an ordinary 1008 steel. However, if die radii, hold-down pressure, draw bead radii, etc., are carefully selected, the required length of line can be generated by replacing the stretch forming component by pulling metal from the flange. Thus, proper tool design can optimize the forming modes for which high strength steels are most suited.

In one study (Ref. 5), 50 and 80 ksi (345 and 632 MPa) yield strength steels were directly substituted for a 30 ksi (207 MPa) yield strength 1008 steel. Without any tooling, lubricant, or press adjustments, the 80 ksi (632 MPa) yield strength steel resisted stretching over the punch (because of the high yield strength) but compensated for this by pulling more metal into the die from the flange. Any modifications by the part designer or toolmaker which can provide this replacement of stretch forming by bend and straighten deformation of flange metal is encouraged.

The four flange corners are more susceptible to wrinkling or buckling if the yield strength of the metal is increased or if the sheet thickness is reduced. This greater tendency of a high strength steel to wrinkle can be compensated by increased hold-down pressures. However, increased hold-down pressures result in increased binder (flange) forces unless the lubricant coefficient of friction is reduced to offset the increased pressures.

The increased forces required to deform the higher strength steels generate higher interface pressures. Lubricants may have to be upgraded to withstand these increased interface pressures without lubricant breakdown. Further, careful lubricant selection is required to avoid increased tool wear.

The reduced work hardening exponent, n, of the high strength steel indicates a reduced ability to resist localization of strain in the presence of a stress gradient. Therefore, part and tool designs should compensate by reducing stress gradients. Included in a long list of possible modifications are: increasing punch and die radii, selecting lubricants to encourage uniform distribution of deformation, avoiding plane strain stretching, reducing flange loads, reducing depths of embossments, bringing more metal in from the flange, etc. Proper part and design changes can help compensate for any reduced stretchability of the new, higher strength-to-weight ratio metals.

Other deformation parameters are less well defined. One example is springback. In bending, springback increases for increased yield strength. This can be corrected by appropriate overbend, overcrown, or subsequent restrike. In stretch forming, however, the interaction of deformation with metal properties is very complex and predictions of springback are difficult.

Some considerations also must be taken into account in the secondary forming operations. These include increased press loads to blank, punch, and shear; blanking clearances; edge cracking during flanging; etc.

Weldability. Low carbon sheet and strip can be welded by all the usual resistance welding processes (seam, spot, and projection) and by shielded metal arc, submerged arc, and gas metal arc processes. One of the prime reasons for the widespread use of low carbon sheet steels is their excellent weldability. The weldability of these steels is the standard against which weldability of alternative materials is compared.

With proper procedures, higher strength structural quality and high strength low alloy steels can be welded by all of the popular production welding processes — shielded metal arc, submerged arc, gas metal arc, and resistance welding. It should be recognized, however, that the ductility of welds in these steels decreases as the carbon and alloy contents increase. Some of the newer high strength low alloy grades are specifically designed to overcome this problem, and their weldability may approach that of low carbon steel. It is important to make certain that the welding practices planned for any particular grade of sheet steel provide the required properties. The steel supplier should be consulted for his recommendations on welding individual products.

6. COMPARISON OF WELDS in 0.120" cold rolled steel with welds in 0.082" HSLA (80 ksi minimum yield point, vanadium bearing).

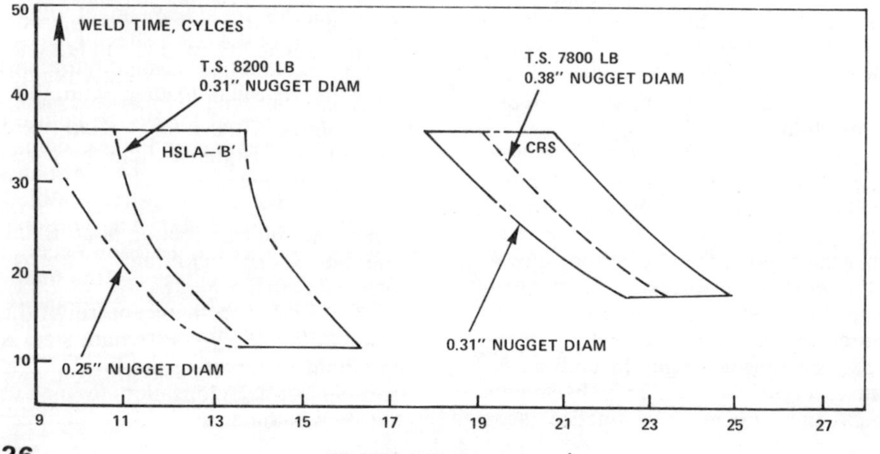

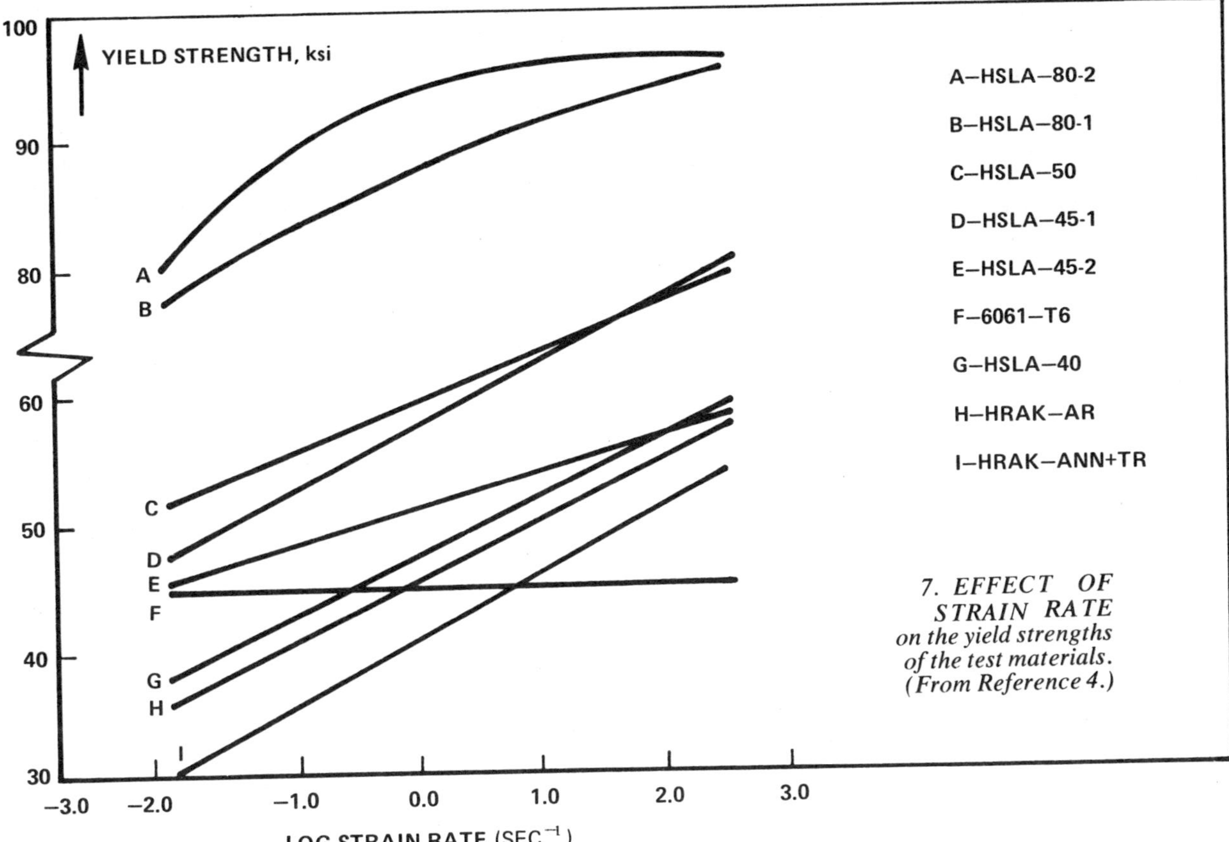

7. EFFECT OF STRAIN RATE on the yield strengths of the test materials. (From Reference 4.)

Resistance spot welding is of primary concern in automotive manufacture. Industry experience with resistance spot welding of high strength low alloy grades on existing equipment has been good. However, changes in electrode forces, weld times, and weld current in comparison to low carbon steels may be required. A welding lobe diagram is the best indicator of the resistance spot welding characteristics of a given sheet steel. It shows the combinations of weld time and current that will produce satisfactory welds under specific conditions. These helpful diagrams can be provided by steel suppliers for individual grades.

Many high strength low alloy steels are compatible with existing automotive welding equipment. This is shown in *Figure* 6. In this example, 0.120″ (3.05 mm) thick carbon steel was replaced by 0.082″ (2.08 mm) thick minimum vanadium treated high strength steel with a yield strength of 80 ksi (632 MPa). Weld strength as measured by tensile shear tests was increased and a larger weld lobe was developed at reduced current levels. In addition weld force was decreased from 2000 to 1600 lb (17 792 to 14 234 kN).

Strength in Finished Part. All steels are characterized by an ability to work harden and strengthen from strain induced during part forming. In addition, many steels will age harden at ambient temperatures or at elevated temperatures such as those incurred during paint-bake cycles. These two properties are important in imparting additional strength to the finished part and should be taken into consideration when comparing steel to other materials. Strength increases in the finished part due to straining and aging of 20-30 ksi (207-138 MPa) are not uncommon. Most steels have unique strain aging characteristics and the purchaser should consult with the steel producer for specifics.

Many steel companies are currently investigating a new family of dual phase steels which are characterized by very rapid work hardening characteristics. Increases of 20,000 psi (138 MPa) in yield point can be obtained in areas of a part which have seen less than 3% strain. This characteristic enables one to start with a relatively low strength steel in producing high strength parts that required complex forming.

Strain Rate Sensitivity. Of increasing importance to the design engineer is the effect of impact loading, controlled crush and energy absorption on vehicle components. A knowledge of the change in mechanical properties of a material with changes in strain rate (strain rate sensitivity) is paramount in understanding and designing for crash protection. Recent studies (Ref. 4) have shown that steel offers advantages to the designer in this area. These studies show that for both low-carbon steels and high strength low-alloy steels, yield and tensile strengths increase with increasing strain rate. The total elongations remain constant. Absorbed energy tends to increase with increasing strain rate. No such increases were found for typical automotive aluminum alloys under the same test conditions. *Figure* 7 shows examples of relative increases in yield strength with strain rate for a number of steels and an aluminum alloy. In a practical sense, ferrous alloys will be stronger at high loading rates than expected from ordinary mechanical property measurements. This provides improved dent resistance, impact loading resistance and energy absorption.

Fatigue Behavior. Designing to avoid fatigue failures in a component of a vehicle is one of the more difficult tasks an engineer faces. Fatigue failure is caused by repeated loading with the number of cycles of loading to failure varying with load range. Fatigue damage occurs gradually over the life span of cyclical loading, with detectable changes taking place in a small portion of the total life, crack initiation occurring at varying percentages of the final life, and crack propagation continuing until final fracture takes place.

Designing to avoid fatigue failures requires knowledge of the following:

▶ Expected load-time history, or better, the local stress-time and strain-time history at the most critical locations.

▶ The nature of the environment in which the component is operated.

▶ The properties of the material as it exists in the finished component at the most critically stressed locations.

▶ The geometry of the component and its notches (stress concentrations, surface finish, manufacturing variability, etc.).

Scatter must also be considered in fatigue life evaluation and prediction. This often calls for statistically based analyses. Circumstances dictate the degree of sophistication required for any part.

Perhaps the most widely used indicator of fatigue resistance is the concept of fatigue limit — the stress level below which fatigue is highly improbable. The high cycle fatigue endurance limit of steels is related to the tensile strength; fatigue endurance limit increases as the tensile strength increases. Thus, the fatigue limit of the higher strength steels increases as the specified yield point and its associated tensile strength increases.

Because tensile strength is a primary factor in fatigue, very little distinction can be made between the fatigue limits of steels within the same tensile strength category.

Many other parameters of the fatigue-resistant properties of a material may be needed depending on the sophistication desired and the criticality of the component. Suppliers should be consulted for specific guidance on this point. ∎

References

1. Chang, D.C. and Justusson, J.W., "Structural Requirements in Material Substitution for Car-Weight Reduction," Paper 760023, SAE Automotive Engineering Congress, Detroit, February 1976.

2. DiCello, J.A., and George, R.A., "Design Criteria for the Dent Resistance of Auto Body Panels," Paper 740081, SAE Automotive Engineering Congress, Detroit, February, 1964.

3. Adams, D.G., Dinda, S., George, R.A., Karry, R.W., Kasper, A.S., Pogorel, J., Swenson, W.E., and Weeks, W.L., "Charger XL: A Lightweight Materials Development Vehicle," Paper 760203, SAE Automotive Engineering Congress, Detroit, February 1976.

4. Chatfield,D.A., and Rote, R.R., "Strain Rate Effects on the Properties of High Strength Low Alloy Steels," Paper 740177, SAE Automotive Engineering Congress, Detroit, February 1974.

5. Keeler, S.P., and Brazier, W. G., "Relationship Between Laboratory Material Characterization and Press-Shop Formability," Proceedings of Microalloying 75, Washington, D.C., October, 1975.

6. Heimbuch, R., Maloney, W.D., and Rose, L.J., "HSLA Steels for Auto Parts," *Automotive Engineering*, Vol. 82, No. 7, July, 1974.

7. Graham. J.A , *SAE Fatigue Design Handbook*, Publication AE4, Society of Automotive Engineers, Warrendale, PA, 1968.

8. Tucker, L.E., Proposed Technical Report on Fatigue Properties, Publication 740279 Society of Automotive Engineers, Warrendale, PA,1974

Art vs Science   By Jack Gilchrist

While most of the data required on the material specification forms can either be developed from engineering input or is adminstratively provided, some of the data particularly in the area of physical characteristics, require judgemental decisions.

Unfortunately, good sense is not necessarily a prerequisite for judgemental decisions. As a result, much of the savings achieved through material utilization economies can be dissipated through the specification of cost extras whose needs are not supported by either fact or experience. For example . . .

The point in time was a late summer evening in 1952, the kind of evening made for outdoor barbecues, twilight golf or swimming parties. But, in the Sterling Stamping Plant there was no time for any of these. Rather, presses stretching the quarter-mile length of the building, were pounding out stampings for the new 1953 model cars soon to be introduced. Conspicuous by its non-contribution to this effort was a 150-ton OBI press in the far corner of the building. It sat there dead and silent, and the four of us crowded around it showed little more activity. There was the plant metallurgist, the die engineer, a die maker and myself, and we were completely perplexed. From early morning we had been attempting to stamp a hot-rolled steel retainer that, when completely formed, was no bigger than the palm of an average man's hand. From the moment the die had been put in the press and the inital hit made, a round extruded flange in the center of the part had been cracking. It had seemed a simple problem at first and the die maker had taken the normal corrective action, checking die clearances, adjusting spring pressures, and finally honing the extrusion punch to remove any semblance of sharp edges, but all to no avail. A call to Engineering brought to the scene the die engineer responsible for the tool design; he came up with several more suggestions with the same lack of success. Finally, when all else had failed, it was decided the problem must be the steel, and samples had been sent up to the lab to verify its drawing quality. Now the day shift production force had gone home, the night shift was in, the second round of the day was underway, and still the press sat unmoving; the retainers, badly needed if cars were to be assembled in the morning, were still in the form of coiled steel. As though to emphasize the seriousness of the problem-the need to get production immediately-the metallurgist was making one of his rare appearances to the pressroom by hand-carrying his findings back down to the press. No, there was nothing wrong with the steel. It was HR DQ 1010 as specified. Ductility was good; it ought to make the part. The die engineer was equally positive about his die. There was no reason why the extrusion shouldn't make. The die maker said he had done all he could. It beat him what the problem was.

Off to one side the press operator, optimistically assigned to the job on the assumption that production would be run, quietly watched the scene. He was one of the older employees on the night shift and not overly impressed by all the brass that the problem had attracted.

He sat on the edge of a stock tub resting his feet until the time when either they got the job going or gave up and put him somewhere else. His imputability in the midst of all the dissary drew me toward him. "We're having a terrible time," I said.

"Yeah."

"It's been like this all day, running a couple of good parts and then that extrusion starts cracking."

"Yeah."

"As a matter of fact, since this new tool came in a week ago, we've only been able to stamp four or five thousand pieces that were any good."

He shifted his weight on the stock tub and looked straight at me. "Yeah, we ran those the night before last, but you'll never stamp them out of that coil of steel you got there."

I stared hard at him. "What do you mean?"

"That steel," he said, "the other night when I ran the job, I couldn't make it with that steel either. Had to use the other stuff."

"Show me what you mean."

He pushed himself off the stock box and lumbered over to the steel storage area. I followed. All of the coils for the retainer stamping were stored in one area, and he looked them over carefully before pointing to one particular coil of steel. "Put that one in," he said, "and we'll run production."

The die maker, the die engineer and the metallurgist all stood by in puzzled amusement as the coiled steel at the job was rebanded and taken away and the new coil put in its place.

"You know," my metallurgist friend pointed out, "all of that steel is from one heat; it's all the same."

"May be, but we're not getting anywhere and I'm desperate. If the old man thinks he can run the job with that particular coil, I'm willing to give him the chance."

The new steel was in place and the operator inched the coil into the die and turned over the press. The part that ejected out was picked up hurriedly by the die maker. He wiped the lubricant off the extrusion surface. "It's cracked," he said with just a slight hint of satisfaction in his voice. The operator wasn't paying any attention. He started his press running on the hop and the parts started to shoot out forty-five to the minute. As they came out, the die maker was picking them up and throwing them in the scrap barrel. "They're all cracked," he said, "but it's getting better." Then the crack disappeared entirely and the press was running good parts.

The four of us looked at each other and then at the press operator. He obviously didn't share our surprise.

"When you get this coil run out," the metallurgist said to him," put the other one back in—the one that was in before."

"If I do," the old man said, "it will just start cracking again."

"Why are you so sure?"

"Because I know that's what happened the other night. When I put that crappy stuff in, they all cracked; when I put his stuff in, it looks good."

"But it's all the same," our metallurgist protested.

"Like hell it is the same," the old man said.

"Can you show us a difference?" I asked.

He stopped the press and we all walked back with him to the coil storage area. He started to point out the good from the bad. "That coil will stamp, that one won't; didn't you see the difference?"

"The stuff you're pointing out," the die maker said, "is a little discolored."

"That's right," the old man answered. "The darker stuff stamps good. The rest of it won't make a piece."

"Okay, make sure that you run only the stuff that you know will stamp, and pick out a coil at the end of the shift for the day shift to run to get them started."

The metallurgist wasn't saying much, but before he left he had the steel man cut off a strip from each coil, good and bad, and these he took with him back to the lab.

Later on, I talked to the metallurgist about his findings. "There was no perceptible difference." he said, "between the coils of steel. They checked to specifications."

I have no explanation to offer as to why one coil produced good parts while a second supposedly from the same heat of steel did not. I do know that this rather traumatic experience reinforced my awareness of the unpredictability of steel once plastic deformation takes place.

Now, many years later, after numerous visits to mills, much reading, and continuous observation of the deformation of sheet steel in the stamping process, I must admit a lack of confidence in predicting how steel will react to any given set of circumstances. I have listened to knowledgeable steel men talk with authority on their product—but their subject is usually on that of making steel, rarely on using it.

No less reticent are die designers. The designer, who speaks with conviction and authority when describing how his die design will stamp a
particular product, seems to lose his air of confidence standing in front
of a double or triple action toggle press awaiting the results of the
first hit.

Both the making of steel and the building of dies may be based on
sound engineering fundamentals, but the stamping of a deeply drawn, irregularly shaped object is as much an art as it is a science; the responsibility of producing the finished product usually is in the hands of
someone who is neither a scientist nor an engineer, who uses the cut and
try approach, and who will readily reverse his direction when he learns
his results are wrong.

As a young die engineer, I worked with a man who made his living
on his ability to trouble-shoot stamping problems. Willie was paid not
for what he did, but for what he knew. If one of the stamping plants
was having difficulties producing a stamping, Willie would be called
down to look over the situation. By the time Willie was called to the
scene, things were usually in a pretty bad shape. The sense of urgency
that usually surrounds a stamping line that cannot produce good parts
had changed to one of panic. Materials of various thicknesses would have
been tried, discarded, and tried again. Everyone's favorite drawing
compound would have had its chance: applied on top of the blank-the
bottom of the blank-the top and bottom-or desperately-in strategic globs.
Likewise, adjustments to air cushions and press rams, as well as to die
ring hold down pressures, had been many and varied. Into such a scene
Willie would appear, calmly puffing his pipe and asking in his heavy
German accent, "What's the problem?"

Usually just his appearance and the sound of his voice-German
accents seem to generate confidence, or at least awe, in stamping plants-
were enough to bring some semblance of sanity to the scene. Willie would
listen carefully to the problem, his head slightly bowed, and his hand on
the bowl of the pipe even though the stem was clenched between his teeth.
His eyes would blink as though in concentration as he listened, and every
once in a while he'd pull the pipe from his mouth to nod vigorously at
some particular point in his listening. When the explanations were over,
he would look at all the torn, widely split stampings and inform the
plant people that what they were doing was the wrong thing. He would then
suggest several alternative corrections, rarely specific, but rather
general in nature, and walk away. Willie's success was based on the in-
disputable fact that after his visit, the problem seemed to shortly disappear, though the people involved could never remember just what Willie
has specifically suggested that had effected the cure.

One time when Willie was called down to help in a chronic oil pan
draw operation problem, the die maker thought he had Willie on the spot
when he made the specific suggestion that the draw radii in the die ring
be increased. The die maker, trying to get Willie to be even more specific, asked him how much the radii should be ground away. Willie stared
at him, blinked his eyes once or twice, and then waved in the direction
of the lower half of the die and said "just steal a little."

Several hours later when Willie came back to see the results of this remedy, he found that the slight crack in the side wall of the oil pan had become a gaping hole. He took one close look at the die in the press, nodded vigorously, and turned to the die maker standing by awaiting his reaction.

"Well," he said, "you stole too much!"

The point I am trying to make is that there is reason to doubt that either the steel maker or the die engineer knows just what will happen when metal is stretched into a state of plasticity in a stamping operation; for this reason they will often specify material restrictions which are expensive to the stamper, and these extra costs, if accepted without question, can add significant costs, if accepted without question, can add significant costs to the product without adding any appreciable benefits.

Not long ago, at a management dinner, a friend and associate explained to the wives present that the dies used in the stamping process were a lot like cookie cutters. He took some ribbing from some of the die engineers present, who felt that this was a slight over-simplification of a highly technical subject.

I'll risk a similar reaction from steel making friends by confessing that there is a striking similarity in my mind between casting an ingot of steel and baking a loaf of bread. What comes out of the process is dependent on a number of factors that are not always predictable. In both cases, whether it be a white-haired, kindly old lady or hard-hatted, jut-jawed steel worker, some tender care goes into the process if the desired results are to be achieved-for steel making, like bread baking, is as much an art as well as a science.

As a result, there is an aura of mystery about steel that is not dispelled by the vast technical books, periodicals or articles that are available to the reader. Opinions and attitudes are so polarized on the subject that extermely competent tool makers will take diametrically opposite views on such subjects as age hardening of steel, the significance of draw quality or killed steels, and the importance of grain flow to the forming process.

I do not mean to imply that the steel industry has induced confusion or mystery or been secretive about the steel making process. Just the opposite is true. The attitude of the steel industry regarding dissemination of technical information is well illustrated by the work of the American Iron and Steel Institute. The Institute. which has as its members the manufacturers of steel, conducts extensive research programs at universities, sponsors work shops for teachers, and publishes numerous manuals and pamphlets. The manual entitled Carbon Sheet Steel is particularly valuable to anyone involved with the stamping of steel. This manual provides a comprehensive explanation of the various steel making processes. It gives a wealth of data on chemical composition and physical characteristics, and documents inspection and testing techniques, and dimensional tolerances of all steel sizes of sheet steel.

The information is invaluable to the stamper, and we are indebted to the American Iron and Steel Institute for making such a manual available.

Still, in spite of this information, or perhaps to some extent because of it, stampers find themselves specifying extras on steel, which, when pressed for reasons, they cannot really justify. Not that reasons aren't given-and with conviction!

"You can't form a part like that without draw quality steel."

"Our steel sits in inventory for a long time between operations, so if we didn't use aluminum killed steel, it would age harden."

"We have to stamp with the grain to get an acceptable part."

"Hot-rolled steel just won't work on this job."

All of these comments are reasons, supposedly, for justifying steel cost penalties on the basis of product quality or manufacturing feasibility. It has been my observation that for the most part, these are really assumptions that could be quickly dispelled if the user took the approach that he would add extras only after standard cost material had proven unsuccessful-and that, only after every possible fix had been tried on tools, equipment, or product design.

On the surface this may seem rather rash and impractical, so let's look at some of the extra costs which in my mind are most frequently ordered ill advisedly.

The D. Q. Myth

For years die engineers have been insisting on draw quality steel and paying in the area of forty-five cents a hundred weight for the privilege without ever knowing exactly what it was they were receiving. In fact, I have yet to find a die engineer, or for that matter, a material quality control inspector who could pick out either visually or by test draw quality from commerical quality steel. I believe many mills would agree that draw quality is, in actuality, an insurance policy which allows the user to return steel to the mill if the steel does not make the part, and stampers for years have had a field day doing just that. Fractures in the forming of a stamping often are an automatic rejection, and the return of steel in some cases becomes so prevalent that the mills have allowed the material to be used with a rebate on the percentage of parts scrapped. The unfortunate part of this is that, in many cases and probably in the great majority of cases, the cause of fractures in the stamping is not found in the steel but in the tools or equipment used in the stamping process.

A short time ago one of our plants was rejecting steel on a job that was drawing two parts at a time on the basis that the material would not produce a part, even though one draw came out consistently good and the other consistently cracked. When the illogicality of this finally became apparent, the die maker corrected the draw station in his die that was

producing the bad part, and the one hundred thousand pounds of steel
which had previously been rejected was rerun through the die with minimum scrap. But even then the steel was used not because of the acknowledgement that there was a die problem rather than a steel problem, but
only because the inability of replacment steel required that the rejected
steel be tried once again. The quality and consistency of steel making
today is such that the material engineer has the right and responsibility
to insist that commerical quality material be the standard, and that the
use of draw quality be specified only when the severity of the draw results
in scrap in sufficient amounts to make the extra cost of draw quality
steel economically feasible.

Over the years, particularly when we've had a buyer's market, steel
users have submitted prints or prototypes of the stampings to be produced
to the mill and received assurances that the steel sold would produce the
part in question, with the user having the right to reject material which
did not. Such rejected material was usually then held for a steel field
representative to come in and approve its return to the mill. All of this
had resulted in abuses to the detriment of the steel companies, abuses that
rapidly get corrected in times of steel shortages. Steel will fracture or
tear for many reasons other than the physical characteristics of the material. The die may be designed so as to require a greater elongation than
the material is capable of giving, or the die construction may be the cul-
prit due to insufficient clearances, so that in the draw process the material locks rather than flows. Or the problem may be die setting or the
condition of the press in which the tool is used. At times, a change in
the type of drawing lubricant used to aid the drawing or forming process
will be enough to cause part fractures. A great deal of steel waste will
be eliminated when our stamping plant personnel develop the philosophy
that the physical qualities of the steel is the last, rather than the
first, thing to be questioned when stamping problems occur. Not only will
this conserve steel, but it will reduce the enormous waste of labor and
press time that often develops because proper die maintenance or press
repair is postponed until the conditions are so bad that even the laziest
and most short-sighted staff must recognize the problem. Frequently,
rejected material, thrown out because it would not make a proper part, has
been put back into production and runs successfully after minor die changes
had been finally made.

Now, if, in fact, we can't tell the difference between draw quality
and commmercial quality steel, why then buy draw quality and pay the extra
cost. If the only reason is to make it easier to return rejected steel to
the mills, the cost of this insurance, except on extremely difficult draws,
far outweighs the savings. Much better to purchase commercial quality
steel and place on your plant personnel the responsibility of making the
material work. If this is done, on those occasions that poor quality steel
can be proven, such as a lamination, pitted surface, or brittleness, a plant
that has a good track record with the mill should not expect difficulty in
obtaining redress. If the mill feels that draw quality or draw quality
killed steel is required, it will insist that such be ordered or refuse
responsibility.

How Important is Age Hardening

What is age hardening? There are those who claim that there is no
such condition. Yet, many steel mills specify in their manuals that steel
should be fabricated within sixty days of receipt in order to minimize the
effect of age hardening. The AISI Steel Manual previously referred to
suggests that special killed steel be specified on difficult draw jobs where
age hardening may be a problem. Obviously, those who claim that steel does
age harden with a negative effect on material ductility have much distin-
guished support.

Let us not, therefore, be too equivocal in denying the existence of
age hardening. Rather, I would suggest that from a commerical point of
view, we just ignore its effect. This is not a revolutionary idea. Many
stampers in the not too distant past, when foreign mill prices were considerably lower than America's, brought in steel from Japan or Europe,
which in travel time alone, used up the sixty days our mills often specify
as the limit for steel to sit prior to age hardening, and I have not heard
that age hardening on foreign steel purchases causes a problem. Also,
many stampers in the overcapacity years of the U.S. mills bought from
warehouses at reduced prices mill surpluses, that had sat for unknown
lengths of time without any apparent age hardening.

I've often wondered why it is that manufacturers, who are concerned
about age hardening of steel and pay premiums to counteract the problem,
are able to put the subject completely out of mind when the opportunity
presents itself to buy bargain material at other than mill. The answer
that I finally settled on is that they are not the same people in both
situations. Those who fuss about age hardening are likely to be operations people, die makers, or production personnel who use the steel and
must answer for its deficiencies, but purchasing decisions rarely rest on
them. These people may be able to influence steel specifications and the
adding of quality extras, but such decisions as the purchase of foreign
steel are usually made at a level of management where cost is paramount,
and where the measuring stick is the cost of the part, not ease of production. The age hardening theory is a luxury that is quickly discarded
when it gets in the way of profit.

Lubricants and Drawing Compounds

How best to lubricate or coat production material in order to facilitate stamping operations and preserve die life is a constant concern of
die engineers and press shop personnel. Just about everyone in the pressroom has his own pet drawing compound, and I must confess sheepishly, that
I, too, as a young die engineer kept on had a small supply of a drawing oil
that I applied on those jobs that defied every other solution. In retro-
spect, that drawing compound, in which I placed so much confidence, was a
security blanket that probably worked no better in correcting sheet metal
fractures than a number of other commercially sold products. Looking
back on my experience in the pressroom and observing the experiences of
many others, I am prepared to make only one sure conclusion and that is,
to a great extent, drawing compounds of all types are both misused and
overused.

The misuse, no doubt, is partially caused because of the variety
that is available in the market place. There are dry compounds and wet
compounds. Some have the appearance of mineral oil, others the texture
of soap, and still others look and feel like glue. Not only that, there
is a salesman not too far removed who is ready to visit you at a moment's
notice to offer convincing evidence that his particular lubricant is the
right one for your job. No wonder the poor fellow who is in charge of
purchasing such lubricants is confused. On a difficult stamping operation, where sheet metal fracturing is a chronic problem, it is not unusual for the specified drawing lubricant to be changed several times over.
When production is needed and the press is producing nothing but scrap,
it is easier to change drawing lubricants than to fire the die engineer,
though that, too, might eventually happen. Often, the drawing compound
that stays with the job is the one in use at the time the press problem
is resolved, even though the solution to the problem may be something
completely unrelated to the compound in use.

An expensive and annoying problem is excessive use of drawing compounds. To begin with, the material itself is costly. Application costs
may be high, either in expenditures for applicators or in the labor costs
involved. Then, after the stamping has been produced, there is still the
cost of removing the lubricant, as is so often required prior to finishing
operations. In addition, there is the housekeeping problems caused by the
dripping compound, and the maintenance and clean-up of dies, presses and
stock containers. All in all, the use of drawing lubricants is an expensive mess, which the pressroom would be better off without. Unfortunately, in many cases they are a necessity and mean the difference between
a completed stamping and a torn piece of scrap. The point is not the
abolishment of drawing compounds, but their controlled usage.

All cold-rolled steel and hot-rolled pickled steel are oiled at the
mill unless the user specifically requests otherwise. This oil, in itself,
not only preserves the surface of the material but acts as a lubricant.
In addition, it is possible to order from some mills, at an extra cost,
a dry coat application of drawing lubricant. The cost, however, runs
fifty cents or more per hundred weight; it coats uniformly all of the
steel in the coil, both offal and blank, and, therefore, is wasted to a
considerable extent.

If you are like most stampers, you purchase your own drawing lubricants to be used as required and to be applied in the areas needed. My
only quarrel with this is that over a period of time there is a tendency
to be indiscriminate in the purchase and use of drawing compounds, so
that on difficult stamping jobs the cost of the lubricant continually
rises as more and more expensive applications are purchased in an attempt
to correct severe die problems; the result is often high-cost greases
dripping out of tote boxes and smearing over dies and presses in a colo-
ssal mess, while the real solution to the problem may be the ram speed
or alignment of the press, or die design considerations.

At one time, I had the responsibility for a stamping plant which
was spending a small fortune on a sulphur-based compound for use in a
high-speed press department. Evidence of its use was everywhere. Dies
came out of the presses coated yellow, as were the presses themselves.

The stuff seemed to permeate the whole plant. Even so, when we attempted
to eliminate its use, our die engineers and press personnel were dogmatic
in their insistence that the sulphur was a necessity in order to maintain
die life and assure high quality stampings. The battle went on for several years with no solution until a third force inadvertently interjected
itself into the fray. A phosphate coating plating system was installed
to meet new surface requirements on many of the stampings furnished by
the pressroom. Shortly after installed, the system became completely
inoperative, unable to phosphate coat in the prescribed manner. Plating
experts finally determined that the cause of the problem was the sulphur
die lubricant, which was contaminating the plating solutions. By necessity, sulphur was removed as an ingredient in the die lubricants. The
negative results in the pressroom were imperceptible.

In this case, and I think in most others, die engineers and pressroom
production personnel are not knowledgeable on the subject of drawing
compounds and lubricants, and they parrot the spiel of whichever salesman
catches their ear.

Many plants could generate significant savings in operating costs by
cancelling out all of their perssroom production lubricant purchase orders,
and re-establishing use and grade according to proven necessity. While
lubricants, to some extent, will still be required, the variety would
probably be greatly reduced, as would the amount and use. The results
would be cleaner plants, lower costs, and reduced headaches in the
Finishing Department.

Length, Width and Grain Direction: The Search for Alternatives

The pricing books published by the steel mills include a notation
that when cut length sheets of steel are specified or required within the
rolling direction parallel to one dimension, that dimension will be considered the length, and the other dimension the width, for the application
of dimension extras. The significance of this is important to cost-con-
scious steel buyers, since the cost of steel increases rapidly as the
width increases. For example, if the material ordered for our hypothetical
part, which has a thickness of .100 and is produced with hot-rolled steel,
was purchased at a width of over thirty inches, there would be no width
cost extra. The same material purchased between twenty-four to thirty
inches would have a width extra of twenty-five inches, it would have a
seventy-cent cost extra per hundred weight. The same material purchased
at ten inches wide which is the coil width coming to the press would have
a ninty-cent per hundred weight penalty. Many steel users buy the most
economical width and have ther material slit to press size other than at
the mill in order to reduce the width size penalty, but no matter where
the slitting is done, it is an expensive process; therefore, it is always
to the stamper's advantage to use the widest possible material consistent
with the most economical nesting layout. This option, however, is lost
when the grain direction becomes a prerequisite either by the product
designer or die maker. The product designer specifies that the grain in
the steel should run in one particular direction on the part in order to
establish part strength. The die designer may specify the same thing to
establish ease of draw or forming.

Often, neither is aware of the cost penalties inherent in such decisions.
Specifying grain flow may very often add considerable material usage to
the part in that it does not allow for freedom in nesting the part in
the blank layout. Often a part must be run out of a strip stock with no
nesting because of grain flow, that otherwise could be run off of sheet
with nested blanks at a considerable savings in material cost. Unfortunately, such a decision is often made without considering the cost
penalty involved.

Not too long ago, an automotive supplier found himself purchasing
steel two inches wide to produce a lever of that same width, but twenty-
nine inches long. The ideal stock purchase size, of course, would have
been to use the twenty-nine inch dimension as the coil width, but this
was not possible because of more than one dollar per hundred weight, plus
a sizable increase in press time resulting from having a feed progression
of twenty-nine inches rather than two inches. An inquiry as to the need
for grain flow specification revealed that this was necessitated to meet
strength requirements. This problem was resolved by designing strengthening beads into the part at no material cost penalty.

No one purposely designs a material penalty into the product. It
is the responsibility of the material engineer always to question the
necessity of grain flow specifications. Then the product or die engineer
is faced with the costs involved, he will very often determine that grain
flow specification is not that essential after all, or will find another
means of attaining the same result.

Steel Surface Condition: Hot-Rolled vs. Cold-Rolled

Mills will not supply hot-rolled steel or strip below the thickness
of .056. It is surprising, however, how often stampers will order cold-
rolled steel at heavier gauges in spite of the cost penalty. The reasons
given are the need for surface quality or stamping feasibility. A third
reason given is the closer tolerance that can be achieved on cold-rolled
steel. These are all logical reasons. But, in practice, they don't
always hold up. Specifying cold-rolled steel for stamping feasibility
is often a cop-out. The surface condition and thickness tolerance argu-
ments used to support the use of cold-rolled steel look better on paper
than in actual practice. For years the automtoive companies specified
cold-rolled steel for the manufacture of seat adjusters for just these
reasons, only to find that hot-rolled steel could produce a part equally
acceptable at less cost. A good material engineer who questions the
specification of cold-rolled steel at all times can achieve a substantial
cost savings for his operation with very little out-of-pocket costs.

We have mentioned here some of the hang-ups that we have found to
be expensive in the stamping business. You can probably think of others
in your own operation that should be challenged. A good ground rule for
a material engineer is to challenge every precept or principle, real or
imagined, that adds cost to the material he specifies. Every once in a
while he will find a tradition that must be honored, but more often he
will discover that the specified extra is based on a theory that does not
hold up when put to the test of actual practice.

Stampable Fiberglass Reinforced Plastic

Azdel sheet is light, tough, and temperature resistant.
Here's what you need to know about
part design, and the tooling and presses involved

Reprinted from Manufacturing Engineering, April 1979

ROBERT D. CAMPBELL
Azdel Product Group
Fiber Glass Div.
PPG Industries, Inc.

THE STAMPING OF PLASTICS is creating considerable interest worldwide because of advantages both in process and product. From the processing point of view, the rapid cycle times for stamping operations and the capability to handle complex shapes allow stamped thermoplastics to compete effectively with metal assemblies requiring multiple operations. In the plastics industry itself, thermoplastic stamping is one of the most productive processes for large, complex parts, such as that shown in *Figure* 1.

The pioneer material of this type in terms of technical development and actual production applications is generally recognized to be Azdel® thermoplastic sheet, a product of PPG Industries' Fiber Glass Div. This material has been used in a variety of unusual and demanding applications, including automotive seats, retainers supporting the elastomeric or "soft" fronts of several car models, battery trays, tractor instrument panels, musical instrument cases, and rotary mower shroud bases, *Figure* 2.

Advantages. Made from 40% continuous fiberglass strands in a polypropylene matrix, Azdel sheet is the only molding material in which the glass fibers flow considerable distances in the mold, yet retain essentially isotropic part properties. The length of the reinforcements and efficient coupling of glass and resin result in properties considerably higher than similar grades of short glass reinforced polypropylene molding compounds and other more expensive engineering resins. Primary advantages of the material include light weight, toughness, and resistance to temperature effects.

With a specific gravity of only 1.19, finished parts weigh 30% less than common reinforced polyester compounds, and offer weight savings of

1. THE STAMPING PROCESS for Azdel thermoplastic sheet provides high productivity in turning out large, complex parts.

Type of Material	Cost Comparison (% of Azdel Cost)		Flex Strength (psi x 10³)	Flex Modulus (psi x 10⁵)	Notched IZOD Impact (ft-lb/in.)	Heat Distortion Temperature at 264 psi (F°)	Specific Gravity	Weight lb/in.³
	Per lb	Per in.³						
Azdel P-100 Series stampable sheet	100	100	22	0.8	10.0	310	1.19	0.0429
G.R.P.P.-Inj. molding (30% glass)	81	70	9	0.5	1.3	275	1.12	0.0404
G.R. Nylon-stampable	115	135	23	0.8	0.8	390	1.37	0.0495
Nylon-Inj. molding (30% glass)	185	248	30	1.1	1.6	410	1.59	0.0574
Polycarbonate-Inj. molding (10% glass)	84	210	19	0.8	2.0	288	1.35	0.0487
Polycarbonate structural foam (unreinforced)	128	91	10	0.3	—	270	0.80	0.0289
Sheet molding compound (25-30% glass)	57	86	25	2.0	11.0	450	1.80	0.0650
Aluminum	65	151	—	105.0	—	—	2.76	0.0996
Steel	30	201	—	300.0	—	—	7.83	0.2827
Zinc die casting	47	262	—	130.0	—	—	6.60	0.2383

more than 50% compared to sheet metal. In terms of toughness, continuous fiberglass reinforced polypropylene has the capacity to withstand impact forces which can permanently deform sheet metal or crack diecast parts. In addition, this unique thermoplastic sheet remains stable at temperature extremes from −40°F (−40°C) up to +280°F (+138°C). A comparison of Azdel sheet properties with those of other materials is given in the accompanying table.

Advantages also exist from the fabricator's standpoint, particularly in terms of parts consolidation. A single reinforced plastic design can often replace a metal assembly made from multiple separate parts.

An analysis of stamped steel versus molded Azdel sheet on an automotive front fascia retainer showed that the steel parts for the component would have required 27 separate machine operations. The Azdel sheet part required one 50-second molding cycle and two secondary operations for hole piercing. Despite the faster steel stamping rates, total press hours required for a one-piece Azdel sheet design were only 40% of those required for the multipiece steel alternative. In addition, dies for the hydraulic and mechanical stamping of this part cost considerably less than dies required for the metal stampings.

Processing Steps. The thermoplastic material is produced continuously in a sheet 48″ (1219 mm) wide and 0.150″ (3.81 mm) thick. The sheet is sheared and slit in-line into blanks as specified by the fabricator so that the weight and area requirements of his production parts are met with blanks cut and ready for preheating.

As a material, Azdel thermoplastic is specifically designed for processing by flow forming in the melt phase. In the actual molding process, precut blanks are heated to near 400°F (204°C) in an infrared oven. The hot blanks are then transferred to water-cooled, matched metal dies, *Figure* 3, in a hydraulic or mechanical stamping press. The material is molded at pressures in the range of 1400 to 2000 psi (9653 to 13 790 kPa), depending on flow geometry. Unlike metal stamping, where the part is shaped by deformation, Azdel sheet stamping involves flow forming to produce complex shapes.

Ovens best suited for heating the sheet are the infrared type consisting of opposing banks of heaters in a sheet metal enclosure. Drafts inside the oven should be avoided. Radiant heat dissipation, however, should be allowed to occur. The sheet is best heated by absorption of infrared radiation, not ambient heat in the oven.

Blanks are heated from both sides as they move through the oven. Since heating takes longer than the stamping cycle, blanks should be processed through a multiple zone oven. The oven should be designed and run to be slave to the molding operation. Normal heating time for Azdel sheet is 90 to 120 seconds. An oven with three heater zones should support a 30 to 40-second part cycle time.

When the blanks exit the oven they are two to three times their original thickness, and are limp and flexible like a thick wool blanket. Although the resin is molten, the glass fiber rein-

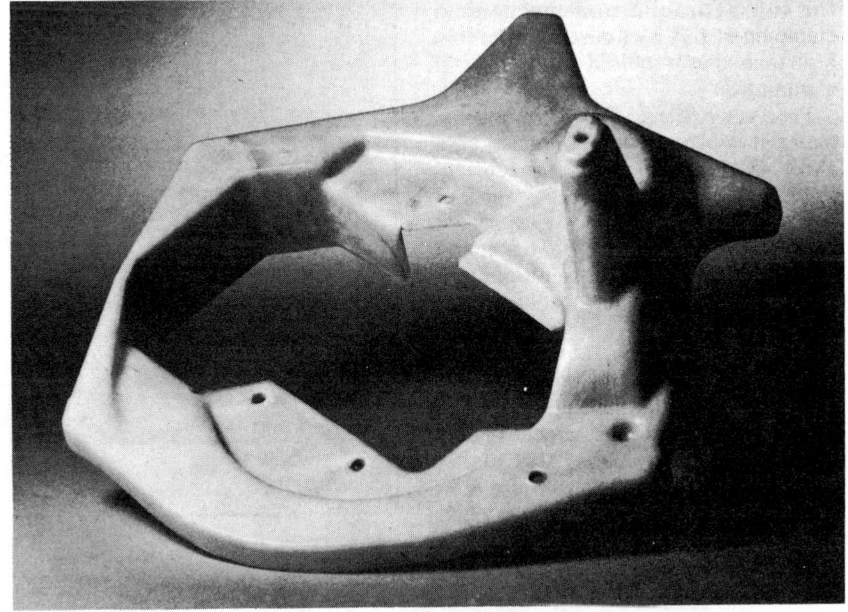

2. THIS SHROUD BASE for a power mower is stamped in one piece from a fiberglass-reinforced thermoplastic sheet.

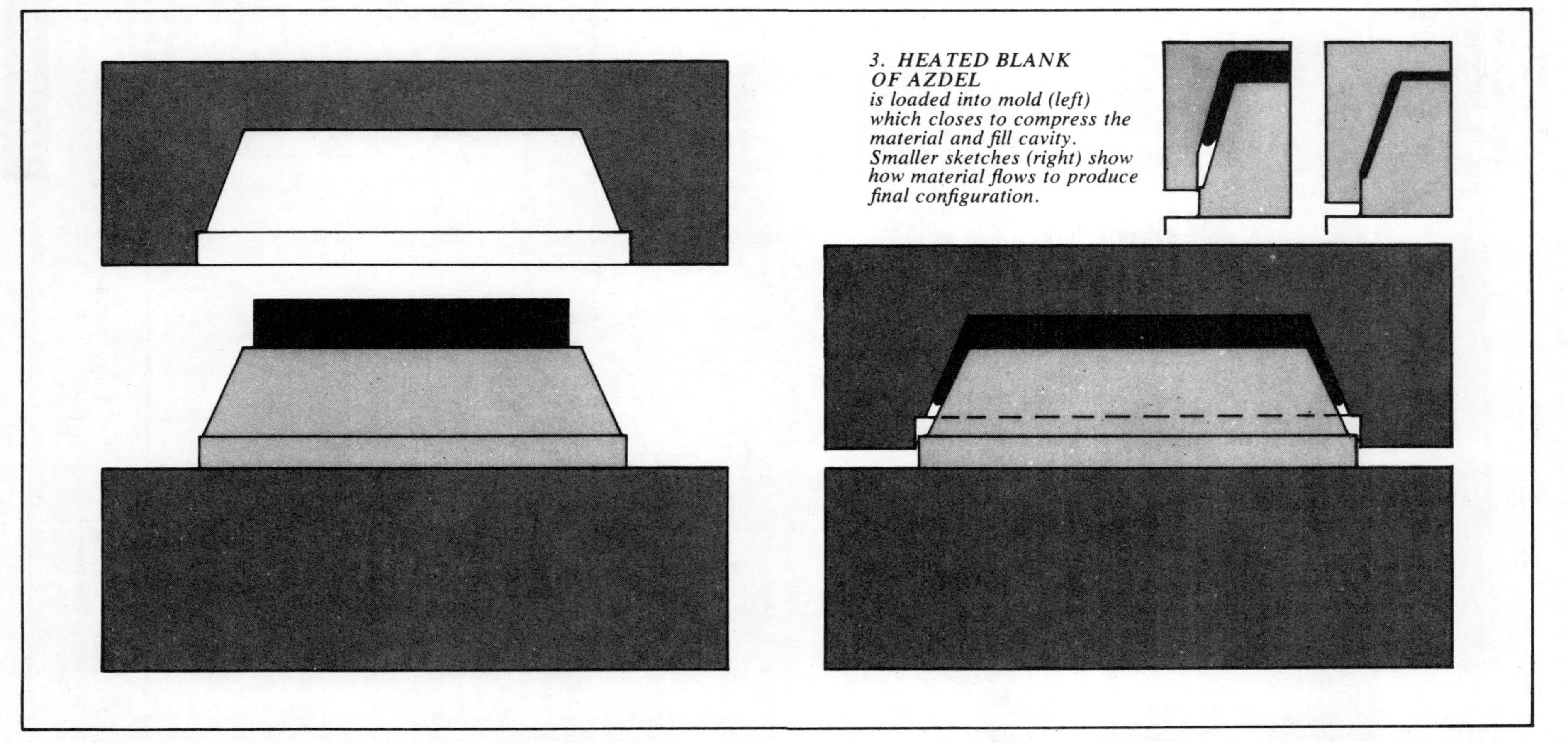

3. HEATED BLANK
OF AZDEL
is loaded into mold (left)
which closes to compress the
material and fill cavity.
Smaller sketches (right) show
how material flows to produce
final configuration.

forcement prevents the blanks from dripping or pulling apart.

Tooling. A prime ingredient for trouble-free production of any Azdel sheet part is a well-designed and constructed die. Each tool has a unique, optimum design, although certain standard features are common to almost all of the dies. The dies combine features from sheet metal dies, injection molds, and compression molds. In principle, the Azdel sheet die is a compression mold equipped with vertical telescoping shutoff (sealing) edges, *Figure* 4, and designed with a heavy duty guidance system.

Azdel sheet dies differ from injection and matched metal molds in two ways. First, injection molds are guided to a large degree by the machine. An Azdel sheet mold needs substantial built-in guidance because it closes onto the material, which flows with the increase in viscosity as the tool touches it.

Second, a mold is a constant volume device. It is either closing against stops, or forms a closed, fixed cavity by virtue of metal-to-metal shutoff. Azdel sheet dies are volumetrically variable — they depend upon the material charge for their part volume and should not be built with stops to control part thickness. The hot Azdel material must be the only stop the die encounters while closed. Stop blocks on these dies are

safety devices only to prevent the two die halves from touching when no material is in the tool. All safety pins, punches, and shutoff devices must follow the same rules.

Like injection molds, Azdel sheet dies are water cooled. Normal mold temperature is in the range of 80 to 140°F (27 to 60°C).

Since the material is not abrasive as it flows, it is not mandatory to harden the entire plug or cavity block. However, the telescoping shutoff edges should always be heat treated to a range of R_C55 to 60 to prevent damage and galling. Azdel sheet dies need only conventional tool steels, preferably those that can be selectively flame hardened. The following AISI steels have given excellent results: P20, 1060, 4140, 6150, D2, and A2.

The telescoping mold seals at the edges before pressure is applied to the sheet blank. For a flash-free part, these edges should be hardened to R_C55 to 60 and fit to a clearance of 0.001 to 0.002" (0.02 to 0.05 mm). The edges seal off the flow of hot material only and are not used to align the die halves or cut any of the hot sheet. As shown in *Figure* 4, allow ½" (12.7 mm) minimum engagement at part thickness. Most important, the edges must be vertical.

Mold Surface Finish. Part design, function, and secondary finishing re-

quirements determine mold surface finish. The higher polished surface promotes material flow and part removal. It is recommended that most Azdel sheet molds be polished to at least an SPI/SPE No. 3 finish, although parts have been successfully molded on rougher finished and grained surfaces. Painted parts sometimes require a matte surface finish obtained by vapor etching the mold.

Chrome plating is frequently used to protect the mold from corrosion. Molds with grained patterns often have a hard flash chrome finish to prevent washout. As many as one million parts have been produced on an unhardened mold with no plating, and very little wear could be detected. Obviously, Azdel sheet is not an abrasive material.

Sometimes trapped air in a part will cause rejects for surface finish, distortion, and unfilled areas. With deep blind ribs, air entrapment often causes diesel burning. Thus, most Azdel sheet molds are vented in some manner. Standard injection mold venting techniques apply. It is common to put air-escape grooves 0.001 to 0.003" (0.02 to 0.08 mm) deep on all ejector pins and along the telescoping shutoff edge.

Standard injection-molding ejector systems and components are used, but caution must be taken about pin placement. Although the Azdel sheet part is

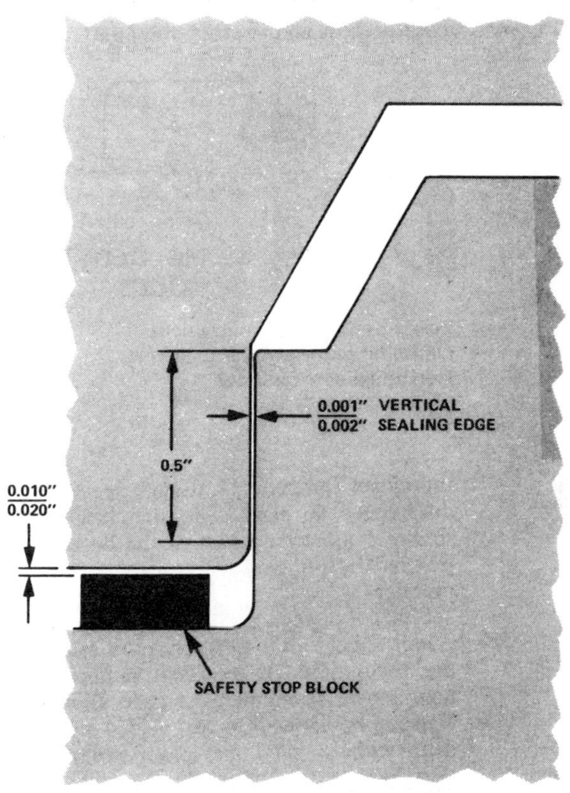

4. TELESCOPING MOLD
seals at the edges before pressure
is applied to the thermoplastic sheet.
Hardened edges are fit
to 0.001 to 0.002" clearance.

solidified when ejected, it is still somewhat warm and soft in some of the thicker sections. Pins located under these points can cause part distortion.

All vertical surfaces should have draft. This draft can be held to a minimum if the design demands it, but the reduction in taper causes more restricted processing techniques and could become a source of molding problems. Draft angles of 1° are suggested for all walls to 3″ (76.2 mm) in length and 2 to 3° for anything longer. An additional 1° of draft is recommended for each 0.001″ (0.02 mm) of depth for a textured surface.

Part Design. Design considerations for Azdel thermoplastic sheet parts are similar to those for compression-molded are possible in most cases, and metal inserts, including edge stiffeners, have been successfully molded in. In addition, various mechanical fasteners, such as self-tapping screws and rivets, can be inserted into a part during postmold operations.

Much of the increased design freedom with Azdel sheet can be attributed to the crack and craze insensitivity of polypropylene as compared to polyester. Polypropylene is also a natural release surface.

While maximum wall thickness is unlimited, the minimum practical wall thickness is 0.060″ (1.52 mm) on horizontal surfaces and 0.070″ (1.78 mm) for vertical surfaces — to a maximum depth of 3″ (76.2 mm). Undercuts are not possible without complex tooling. press. This design is such that the strain or load is imposed on the frame vertically, and the frame is thus capable of resisting any normal press loading without excessive deflection. Gap frame or open-back inclinable presses are not applicable for Azdel sheet because of the angular deflection incurred under load.

Most Azdel sheet work on mechanical presses has been with crankshaft-type press drives, although eccentric gear, eccentric shaft, and rocker arm press drives also have been used successfully. With crankshaft press drives, the nongeared type is not recommended.

Stopping the press at bottom dead center (BDC) is the heart of Azdel

5. DESIGN GUIDELINES

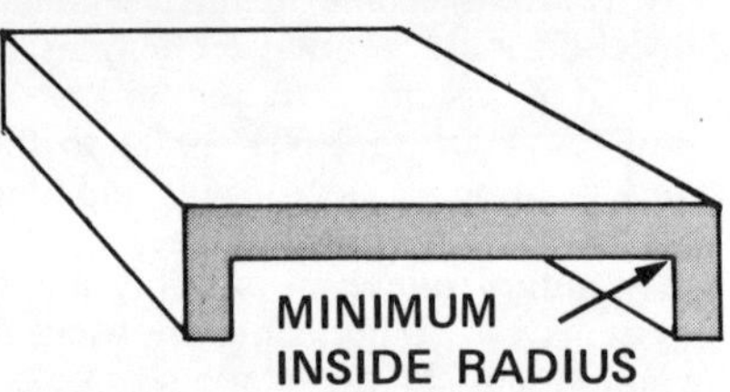

MINIMUM INSIDE RADIUS

No minimum. Azdel parts with square corners have been molded. Generous radii are recommended to assist material flow and to strengthen part.

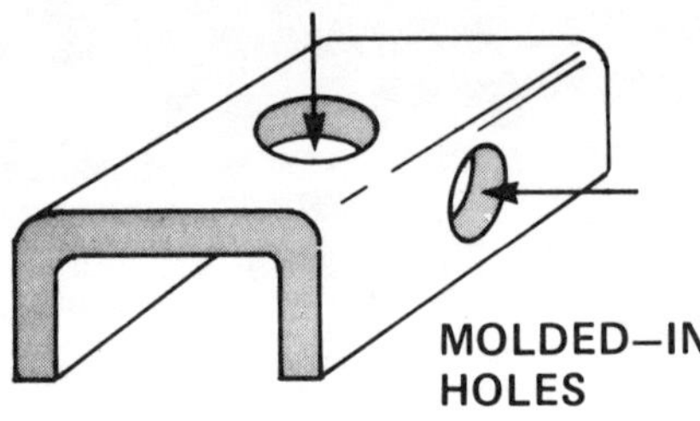

MOLDED—IN HOLES

Yes, if parallel to die movement. Limited if perpendicular to stroke. Horizontal core pulls are not recommended.

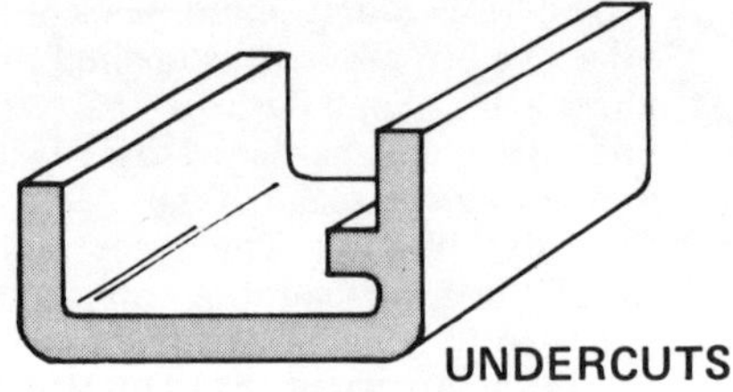

UNDERCUTS

Yes, but limited in depth.

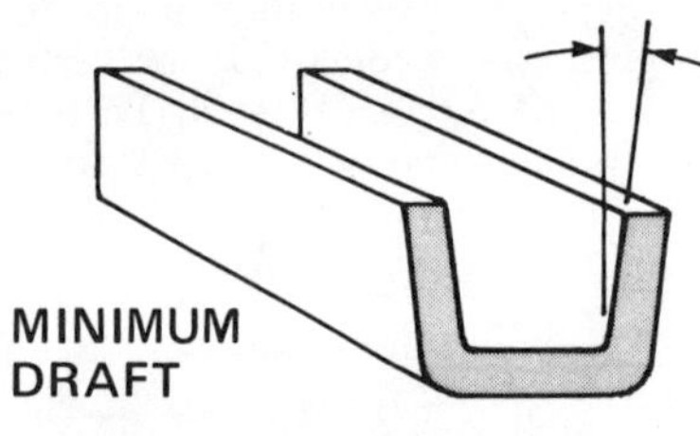

MINIMUM DRAFT

1/4″ - 3″ Depth: 1/2° - 1°
Over 3″ Depth : 1° - 3°

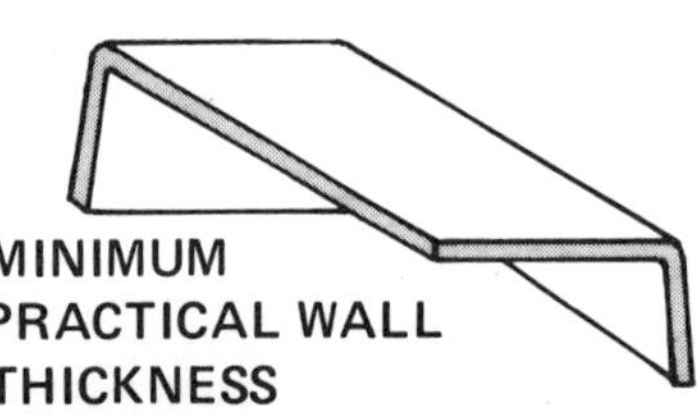

MINIMUM PRACTICAL WALL THICKNESS

0.060″ Horizontal
0.070″ Vertical to 3″ maximum depth

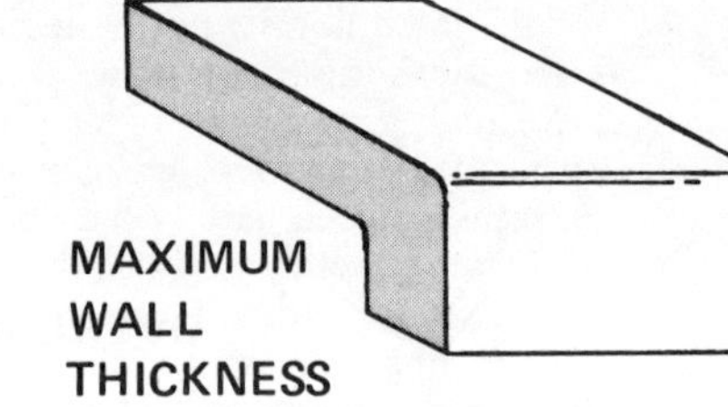

MAXIMUM WALL THICKNESS

Unlimited

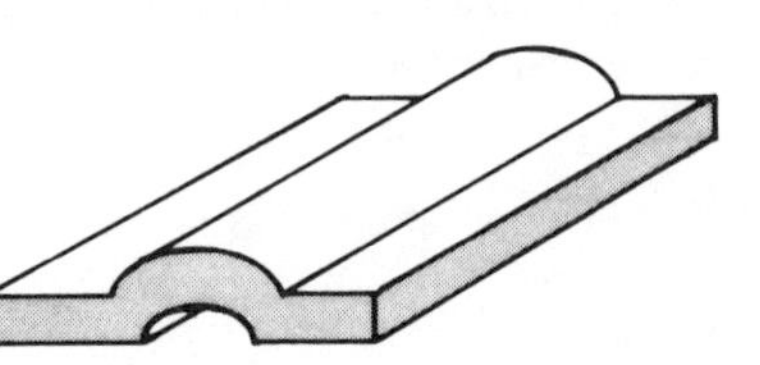

CORRUGATED SECTIONS

Yes

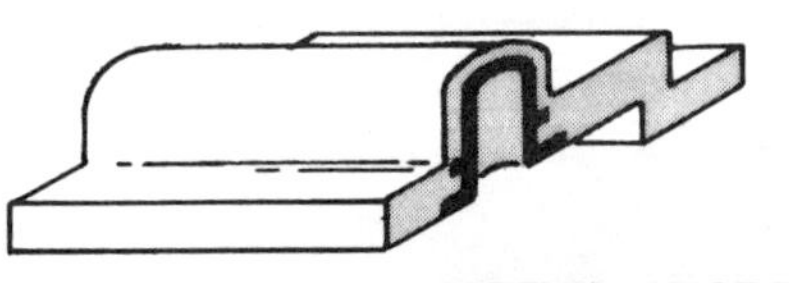

METAL INSERTS

Yes. Can be molded into part or installed later.

thermoset components, though specific differences do exist. The sketches in *Figure* 5 illustrate some of the design possibilities and recommended practices.

Sharp corners can be molded in Azdel sheet, although generous radii are recommended to assist material flow and to strengthen the part. Blind ribs can be formed up to 1″ (25.4 mm) deep without glass and resin separation. Bosses can be formed as small as ⅛″ (3.17 mm) in diameter and ½″ (12.7 mm) high. However, large bosses should be tied to a wall or rib for strength and for proper mold filling.

Molded-in holes in the line of draw sible without complex tooling.

Because Azdel sheet is formed on cool dies, other materials, such as vinyls (with a fabric backing) and carpet facings, can be placed in the die to produce parts from combinations of materials. An example is the compartment doors on the lower rear deck inside one the newer car models. These are made from Azdel sheet, a textured ABS plastic ring, and a die-cut carpet facing.

Presses Used. Azdel sheet parts can readily be formed at high production rates using conventional metalworking equipment. A mechanical straight-side press is similar in frame design to a closed-side, housing-type hydraulic sheet stamping and the major difference from metal stamping. Metal stampers, of course, try to avoid stopping the press at BDC. Deliberately doing this, however, as is done with Azdel sheet, entails a different concept. When the press is set to stop on BDC to mold an Azdel sheet part, it is not locked on bottom as would be the case with a double hit in sheet metal stamping. An adjustable position limit switch is used to stop the press.

It is set to be activated, usually by the moving platen, near the bottom of the stroke. When this switch is triggered, it causes the clutch to be disengaged and the brake to be applied. The switch should be wired into an adjustable timer which is used to open the press automatically after a predetermined time interval. When the press closes on the blanks and fills the mold, the hot sheet becomes an effective shock absorber. Thus, the forming energy used to flow and compress the plastic becomes, for the most part, the brake.

The greatest concern when using mechanical presses is the ability to stop the press at the same place every time. As a rule of thumb, a press with a 20″ (508-mm) stroke should stop within ±2° of BDC. In general, the mechanical press should have a pneumatically or hydraulically operated clutch and brake, a housing-type frame, tight guidance of

rapid approach to the material without contacting it at this speed. An adjustable position switch is used to change from this fast advance to an intermediate and/or pressing speed.

This switch is set so the press changes to its slower pressing speed about 1″ (25.4 mm) before contacting the hot sheet. When the moving half of the mold contacts the material, it must flow the material in one continuous and smooth motion to fill up the cavity.

This swift and continuous motion is important to successful molding of Azdel sheet. The moving mold half should never be allowed to stop and hesitate on the hot material while the speed change takes place. Once material movement starts, it should be com-

cases where only a few simple parts were needed.

The basic material for prototyping of Azdel sheet parts is Kirksite, a zinc-based alloy containing aluminum, copper, and magnesium. It is easily ground, takes a brilliant polish, machines readily, and is easily welded. Cast Kirksite also reproduces pattern detail very well, making it suitable for a grained surface mold. It is relatively inexpensive and can be reprocessed into other molds or reclaimed.

Postmolding Operations. Azdel sheet parts are easily adapted to most cutting and assembly techniques. On low-volume runs, steel rule dies have been used to cut the sheet stock with excellent results. Standard sheet metal tool-

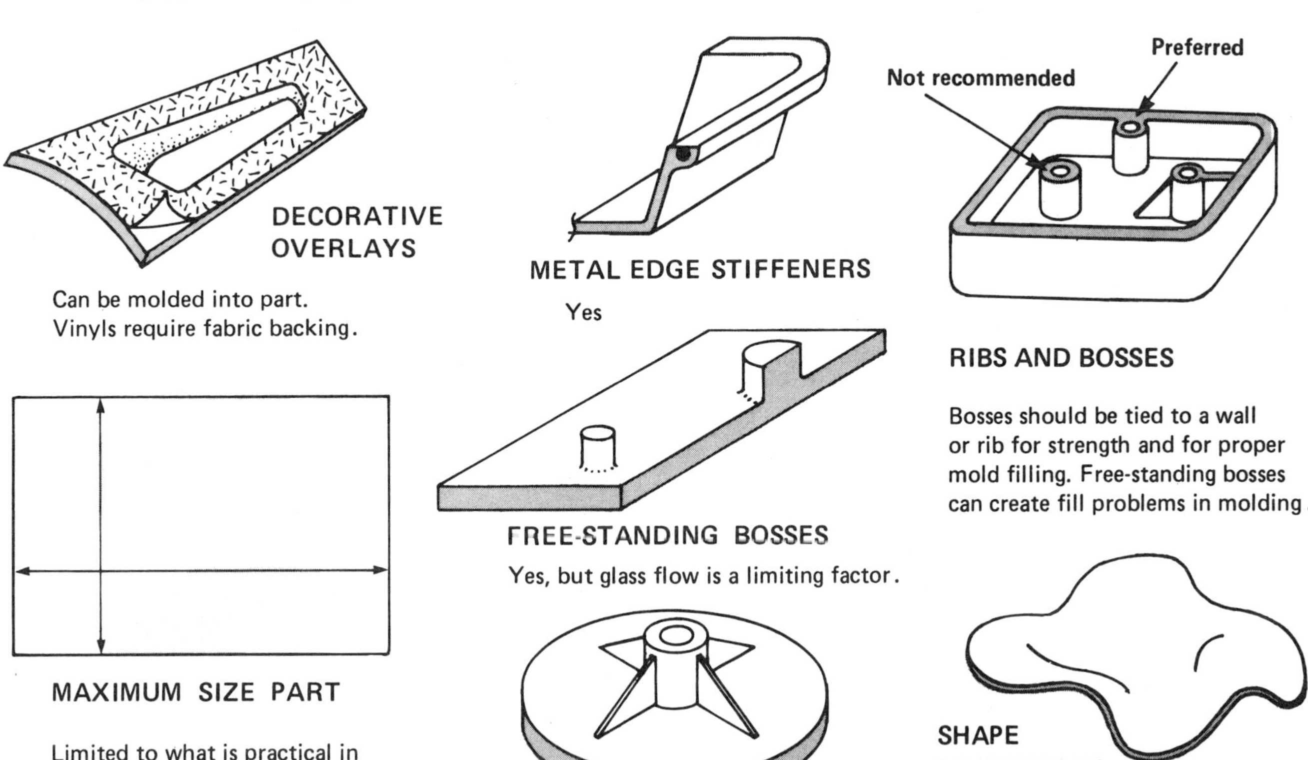

DECORATIVE OVERLAYS

Can be molded into part.
Vinyls require fabric backing.

METAL EDGE STIFFENERS

Yes

RIBS AND BOSSES

Bosses should be tied to a wall or rib for strength and for proper mold filling. Free-standing bosses can create fill problems in molding.

FREE-STANDING BOSSES

Yes, but glass flow is a limiting factor.

MAXIMUM SIZE PART

Limited to what is practical in terms of processing—oven size and ability to handle heated blanks.

RIBS AND CORED HOLES

Yes, but glass flow is a limiting factor.

SHAPE LIMITATIONS

Compression moldable

the slide, and a tight crankshaft bushing.

Hydraulic presses are best suited for Azdel sheet forming by virtue of their versatility, adjustment capabilities, and ram movement features. Generally, they are less expensive than mechanical presses, less time is required to set a mold in them, and no ram adjustment is required. They can be designed to open slowly during breakaway and then rapidly lift fully open. They have better ram guidance and thus are less susceptible to side thrust movement.

Forming Azdel sheet parts on hydraulic presses is somewhat different than on mechanical presses. An automatic slow-down valve allows a very

pleted in about three seconds, or less. After the mold cavity fills out and ram movement stops, a preset timer delays opening of the press while the part solidifies. When the press opens, it has a slow initial movement, then a rapid speed to full open. Both top and bottom drive presses are acceptable.

Prototype Parts. It is often desirable to produce prototype parts for part evaluation and/or tooling and processing investigation. A variety of materials is used today in the production of tooling for short runs of parts. Those used on Azdel sheet have been Kirksite, aluminum, wood, and epoxy. Wood and epoxy have been used in very special

ing is used to blank, pierce, cut, drill, or tap the sheet. When piercing holes, the punch should be sized to the high limit dimensions, since pierced holes tend to close slightly. Azdel sheet has a shear strength of 9000 psi (62.05 MPa).

From the materials engineering standpoint, stamped fiberglass reinforced components offer economical structural performance and light weight. Continuous glass reinforcement provides improved toughness and extended temperature ranges over equivalent molding materials. As a result, designs can be simplified to provide superior performance with fewer separate components. ■

CHAPTER 2

COST ESTIMATING

Die Estimating

By Jack Bradley

PREFACE

Production tooling is one of the most difficult of all industrial operations
to standardize and evaluate for cost estimating.

The fact that die building demands a high degree of accuracy; is seldom
repetitive, rarely more than one of a design; requires highly skilled
labor to perform the work involved; it is impossible to standardize
enough of the work to assign elemental time factors. However, if we
group the operations required to machine or assemble the die components,
standard data factors and guides for construction can be used effectively.

The basic tool feature classification charts included in this system will
provide a basis to determine the various die components required as well as
the unit hour and cost factors assigned to specific types of dies.

Estimating the time required for manufacturing a production die is facilitated by actual experience in the making, plus the ability to mentally
visualize the machine and handwork necessary for construction. An estimator
with these qualifications will be able to easily adapt the charts and tables
that follow.

Estimating factors shown in the charts are based on actual machining and
handwork studies. As a result, die build sources must be rated and the
factors adjusted to satisfy the differences in efficiencies between them.
The widely extended scope of tooling requires that this system of die
estimating embrace only the generally accepted methods of manufacture and
tool complexity.

The "Basic Die" classification pictures shown in this guide serve to
illustrate the components considered to be standard for general use. Basic
Die unit/hour factors are established for these general construction
designs. Any deviation from the design concepts will necessitate
additional unit/hour factors for the specific components required to augment
the design. Specific features and components are listed on one of the
following pages.

Unfortunately, most tooling-cost estimates are required during the product
concept phase of design, long before tool designs are finalized. For this
reason, the system is based on part print information and past experience.

A rough sketch should precede every estimate. Without a complete design,
the sketch gives the estimator a better "feel for the job" and brings

attention to most of the components required in the die.

Elements having the greatest influence on the estimate must be given first
consideration. Using the outline of basic features, in the order shown
below, will encourage a more efficient appraisal of elements required for
specific die types.

    Basic Die Classification

    Blank Complexity
        a. Punch
        b. Stripper
        c. Shedder
        d. Template

    Form Complexity

    Development Factor

    Die Components

    Squaring-Up

    Design Time

    Material Cost

<u>Basic Die Classification</u>. The estimator must determine from a part print,
the type of die considered to be the most feasible to produce the part
(Pierce die - compound, etc.).

The basis for an estimate can be established by selecting one of the tools
shown on the Basic Die Classification pictures.

All of the components shown in the picture have been evaluated for machining
time on the basis of being the largest practical for a given size.

<u>Summary Sheet</u>. Once the estimator is aware of the components in the Basic
Die, he can use the "Basic Die Summary Sheet" (Chart #1) to establish unit
hours and material cost without further investigation.

<u>Blank Complexity</u>. In addition to Basic Die values, the time for machining
pertinent features of the blank must be calculated using the "Blank
Complexity" (Chart #2). The time for machining the blank punch, stripper
opening and compound die shedders can also be determined by using a
percentage of the Blank Complexity factor.

<u>Form Complexity</u>. To complete an estimate for a form die, time for machining
the form die punch and die block can be derived from the predetermined
values listed for each type of form, as shown on the "Form Complexity"
(Chart 3).

<u>Development Time</u>. Part material and the degree of forming accuracy will
influence an add-to factor for development time. See Chart #4 entitled

"Development Time for Formed Angles."

<u>Die Components</u>. Upon determination of the tooling requirements, it may be
necessary to revise the Basic Die design concept in order to manufacture
the part to engineering drawing. Where additional time for added
components is necessary the estimator may select from the predetermined
"Component List" the time factors needed to augment the Basic Die (see
Chart #5).

<u>Squaring-Up</u>. Time required to machine components not included in the
Basic Die or Component List may be calculated using the predetermined
factors on the "Squaring-Up" Chart, #6.

<u>Design Time</u>. "Design Time" is based on a percentage of the tool build
estimate. Average value would be 30 percent. In some areas of tool
classification, this may not be valid, but must be left to the discretion
of the estimator.

<u>Material Cost</u>. Material cost factors used in this estimating system were
established on a $4.00 per pound figure for all steels. This price is
high by plan to offset the cost of such things as screws and dowels that do
not get special consideration.

Die Set prices are considered to be average for a commercial all-steel
shoulder bushing type set.

> Note: Elemental time factors assigned to machining operations
> are based on the use of oil hardening steels. Therefore, when
> air hardening steels are used, an additional 25 percent must be
> added to the total esimtated machining hours.

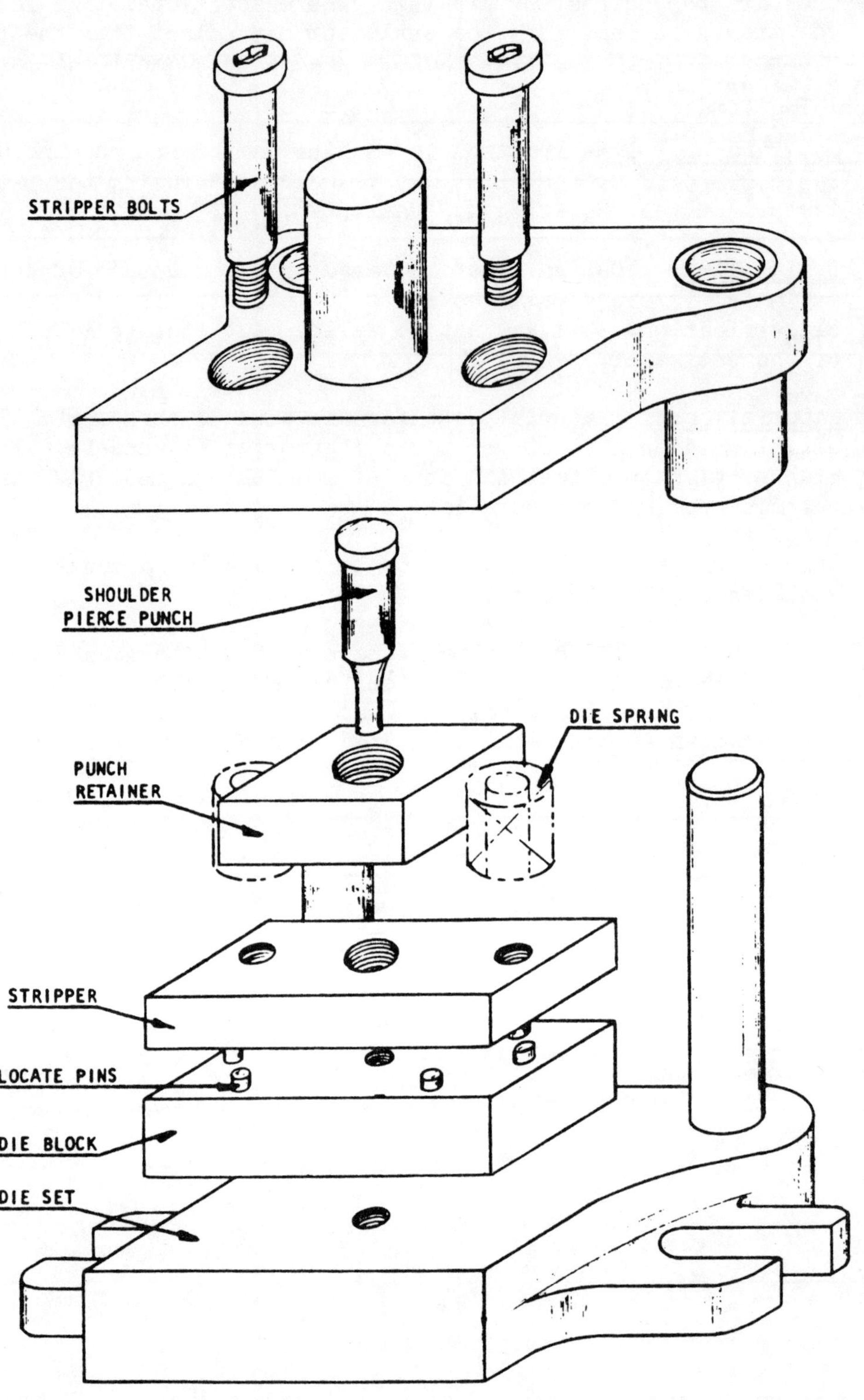

PIERCE OR BLANK

PIERCE DIE
UNIT HOURS & COST FACTORS

FEATURE	DIE SPACE			
	16 - 20 SQ. IN.	21 - 36 SQ. IN.	37 - 64 SQ. IN.	65 - 100 SQ. IN.
Die Block Machining Time	1.6	2.2	3.0	3.3
Punch Pad Machining Time (1/2 Die Area)	.8	1.1	1.5	1.9
Stripper Machining Time	1.4	2.0	2.6	3.6
Locating Pins (6)	6.0	6.0	6.0	6.0
Machining Clearance For Slug in Die Set	2.0	2.5	3.0	3.5
Die Down Stop	2.0	2.0	2.0	2.0
Stripper Bolt Drilling & Tapping	8.0	8.0	16.0	16.0
Drilling & Tapping For MT'G Holes	3.0	3.7	3.7	3.7
Toolmaker Time Ref. MAT-L Layout etc	2.0	2.0	4.0	4.0
Tryout	1.0	1.0	2.0	2.0
TOTAL, UNIT HOURS	27.8	30.5	43.8	46.5
+ Blank Complexity				
Punch Complexity				
Stripper Complexity				
Material Cost	$120.00	$200.00	$300.00	$400.00
Die Set Cost	$ 62.00	$ 74.00	$ 96.00	$135.00
Total Material Cost	$182.00	$274.00	$396.00	$535.00

BLANK DIE
UNIT HOURS & COST FACTORS

FEATURE	DIE SPACE			
	16 - 20 SQ. IN.	21 - 36 SQ. IN.	37 - 64 SQ. IN.	65 - 100 SQ. IN.
Die Block Machining Time	1.6	2.2	3.0	3.8
Punch Pad Machininig Time	.8	1.1	1.5	1.9
Blank Punch Machining Time	.8	1.1	1.5	1.9
Stripper Plate Machining Time	1.4	2.0	2.6	3.6
Guide Rails	4.0	6.0	10.0	16.0
Die Down Stop	2.0	2.0	2.0	2.0
Apron	2.5	3.0	3.5	4.0
Pull Back Stop	2.0	2.0	2.0	2.0
Drilling & Tapping For MT'G Holes	3.0	3.7	3.7	3.7
Toolmaker Time Ref. Layout, MAT'L etc.	2.0	2.0	4.0	4.0
Tryout	2.0	4.0	6.0	8.0
TOTAL, UNIT HOURS	22.1	29.1	39.8	50.9
+Blank Complexity				
Punch Complexity				
Stripper Complexity				
Material Cost	$132.00	$272.00	$420.00	$500.00
Die Set Cost	$ 62.00	$ 74.00	$ 96.00	$135.00
Total Material Cost	$194.00	$346.00	$486.00	$585.00

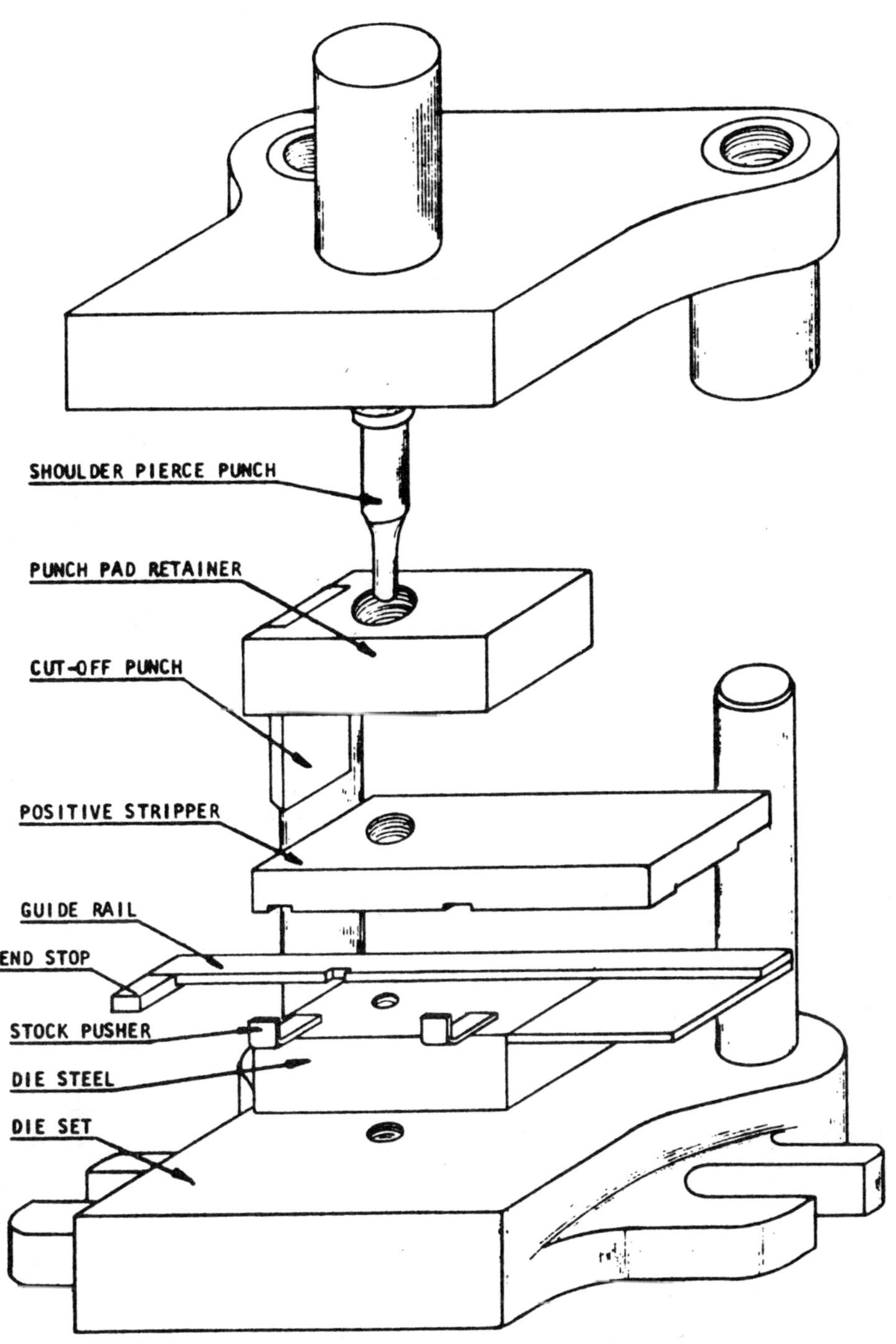

SHOULDER PIERCE PUNCH
PUNCH PAD RETAINER
CUT-OFF PUNCH
POSITIVE STRIPPER
GUIDE RAIL
END STOP
STOCK PUSHER
DIE STEEL
DIE SET

PIERCE & CUT-OFF DIE
UNIT HOURS & COST FACTORS

FEATURE	DIE SPACE			
	16 - 20 SQ. IN.	21 - 36 SQ. IN.	37 - 64 SQ. IN.	65 - 100 SQ. IN.
Die Block Machining Time	1.6	2.2	3.0	3.8
Punch Pad Machining Time	1.5	2.0	3.0	4.0
Stripper Machining Time	1.6	2.0	3.0	4.0
Cut-Off Punch Machining Time	2.2	2.4	3.5	4.5
Apron	2.5	3.0	3.5	4.0
Stock Guide Rails	2.0	3.5	3.8	4.0
End Stop	2.0	2.0	2.0	2.0
Stop (First-Stock)	3.0	3.0	3.0	3.0
Pushers 3.0 Each	6.0	6.0	6.0	6.0
Die Down Stop	2.0	2.0	2.0	2.0
Drilling & Tapping For MT'G Holes	3.0	3.5	4.0	4.5
Toolmaker Time Ref. Layout, MAT'L etc.	3.0	4.0	6.0	8.0
Tryout	3.0	3.5	5.0	6.0
TOTAL, UNIT HOURS	34.4	39.1	47.8	55.8
+ Blank Complexity				
Punch Complexity				
Stripper Complexity				
Material Cost	$130.00	$210.00	$395.00	$475.00
Die Set Cost	$ 62.00	$ 74.00	$ 96.00	$135.00
Total Material Cost	$192.00	$275.00	$491.00	$610.00

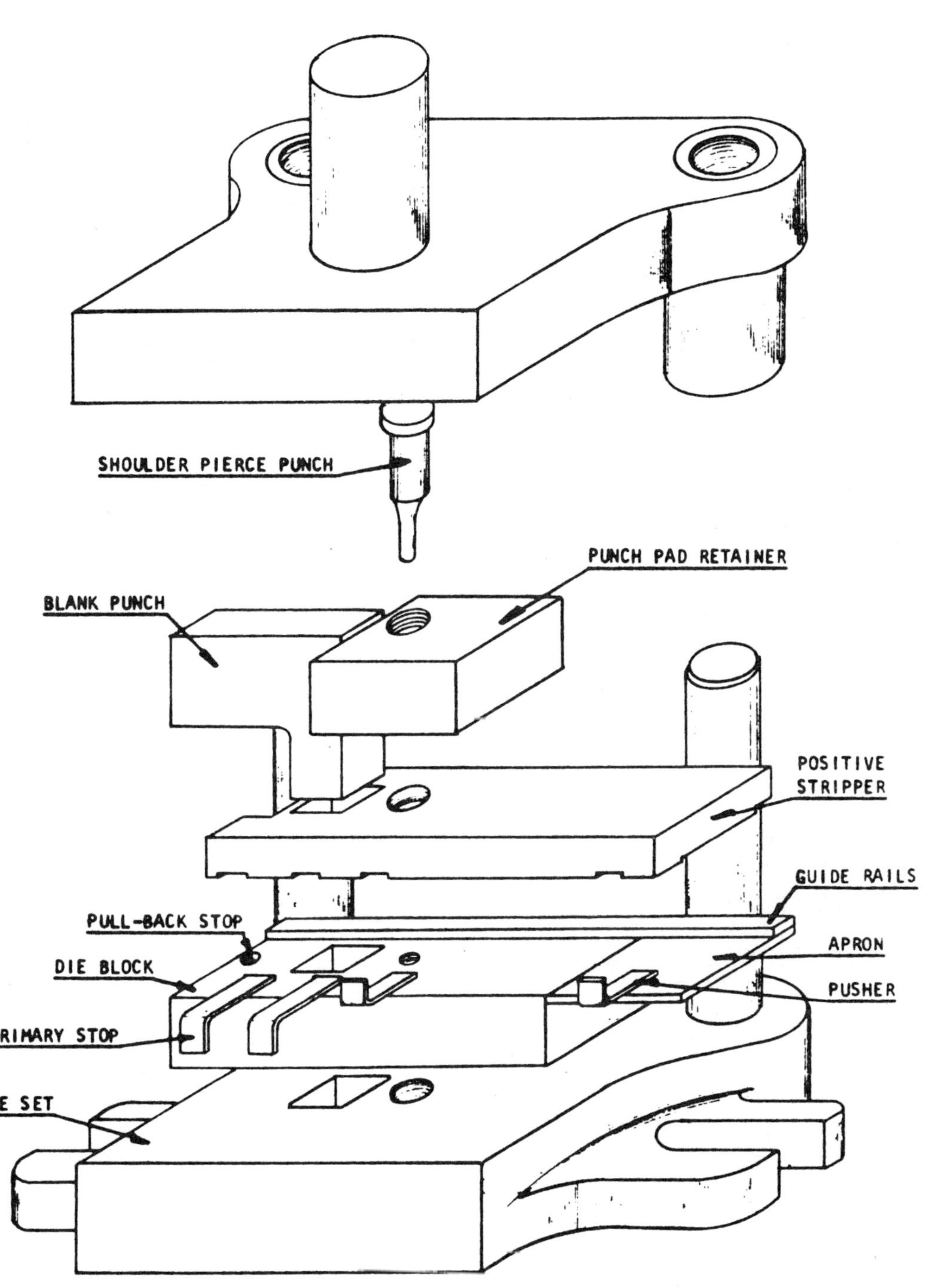

PIERCE AND BLANK THRU DIE

PIERCE & BLANK THRU DIE
UNIT HOURS & COST FACTORS

FEATURE	DIE SPACE			
	16 - 20 SQ. IN.	21 - 36 SQ. IN.	37 - 64 SQ. IN.	65 - 100 SQ. IN.
Die Block Machining Time	1.6	2.2	3.0	3.8
Punch Pad Machining Time	1.5	2.0	3.0	4.0
Stripper Machining Time	1.6	2.0	3.0	4.0
Apron	2.5	3.0	3.5	4.0
Guide Rails	4.0	6.0	10.0	16.0
Pull Back Stop Pin	2.0	2.0	2.0	2.0
Die Down Stops	2.0	2.0	2.0	2.0
Blank Punch Machining Time	1.2	1.5	2.2	4.2
Drilling & Tapping For MT'G Holes	3.0	3.5	4.0	4.5
Toolmaker Time Ref. MAT'L Layout etc.	3.0	4.0	6.0	8.0
Tryout	3.0	3.5	5.0	6.0
Stock Pushers 3.0 Each	6.0	6.0	6.0	6.0
Primary Stops	6.0	6.0	6.0	6.0
TOTAL, UNIT HOURS	37.4	43.7	56.1	70.5
+ Blank Complexity				
Punch Complexity				
Stripper Complexity				
Material Cost	$140.00	$230.00	$400.00	$495.00
Die Set Cost	$ 62.00	$74.00	$96.00	$135.00
Total	$202.00	$304.00	$496.00	$630.00

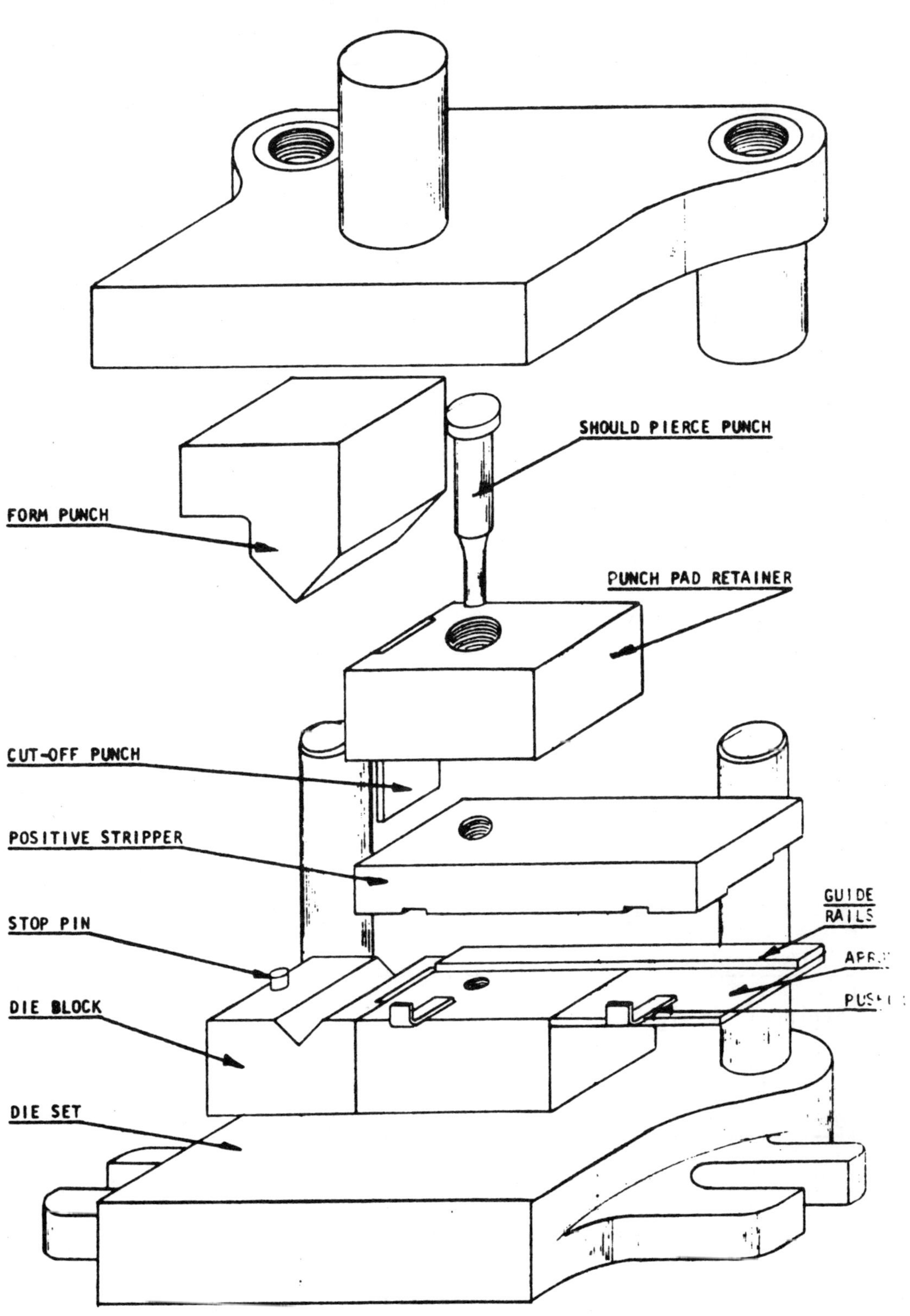

PIERCE, CUTOFF AND FORM

FEATURE	DIE SPACE			
	16 - 20 SQ. IN.	21 - 36 SQ. IN.	37 - 64 SQ. IN.	65 - 100 SQ. IN.
Die Block Machining Time	1.6	2.2	3.0	3.8
Punch Pad Machining Time	1.5	2.0	3.0	4.0
Stripper Machining Time	1.6	2.0	3.0	4.0
Apron	2.5	3.0	3.5	4.0
Guide Rails	4.0	6.0	10.0	16.0
End Stop Block	2.0	2.0	2.0	2.0
Die Down Stop	2.0	2.0	2.0	2.0
Stop Pushers 3.0 Each	6.0	6.0	6.0	6.0
First Stop	3.0	3.0	3.0	3.0
Tryout	6.0	7.0	8.5	10.0
Drilling & Tapping For MT'G Holes	6.0	6.5	7.0	8.0
Toolmaker & Time Ref. Layout, MAT'L etc.	3.0	4.0	6.0	8.0
Form Punch Machining Time	1.2	1.4	3.2	4.6
Cut-Off Punch	1.4	1.5	1.8	2.4
TOTAL, UNIT HOURS	41.8	48.6	62.0	77.8
+ Blank Complexity				
Punch Complexity				
Stripper Complexity				
Material Cost	$140.00	$230.00	$405.00	$500.00
Die Set Cost	$ 62.00	$ 74.00	$ 96.00	$135.00
Total Material Cost	$202.00	$304.00	$501.00	$635.00

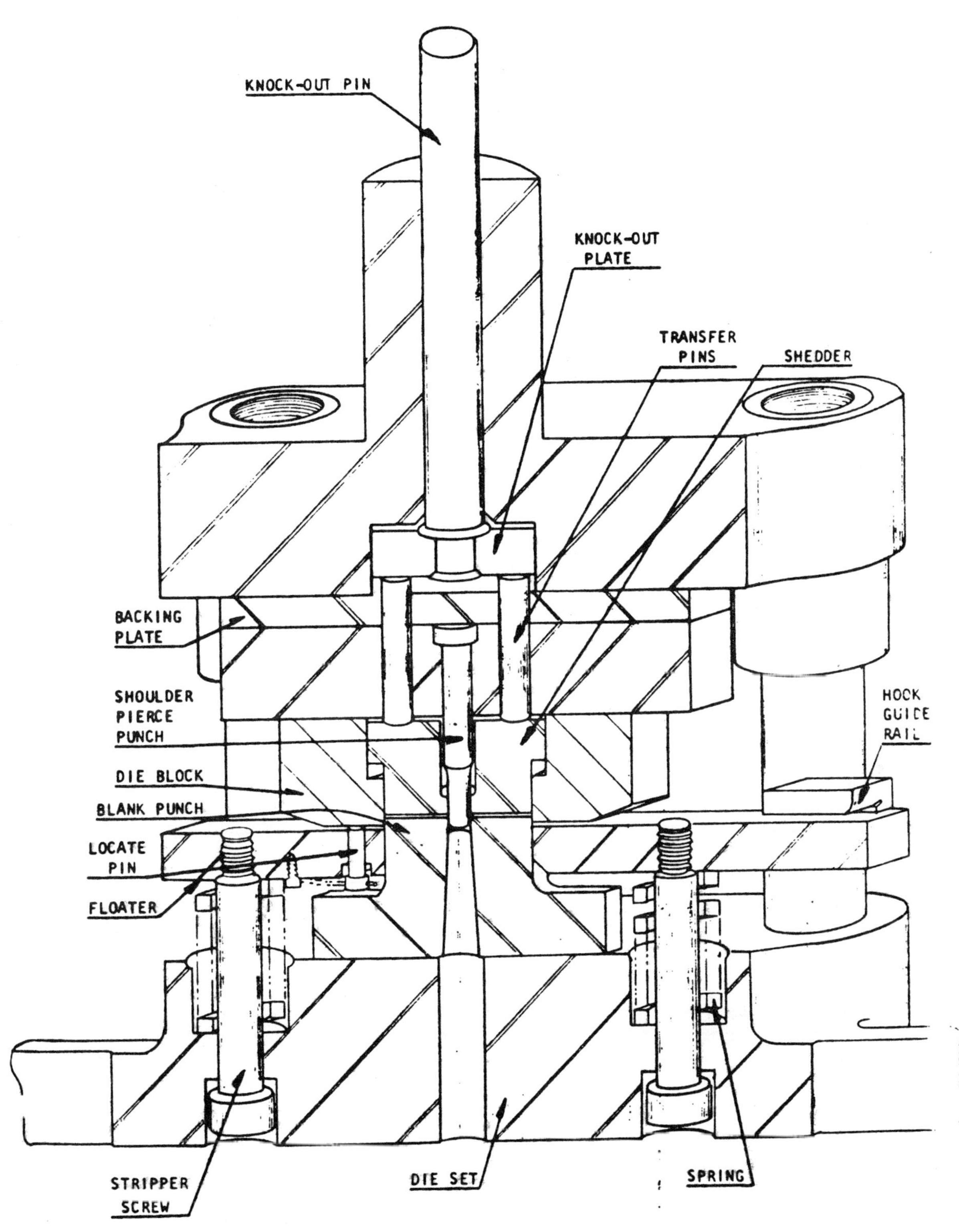

COMPOUND PIERCE AND BLANK

COMPOUND PIERCE & BLANK DIE
UNIT HOURS & COST FACTORS

FEATURE	DIE SPACE			
	16 - 20 SQ. IN.	21 - 36 SQ. IN.	37 - 64 SQ. IN.	65 - 100 SQ. IN.
Die Block Machining Time	2.0	2.5	3.8	4.5
Punch Pad Machining Time	1.5	2.0	3.0	4.0
Shedder Machining Time	.8	1.5	1.8	2.6
Floater (Stripper) Machining Time	1.6	2.0	3.0	4.0
Knock-out Machining Time	1.0	3.0	5.0	7.5
Transfer Pins (1) M. each	2.0	2.0	3.0	4.0
Die Down Stop	2.0	2.0	2.0	2.0
Guide Rails	2.0	2.0	4.0	6.0
Locate Pin (Pull Back Stop)	2.0	2.0	2.0	2.0
Stripper Bolts For Floater (Stripper) Control	8.0	16.0	16.0	16.0
Drilling & Tapping For MT'G Holes	4.0	4.0	6.0	6.0
Blank Punch Machining Time	1.2	1.5	2.2	4.2
Toolmaker Time Ref. Layout MAT'L etc.	12.0	14.0	16.0	18.0
Tryout	4.0	6.0	8.0	10.0
Backing Plate	1.4	1.6	2.2	3.0
TOTAL, UNIT HOURS	45.5	62.1	78.0	93.8
Blank Complexity				
Punch Complexity				
Stripper Complexity				
Shedder Complexity				
Material Cost	$144.00	$259.00	$460.00	$625.00
Die Set Cost	$ 65.00	$ 80.00	$114.00	$148.00
Total Material Cost	$209.00	$339.00	$574.00	$773.00

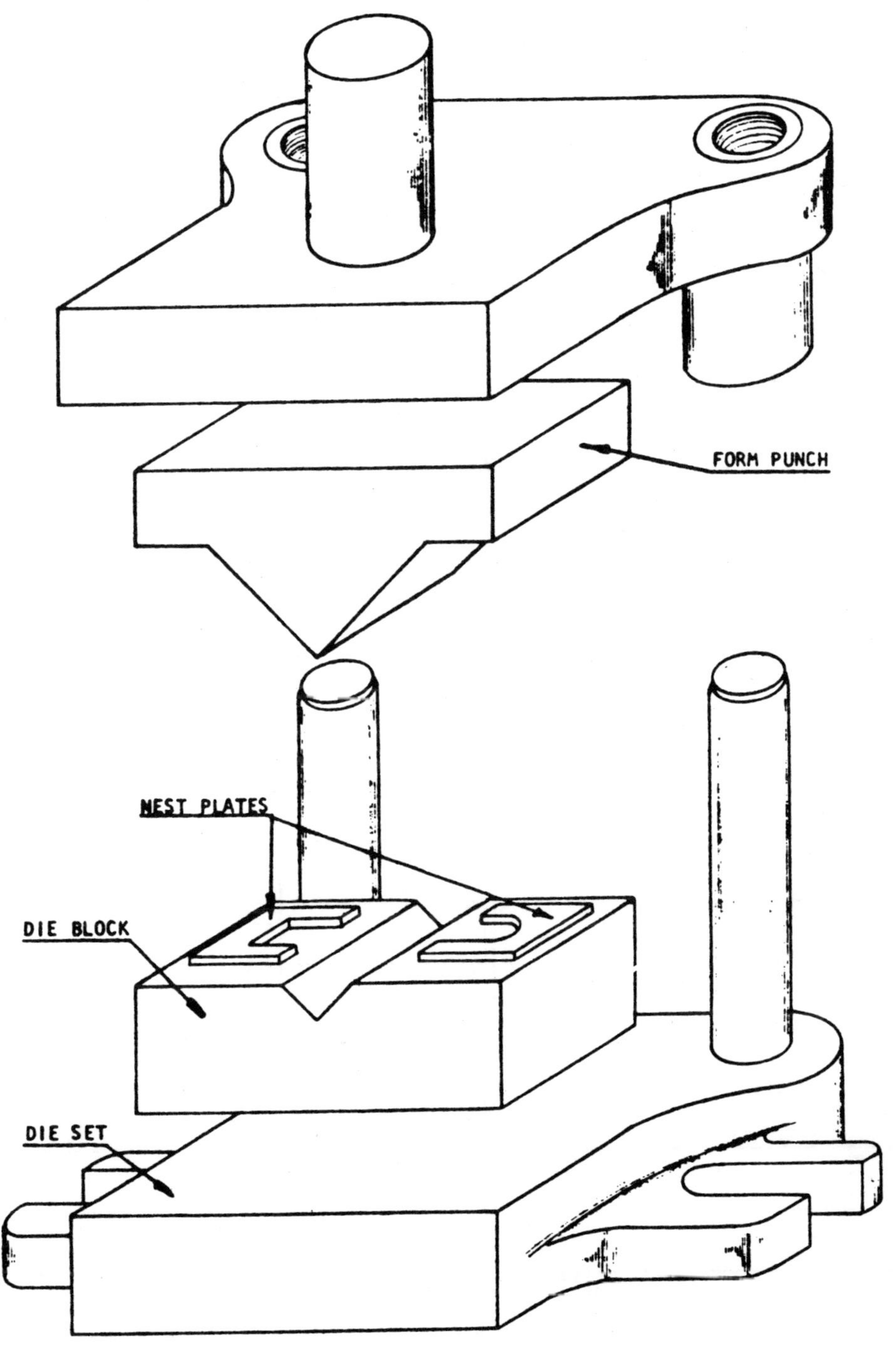

"V" FORM

"V" FORM DIE
UNIT HOURS & COST FACTORS

FEATURE	DIE SPACE			
	16 - 20 SQ. IN.	21 - 36 SQ. IN.	37 - 64 SQ. IN.	65 - 100 SQ. IN.
Die Block Machining Time	1.8	2.4	3.8	4.8
Form Punch Machining Time	1.6	2.4	3.8	4.8
Nest Plate Machining Time	3.0	4.0	6.0	8.0
Die Down Stop	2.0	2.0	2.0	2.0
Drilling & Tapping MT'G Holes	5.0	5.0	6.0	6.0
Toolmaker Time Ref. Layout, MAT'L etc.	2.0	2.0	4.0	4.0
Tryout	8.0	8.0	10.0	10.0
TOTAL, UNIT HOURS	23.4	25.8	35.6	39.6
+ Form Complexity				
Development Factor				
Material Cost	$100.00	$190.00	$240.00	$300.00
Die Set Cost	$ 62.00	$ 74.00	$ 96.00	$135.00
Total Material Cost	$162.00	$264.00	$336.00	$435.00

FORM PUNCH
LOCATE PINS
LOCATE PIN
LOCATE NEST
DIE BLOCK
DIE SHOE

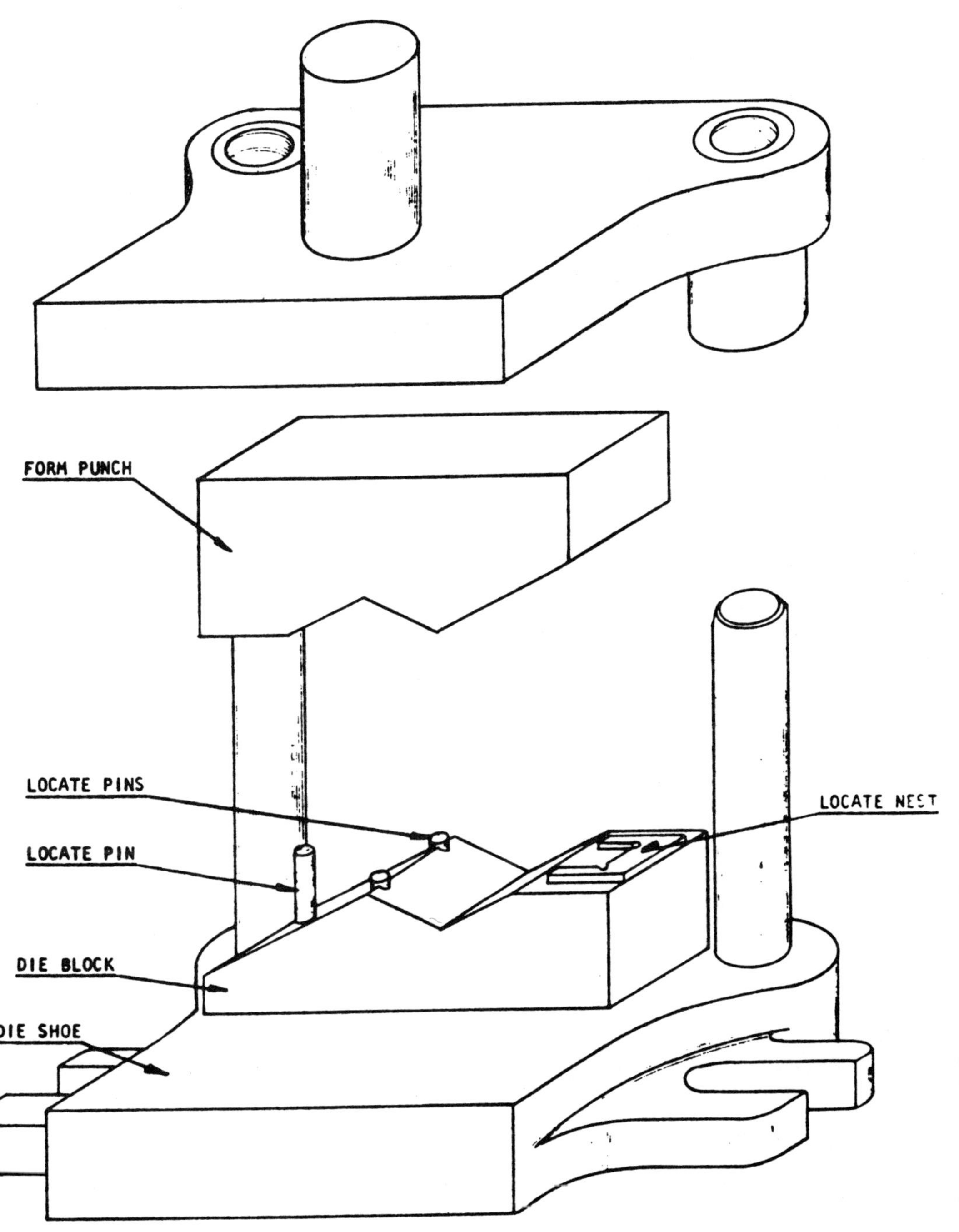

"Z" FORM

"Z" FORM DIE
UNIT HOURS & COST FACTORS

FEATURE	DIE SPACE			
	16 - 20 SQ. IN.	21 - 36 SQ. IN.	37 - 64 SQ. IN.	65 - 100 SQ. IN.
Die Block Machining Time	1.8	2.2	3.2	4.6
Form Punch Machining Time	1.6	1.8	2.6	3.6
Locate Nest Machining Time	1.2	1.2	1.4	1.4
Locate Pins (3) Min.	2.0	2.0	2.0	2.0
Die Down Stop	2.0	2.0	2.0	2.0
Drilling & Tapping etc. For MT'G Holes	3.0	3.5	4.0	4.0
Toolmaker Time Ref. Layout, MAT'L Req.	2.0	2.5	3.0	4.0
Tryout	3.0	3.5	4.0	4.5
TOTAL, UNIT HOURS	16.6	18.7	22.2	26.1
+ Form Complexity				
Development Factor				
Material Cost	$100.00	$190.00	$240.00	$300.00
Die Set Cost	$ 62.00	$ 74.00	$ 96.00	$135.00
Total Material Cost	$162.00	$264.00	$336.00	$435.00

BASIC "L" FORM DIE

STRIPPER BOLTS
DIE SPRINGS
FORM PUNCH
PRESSURE PAD
PILOT LOCATE PINS
HEEL BLOCK
DIE SET
DIE BLOCK

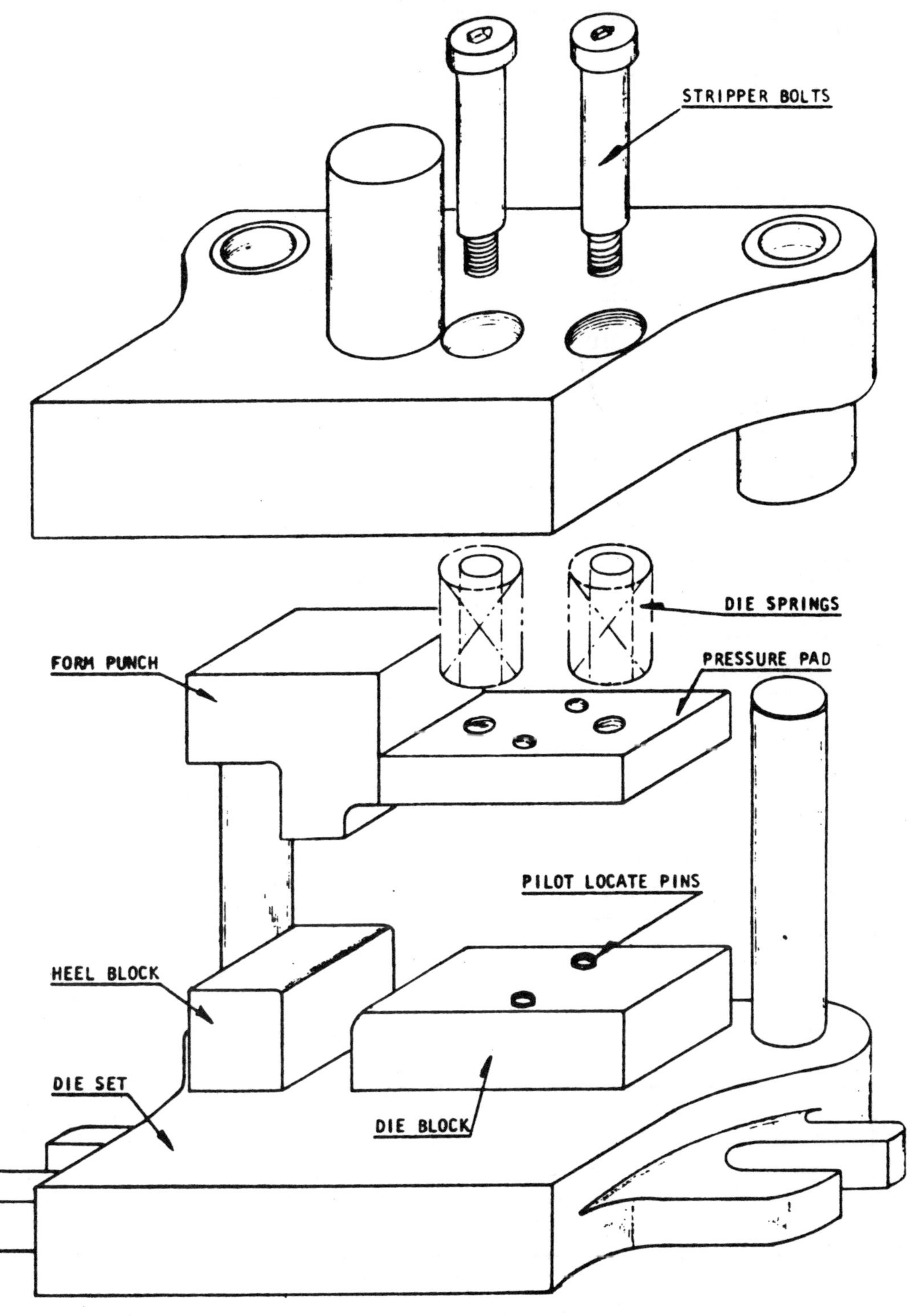

"L" FORM

"L" FORM DIE
UNIT HOURS & COST FACTORS

FEATURE	DIE SPACE			
	16 - 20 SQ. IN.	21 - 36 SQ. IN.	37 - 64 SQ. IN.	65 - 100 SQ. IN.
Die Block Machining Time	1.4	1.4	2.0	2.6
Form Punch Machining Time	1.6	1.6	1.8	2.2
Heel Block For Form Punch	1.2	1.2	1.8	2.0
Pressure Pad Machining Time	1.2	1.6	2.2	2.4
Pilots or Locators (1) Diamond (1) Plain	5.0	5.0	5.0	5.0
Stripper Bolts For Pressure Pads 4.0 Each	8.0	16.0	16.0	16.0
Drilling & Tapping, etc. MT'G Holes	4.0	4.0	5.0	6.0
Toolmaker Time Ref. Layout, MAT'L Req. etc.	3.0	4.0	6.0	8.0
Tryout	3.0	3.5	5.0	6.0
Die Down Stop	2.0	2.0	2.0	2.0
TOTAL, UNIT HOURS	30.4	40.3	46.8	52.2
+ Form Complexity				
Development Factor				
Material Cost	$102.00	$192.00	$245.00	$320.00
Die Set Cost	$ 62.00	$ 74.00	$ 96.00	$135.00
Total Material Cost	$164.00	$266.00	$341.00	$450.00
Springs $.50 each				

BASIC DIE SUMMARY SHEET

CHART # 1

TYPE OF DIE	DIE SPACE 16-20 sq. in		DIE SPACE 21-36 sq. in.		DIE SPACE 37-64 sq. in		DIE SPACE 65-100 sq. in.	
	UNIT HOURS	MATERIAL COST	UNIT HOURS	MATERIAL COST	UNIT HOURS	MATERIAL COST	UNIT HOURS	MATERIAL COST
PIERCE DIE	27.8	$182.00	30.5	$274.00	43.8	$396.00	46.5	$535.00
BLANK DIE	22.1	$194.00	29.1	$346.00	39.8	$486.00	50.9	$585.00
PIERCE & CUT-OFF DIE	34.4	$192.00	39.1	$275.00	47.8	$491.00	55.8	$610.00
PIERCE AND BLANK THRU DIE	37.4	$202.00	43.7	$304.00	56.1	$496.00	70.5	$630.00
PIERCE CUT-OFF & FORM DIE	41.8	$202.00	48.6	$304.00	62.0	$501.00	77.8	$638.00
COMPOUND PIERCE & BLANK DIE	45.5	$209.00	62.1	339.00	78.0	$574.00	93.8	$773.00
"V" FORM DIE	23.4	$162.00	25.8	$264.00	35.6	$336.00	39.6	$435.00
"Z" FORM DIE	16.6	$162.00	18.7	$264.00	22.2	$336.00	26.1	$435.00
"L" FORM DIE	30.4	$164.00	40.3	$266.00	46.8	$341.00	52.2	$450.00

The total unit values taken from this chart will be determined by the blank shape complexity. Features such as, overall blank size, angles, sides and radii must be interpreted as factors influencing the machining time of die blocks, punches, shedders, strippers and templates.

The total of all unit hours derived from this chart will be the estimated time for finish machining a blank opening in the die block.

To estimate the machining time for other components, influenced by the blank complexity, use the following percentages of the die block estimate.

Blank Punch 50%   Stripper 40%   Shedders 40%   Templates 10%

QUANTITIES OR AREA OF THE BLANK	BLANK SIZE "a"	INSIDE ANGLES 90° OR LESS "b"	NUMBER OF SIDES .250 OR LONGER "c"	NUMBER OF TOLERANCED INSIDE RADII "d"	NUMBER OF OUTSIDE RADII "e"	ROUND BLANKS OR PIERCED HOLES "f"	
		UNIT HOURS				DIAMETER	UNIT HOURS
1	5.5	0.7	0.5	0.5	1.5	up to .125	6
2	5.7	1.2	0.7	0.7	1.8	.1260-.250	5
3	5.9	1.7	1.1	0.9	2.2	.251-.500	4.5
4	6.3	2.5	1.7	1.2	2.6	.501-1.000	6.0
5	6.6	3.3	2.3	1.4	3.0	1.001-1.500	7.5
6	6.9	4.0	3.2	1.7	3.5	1.501-2.000	10.0
7	7.5	4.7	4.1	2.0	4.0	2.001-2.500	12.0
8	8.0	5.5	5.0	2.2	4.5	2.501-3.000	14.0
9	8.3	6.2	6.1	2.4	5.0	3.001-3.500	16.0
10	8.5	7.0	7.2	2.6	5.5	3.501-4.000	18.0
12	9.6	8.5	8.5	2.8	6.0	4.001-4.500	20.0
14	11.4					4.501-5.000	22.0
16	13.0					5.001-5.500	24.0
18	15.0					5.501-6.000	26.0
20	16.7						
22	17.0						
24	18.7						

LENGTH OF FORM LINE	TYPE OF FORM			
	"L"	"U"	"Z"	"V"
	UNIT HOURS			
Up to 1.0"	10.0	16.0	16.0	6.0
1.0 to 2.0"	11.5	18.0	18.0	7.0
2.0 to 3.0"	13.0	20.0	20.0	8.0
3.0 to 4.0"	14.5	22.0	22.0	9.0
4.0 to 5.0"	16.0	24.0	24.0	10.0
5.0 to 6.0"	17.5	26.0	26.0	11.0
Each Additional 1 in.	1.5	2.0	2.0	1.0

FORM COMPLEXITY

See Development Factor

DEVELOPMENT TIME FOR FORMED ANGLES CHART #4

MATERIAL HARDNESS	ANGULAR TOLERANCES (TOTAL)	THICKNESS								
		.060	.50	.040	.030	.025	.020	.015	.010	.005
		% OF FORM ALLOWANCE - FORM COMPLEXITY								
HARD	10° or More	-	-	-	-	-	-	5	10	20
	7	-	-	-	-	-	5	10	15	30
	5	-	-	-	5	7	10	15	25	40
	4	-	-	5	7	10	15	20	30	50
	3	-	5	10	20	30	40	55	75	100
	2	40	50	60	70	85	100	140	-	-
	1°	-	-	-	-	-	-	-	-	-
H. H. AND 1/4 HARD	5	-	-	-	-	-	5	10	15	25
	4	-	-	-	-	5	10	15	25	40
	3	-	-	-	5	10	20	30	45	65
	2	10	20	30	40	50	65	90	125	-
	1°	60	120	-	-	-	-	-	-	-
SOFT	5°	-	-	-	-	-	-	-	-	-
	4	-	-	-	-	-	-	-	-	-
	3	-	-	-	-	-	-	5	15	30
	2	5%	10	15	20	25	35	50	75	100
	1°	50%	60	75	95	120	160	-	-	-

<u>CORNER SET FOR FORMING:</u>

1.0 HOUR + 0.6 HOURS FOR EACH ADDITIONAL INCH

COMPONENT LIST

CHART #5

ITEM		UNIT HOURS
1. Air Cushion Pin		0.5
2. Automatic Stop		3.0
3. Carbide Wear Block (in stock guide)		2.0
4. Cut-off punch-open type w/ pressure pad (+1.0 for each addition inch)		7.0
5. Ejector-plunger pin		2.0
6. Equalizer buttons		1.0
7. Extrusions and Embosses	simple	3.0
	complex	5.0
8. Hook Stripper	single	3.0
	double	5.0
9. Knockout pin	single	1.0
	double w/ simple shedder	5.5
	multiple w/ shedder and spider	8.0
10. Locate pin	diamond	3.0
	round	2.0
11. Pilot		2.0
12. Primary Stop		3.0
13. Pusher		3.0
14. Stripper Bolts (for stripper guides)		4.0
15. Toggle clamps		2.0
16. Trasfer pin		1.0
17. Parallels	small	2.0
	large	3.0

3	1/4	1/2	3/4	1	2	3	4	5	6	7	7-1/2	8	9	10	11
2	1/2	3/4	1	2	3	4	5	6	7	8	9	10	11	12	13
1	1	2	3	4	5	6	7	8	9	10	11	12	13	14	15
1/2	2	3	4	5	6	7	8	9	10	11	12	13	14	15	16

1	.8	.8	1.0	1.0	1.2	1.2	1.2	1.4	1.4	1.4	1.4	1.6	1.6	1.6	1.6
2	.8	1.0	1.2	1.2	1.4	1.4	1.4	1.6	1.6	1.6	1.8	1.8	1.8	2.0	2.0
3	1.0	1.2	1.2	1.4	1.4	1.6	1.6	1.8	1.8	2.0	2.0	2.2	2.2	2.4	2.4
4	1.0	1.2	1.4	1.4	1.6	1.6	1.8	2.0	2.2	2.4	2.4	2.6	2.8	2.8	3.0
5	1.8	1.4	1.4	1.6	1.8	1.8	2.0	2.2	2.4	2.6	2.6	2.8	3.0	3.2	3.4
6	1.2	1.4	1.6	1.6	1.8	2.0	2.2	2.4	2.6	2.8	3.0	3.2	3.4	3.6	3.8
7	1.2	1.4	1.6	1.8	2.0	2.2	2.4	2.6	2.8	3.0	3.2	3.4	3.6	3.8	4.2
8	1.4	1.6	1.8	2.0	2.2	2.4	2.6	3.0	3.2	3.4	3.6	3.8	4.0	4.2	
9	1.4	1.6	1.8	2.2	2.4	2.6	2.8	3.2	3.4	3.6	3.8	4.2	4.4		
10	1.4	1.6	2.0	2.4	2.6	2.8	3.0	3.4	3.6	3.8	4.2	4.6			
11	1.4	1.8	2.0	2.4	2.6	3.0	3.2	3.6	3.8	4.2	4.8				
12	1.6	1.8	2.2	2.6	2.8	3.2	3.4	3.8	4.2	4.8					
13	1.6	1.8	2.2	2.8	3.0	3.4	3.6	4.0	4.4						
14	1.6	2.0	2.4	2.8	3.2	3.6	3.8	4.2							
15	1.6	2.0	2.4	3.0	3.4	3.8	4.2								

<u>DIE ESTIMATING WORKSHEET</u>

PROBLEM #1

---

1. CLASSIFICATION PIERCE AND BLANK THRU DIE

Unit Hours

2. UNIT HOURS FACTOR (See Basic Die Summary Sheet, Chart #1)
    Die Space 21-36 sq. in.   43.7

3. BLANK COMPLEXITY, Chart #2

Unit Hours

a.	Blank Size Angles	(2.625 sq. in.)	5.9
b.	Inside Angles 90° or less	(4)x(2.5)	2.5
c.	Number of Sides .250 or longer	(4)x(1.7)	1.7
d.	Number of Toleranced Inside Radii	(4)x(1.2)	1.2
e.	Number of Outside Radii	(0)x(0)	---
f.	Round Blanks or Pierced Holes	(size)-(qty.)x(UH)	---
		(.375)-(1)x(4.5)	4.5
		( )-( )x( )	
		( )-( )x( )	
		( )-( )x( )	
		( )-( )x( )	

Blank Complexity Sub-Total   15.8   15.8

---

4. BLANK PUNCH, Chart #2
    Blank Complexity Sub-Total (15.8)x(50%)   7.9

5. STRIPPER, Chart #2
    Blank Complexity Sub-Total (15.8)x(40%)   6.3

6. SHEDDERS, Chart #2
    Blank Complexity Sub-Total (NA)x(40%)   ---

7. TEMPLATES, Chart #2
    Blank Complexity Sub-Total (NA)x(10%)   ---

8. FORM COMPLEXITY, Chart #3
    Type of Form Factor   (NA) x Length of Form   ---

9. FORM DEVELOPMENT, Chart #4
    Development Percentage x Type of Form Factor   ---

10. COMPONENT LIST, Chart #5

---

UNIT HOURS   73.7

UNIT HOURS	73.7
ADDITIONAL HOURS FOR AIR-HARDENING STEEL (Unit Hours x 25%)	18.42
TOTAL BUILD HOURS	92.12
BUILD RATE x TOTAL BUILD HOURS ($20 x 92.12)	$ 1842.00
DESIGN COST (10% x $1842)	$ 184.00
MATERIAL COST (See Basic Die Summary Sheet, Chart #1)	$ 202.00
COMPLETE DESIGN & BUILD COST	$ 2228.00

REMARKS:

<u>DIE ESTIMATING WORKSHEET</u>

PROBLEM #2

				Unit Hours
1.	CLASSIFICATION PIERCE AND BLANK THRU DIE			
2.	UNIT HOURS FACTOR (See Basic Die Summary Sheet, Chart #1) Die Space 21-36 sq. in.			48.6

3. BLANK COMPLEXITY, Chart #2

			Unit Hours	
a.	Blank Size Angles	(2.656 sq. in.)	5.9	
b.	Inside Radii 90° or less	(0)x(-)	---	
c.	Number of Sides .250 or longer	(2)x(.7)	.7	
d.	Number of Toleranced Inside Radii	(0)x(-)	---	
e.	Number of Outside Radii	(-)x(-)	---	
f.	Round Blanks or Pierced Holes (size)-(qty.)x(UH)		---	
		(.250)-(1)x(5)	5.0	
		( )-( )x( )		
		( )-( )x( )		
		( )-( )x( )		
		( )-( )x( )		
	Blank Complexity Sub-Total		11.6	11.6

		Unit Hours
4.	BLANK PUNCH, Chart #2 Blank Complexity Sub-Total (11.6)x(50%)	5.8
5.	STRIPPER, Chart #2 Blank Complexity Sub-Total (11.6)x(40%)	4.6
6.	SHEDDERS, Chart #2 Blank Complexity Sub-Total ( )x(40%)	---
7.	TEMPLATES, Chart #2 Blank Complexity Sub-Total (N.)x(10%)	---
8.	FORM COMPLEXITY, Chart #3 Type of Form Factor (1.4) x Length of Form	7.0
9.	FORM DEVELOPMENT, Chart #4 Development Percentage x Type of Form Factor	2.8
10.	COMPONENT LIST, Chart #5	---
	UNIT HOURS	80.4

UNIT HOURS	80.4
ADDITIONAL HOURS FOR AIR-HARDENING STEEL (Unit Hours x 25%)	---
TOTAL BUILD HOURS	80.4
BUILD RATE x TOTAL BUILD HOURS ($20 x 80.40)	$ 1608.00
DESIGN COST (10% x $1608)	$ 160.00
MATERIAL COST (See Basic Die Summary Sheet, Chart #1)	$ 304.00
COMPLETE DESIGN & BUILD COST	$ 2072.00

REMARKS:

<u>DIE ESTIMATING WORKSHEET</u>

SAMPLE

---

1. CLASSIFICATION PIERCE AND BLANK THRU DIE

        Unit Hours

2. UNIT HOURS FACTOR (See Basic Die Summary Sheet, Chart #1)
    Die Space   -   sq. in.

---

3. BLANK COMPLEXITY, Chart #2

        Unit Hours

   a. Blank Size Angles   (sq. in.) ______

   b. Inside Radii 90° or less   ( )x( ) ______

   c. Number of Sides .250 or longer   ( )x( ) ______

   d. Number of Toleranced Inside Radii ( )x( ) ______

   e. Number of Outside Radii   ( )x( ) ______

   f. Round Blanks or Pierced Holes   (size)-(qty.)x(UH) ______

                                   ( )-( )x( ) ______

                                   ( )-( )x( ) ______

                                   ( )-( )x( ) ______

                                   ( )-( )x( ) ______

                                   ( )-( )x( ) ______

                    Blank Complexity Sub-Total

---

4. BLANK PUNCH, Chart #2
               Blank Complexity Sub-Total ( )x(50%)

---

5. STRIPPER, Chart #2
               Blank Complexity Sub-Total ( )x(40%)

---

6. SHEDDERS, Chart #2
               Blank Complexity Sub-Total ( )x(40%)

---

7. TEMPLATES, Chart #2
               Blank Complexity Sub-Total ( )x(10%)

---

8. FORM COMPLEXITY, Chart #3
               Type of Form Factor   ( ) x Length of Form

---

9. FORM DEVELOPMENT, Chart #4
               Development Percentage x Type of Form Factor

---

10. COMPONENT LIST, Chart #5

UNIT HOURS

Die Estimating Worksheet Sample Continued

UNIT HOURS	79
ADDITIONAL HOURS FOR AIR-HARDENING STEEL (Unit Hours x 25%)	
TOTAL BUILD HOURS	
BUILD RATE x TOTAL BUILD HOURS	$
DESIGN COST	$
MATERIAL COST (See Basic Die Summary Sheet, Chart #1)	$
COMPLETE DESIGN & BUILD COST	$

REMARKS:

Die and stamping estimating

Estimating a progressive draw die

By **Paul Prikos**, Vice President, Prikos & Becker Tool Co., Skokie, Ill.
and President, X-L Punch Co., Niles, Ill.

Until recent years, the literature on progressive draw dies has been sparse; few diemakers know this work thoroughly and there are even fewer competent draw die designers and estimators.

In my own experience in trying to give a reasonably accurate progressive die estimate, my fortunes have gone against me. Anyone can load the estimate, but to really hit it competitively is an art. I have finally developed a method of estimating which is adequate but far from perfect; however, until a better method to approach a norm applicable to most progressive dies becomes available, the method explained in this article will be used.

The part to be made in a progressive die is shown in Fig. 1. After reviewing the part the estimator must first determine the number of reductions required to produce the shell. This can reasonably be done with any one of several methods in present die design books. I use the percentage method to estimate the number of draws.

Several important things should be considered in estimating a progressive

draw die: (1) Shall we use the hourglass notch method which makes piloting more difficult, yields a distorted strip, but saves appreciable material costs due to a narrower strip, or (2) shall we use the double-circle lance method which allows piloting in the scrap part of the strip because the cup is suspended within the double-circle lance? It is a straight strip, but uses more stock material in the width.

Naturally, there are variations of the above two methods, but in principle these are the two selections. To calculate the price, always make a strip layout after deciding on the above-mentioned two methods and proceed with the estimate (Fig. 2).

Here is the estimate:

Station 1	20.00 hours
Stations 2 to 5 inclusive	100.00 hours
Station 6	20.00 hours
Station 7	30.00 hours
	170.00 hours
Development time	50.00 hours
	220.00 hours

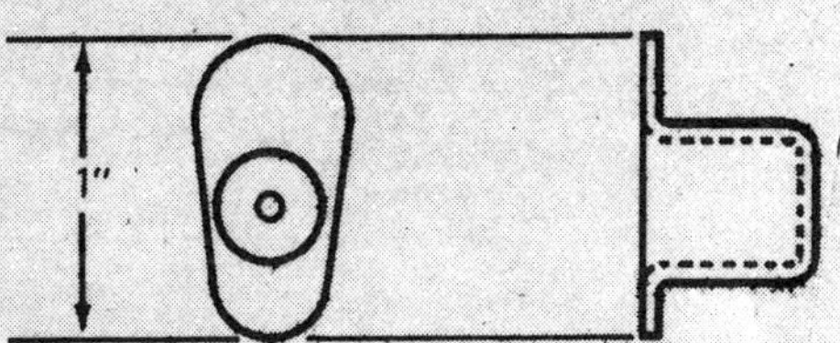

Fig. 1. *Piece-part for which progressive draw die is to be estimated.*

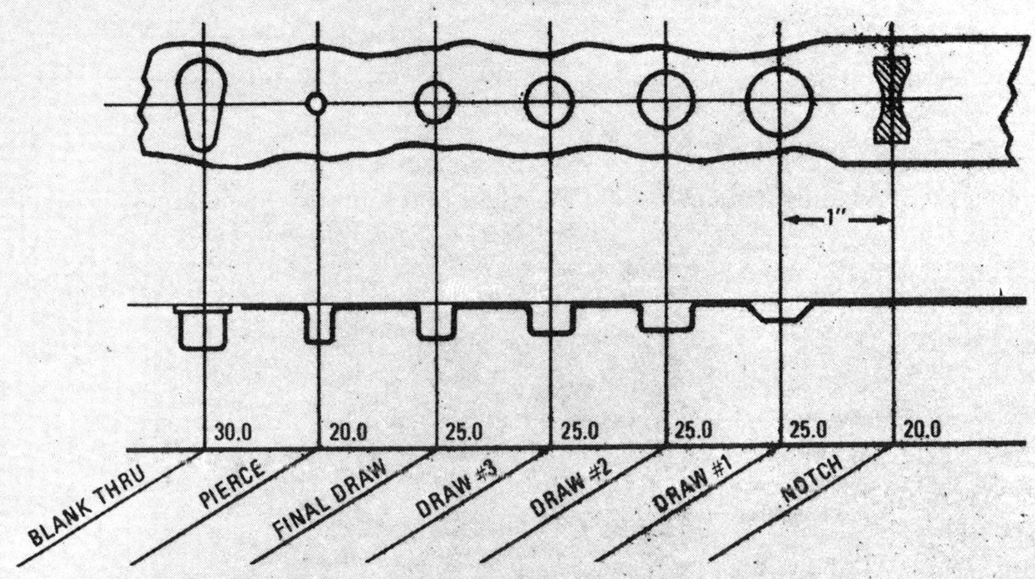

Fig. 2. *Strip layout indicating notching, progressive draw, pierce and blank stations. A progressive draw die quote can be worked up from this layout.*

Extra for spring stripper work, adjustment springs and timing of die .... 30.00 hours

Total .... 250.00 hours

Total die cost in dollars:

Material and die set...... $ 365.00

Layout (design) 475.00

Inspect 50.00

Build, 250 hours @ $16.00 .... 4,000.00

Total cost of die $4,890.00

Because most progressive draw dies are difficult to handle we seldom find parts that exceed a coffee cup or water

CUSTOMER _VI_   PART NO. _PROG. DRAW DIE_   DATE _1-1976_

		NOTES:
1. Prog. ✓ Blank___ Comp.___ Die $	_SEE_	
2. 1st Form Die	_DIE_	
3. 2nd Form Die	_QUOTE_	
4. Other (Pierce) Die		
5. Mill/Drill Fixt.		
6. Other		
Total $.		

PRODUCTION MATERIAL

.04 X / X / X .1 (ALUM) X 1000 = 4 LBS PER M

QUANTITY LEVEL	1.00 10 M	.85 25 M	.80 100 M	.75 250 M
Material Costs	$ 4.00	$ 3.40	$ 3.20	3.00
Inspection, Check, Set-up	2.50	1.00	.25	—
Blank	14.00	10.00	8.00	6.00
Deburr	3.00	2.50	2.00	2.00
1st Form				
2nd Form				
Other				
Mill.				
Grind.				
Drill/Ct'sink				
Sand/Deburr				
Tapping				
Plating	NO			
Inspection	3.00	2.00	1.50	1.00
Die Maintenance	3.00	3.00	3.00	3.00
Ship	3.00	2.00	2.00	1.50
	$ 32.50	$ 23.90	$ 19.95	16.50
Plus 10 % markup and/or (scrap, material Sales comm.)	3.25	2.39	2.00	1.65
TOTAL	$ 35.75 per M	$ 26.29 per M	$ 21.95 per M	18.15 M

Del. Tooling ______ Production ______

F.O.B. Our Plant

Estimating

glass in size. Hence, we can resolve each station from between 25 to 40 hours and then multiple by the number of draws and redraws.

The estimator must also give consideration to any additional time required in the press room to adapt the die to a given press. Velocity, tonnage, and accuracy of the punch might necessitate some slight die modifications when in actual production. Draw compound and draw quality of the piece-part material are important qualifications in the successful run of the progressive draw die.

I want to stress the wisdom of the die designer to provide additional idle stations in the die for an emergency extra draw, or sizing, that may be required. Wherever possible, draw punch adjustments should be designed and built into the die to avoid extensive disassembly time by the diemaker. A progressive draw die should run with an automatic feed to help the proper station-to-station advance. ● ● ●

Die and stamping estimating

A.B.C. concept of die estimating

By **Paul Prikos,** Vice President, Prikos & Becker Tool Co., Skokie, Ill.
and President, X-L Punch Co., Niles, Ill.

To further help in understanding the estimating of dies, I have prepared a series of estimating units and hints that are applicable to many piece-part designs. These individual hints are associated with the hours required in die construction. Therefore, the following A.B.C. Concept of die estimating (Table 1) can be intelligently applied.

Tne scope of the estimate given so far pertains to good Class A dies most commonly associated with punch press work. The die estimated could also be built for twice the cost if carbide sections were required; if extensive jig grinding was needed; if special company-standard design features were called for; if tolerance requirements were severe; if extra-thick die blocks were specified to increase the life of the die; and finally, if the dies required extensively detailed designs. Furthermore, some of the more exotic piece-part materials can play havoc with your estimate. Stainless steel, beryllium copper, and the like, are rough at times compared to brass and cold rolled steel. The estimator must weigh and consider all these factors as part of the development time in coming to the proper die costs.

Needless to say, such intricate dies as progressive pierce, notch, shave, form, and blank—that become very involved—have not been investigated. It is best to accomplish the basic estimating procedure, compare it with others and, with practice, develop a knack for thorough estimates.

Table 1. Estimating dies—general A.B.C. Concept.
Figures preceding the letters "a" through "1," inclusive,
represent the number of hours required to perform that particular operation.

5 hours	a	3/4-inch diameter hole, maximum
30 hours	b	Cluster of holes
10 hours	c	Oblong hole
10 hours	d	Square hole
20 hours	e	Horseshoe opening
15 hours	f	Shear form
10 hours	g	Deboss
12 hours	h	1/2 by 1-inch typical notch
13 hours	i	1 by 1-inch typical notch
15 hours	j	1-inch-step typical step notch
20 hours	k	1-inch typical curve notch
30 hours	l	Combination curve and step notch

m   High-carbon, high-chrome dies worth 25 percent more than oil-hardening tool steel

n   Close-fitting dies requiring lapping, etc. (These are more expensive. Remember, it takes only 10 minutes to remove 1/2 inch of material from a 6 by 6-inch block, but it may take 5 hours to remove that final 0.0001-inch.)

o   H.S.S. dies are worth 30 percent more than O.H.T.S.

p   On large dies add hours for special handling

q   Provide for extra die when sizing is a problem

r   Use jig grinding sparingly—it adds to die costs

s   Sectional dies are more expensive than solid dies

t   Spring strippers are more expensive than stationary strippers

u   Remember maintenance and safety in dies

v   Die layouts prevent costly hindsight errors

w   Reducing die block thickness does not necessarily reduce costs because of inevitable heat-treat warpage

x   Remember the diemaker is human and makes mistakes, give him some leeway in the estimate

y   When the piece-part is too difficult to estimate, do not pretend to know the answer with a guess. It is simply a time-and-material situation

z   Sometimes the high estimate may actually be too low. Conversely, the low estimate may be too high.

estimating

Table 2. Stamping quotation—general guidelines following A.B.C. Concept.

Blanking

    1. Large parts................................. 250 per hour
    2. Medium .... 400-650 per hour
    3. Small (auto-feed)................... 1000-4000 per hour

Forming (hand loading)

    1. Large parts........................... 200-300 per hour
    2. Medium .... 300-500 per hour
    3. Small.................................. 500-750 per hour

Tapping

    General .... 250-500 per hour (1 hole)

Spotwelding

    General .... 200-400 per hour (1 spot)

Table 2 contains some general guidelines that I have used successfully for estimating stampings. They complement the A.B.C. Concept of die estimating brought out in the article.

●●●

Die and stamping estimating

Estimating a subassembly

we will follow through a four-step estimating procedure—progressive die, stamping, screw machine operation and assembly.

By **Paul Prikos,** Vice President, Prikos & Becker Tool Co., Skokie, Ill.
and President, X-L Punch Co., Niles, Ill.
and **Paul T. Prikos,** President, X-L Engineering Corp., Niles, Ill.

There has been a noticeable trend in recent years for many corporations to subcontract not only stampings but also subassemblies. Of the many reasons for this shift in thinking the most apparent is one-source responsibility, thus eliminating additional purchasing and inspection costs as well as expediting.

We have observed the gradual shift of many customers toward concentration on these areas that provide maximum utilization of their expertise; namely research and development, purchasing, marketing and final assembly. Subcontracting much of the manufacturing portion of their end products is the net result.

Much of the piece-part content of most assembled products can be purchased at competitive prices from the myriad of companies that service both medium and large corporations. Many of these subcontracting firms have developed among them a network of sophisticated specialties that represent design and manufacturing capabilities too extensive and time consuming for any one company to develop. As a result, purchasing is acquiring a more dominant corporate staff influence.

Requests for an expanding volume of subassembly estimates are handled by employing the facilities of two companies. The stamping and fabrication facilities of Prikos & Becker Tool Co. and the precision machining and grinding facilities of X-L Engineering Corp.

Now, for the estimates! The example shown in Fig. 1 represents the complete subassembly. Indicated is the need of a progressive die which will produce a complete stamped and formed piece-part (right, Fig. 2). Next, a multiple-diameter metal shaft must be machined; apparently a screw machine operation. Finally, there is the marriage of the stamping to the shaft to form a subassembly. Four estimates will be covered in this article:

1. The progressive die estimate.
2. The stamping estimate.
3. The screw machine piece-part estimate.

4. The subassembly estimate.

The Progressive Die Estimate

Start out with a strip layout such as that illustrated in Fig. 3. Using the strip layout as a guide, the following estimate can be made:

1. Perf and pilot....... 30.0 hours
2. Rectangular 1/4-inch hole .... 6.0 hours
3. Long rectangular opening .... 12.0 hours
4. Two small round holes 10.0 hours
5. Large notch .... 25.0 hours
6. Shear form .... 15.0 hours
7. 90-degree form bend 30.0 hours
8. Blank thru .... 20.0 hours
9. Misfeed detector ... 10.0 hours
10. Pads for adjusting of punch heights when sharpening .... 15.0 hours

Fig. 1. *This bracket-and-shaft subassembly will be estimated in this final article of an eight-part series on die and stamping estimating.*

Fig. 2. *Component parts of subassembly being considered. Stamped bracket (right) will be estimated first, followed by screw machine part (left) and, finally, the subassembly.*

11. Spring stripper ..... 30.0 hours
12. Lifters (stock) ..... 15.0 hours
13. Assembly .... 35.0 hours
14. Tryout and debugging.......... 15.0 hours

    Total .... 268.0 hours

Total die cost in dollars:

Die design $650.00
Build; 268 hours @ $16.00................. 4,288.00
Inspection 75.00
Material—die set, tool steel perfs, springs, h.t. screws, other purchases.............. 975.00

Total................... $5,988.00

A few comments are in order to explain the above estimate. Die design costs could vary a few hundred dollars depending upon how involved and detailed one believes his particular die should be. Material purchases are also flexible depending upon the number of items purchased as standard die com-

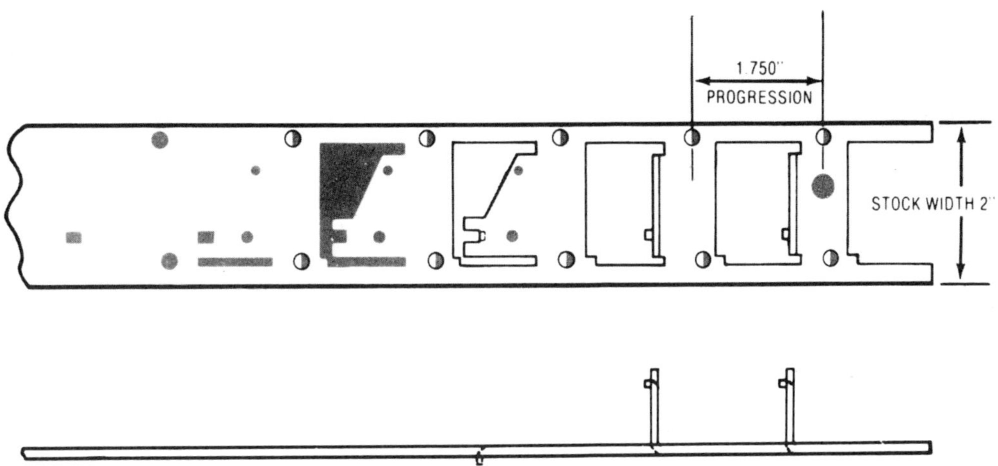

Fig. 3. *Estimating a progressive die is simplified by starting out with a strip layout such as this.*

production volume can justify this extra investment. (An actual scrap strip for this part, including a completed stamping at right, can be seen in Fig. 4.)

The Stamping Estimate

When a stamping is part of an assembly that is joined with other parts, manufacturing tolerances are considered carefully in the area where parts mate. In our case a shaft will be staked to the stamping (Figs. 1 and 2). We have the opportunity to alter slightly the tolerance in the hole to match the shaft for ease of assembly.

The hole that accepts the shaft should have a good "feature size." This means the break in the hole should present as much of a straight-sided area as possible to assure assembly of the pin or shaft in a perpendicular position. Flatness of the stamping is important at the base area where the stamping and shaft are joined.

The part is run in a 75-ton variable-speed press with coil stock, automatically fed. There is a misfeed detector to prevent excessive die damage in the event of a malfunction. Generally

Fig. 4. *Actual scrap strip (including completed stamping at right) produced by progressive die designed on the basis of strip layout shown in Fig. 3.*

Estimating

STAMPING QUOTATION

CUSTOMER   VIII   PART NO.   STAMPING   DATE 11/1/76

		NOTES:
1. Prog........ Blank...... Comp...... Die	$ SEE PROG. DIE	
2. 1st Form Die	QUOTE	
3. 2nd Form Die		
4. Other (Pierce) Die		
5. Mill/Drill Fixt.		
6. Other		
Total	$.	

PRODUCTION MATERIAL

.093 X 2 X 1.75 X .3 X 1000 = 98 lbs. per M

QUANTITY LEVEL	.60 2M	.40 5 M	.30 10 M	.22 20M
Material Costs	$ 59.00	$ 39.00	$ 30.00	22.00
Inspection, Check, Set-up	11.00	4.50	2.25	1.10
Blank	25.00	15.00	12.00	8.00
Deburr	8.00	5.00	4.00	3.00
1st Form				
2nd Form				
Other				
Mill.				
Grind.				
Drill/Ct'sink				
Sand/Deburr				
Tapping				
Plating				
Inspection	15.00	5.00	3.00	2.00
Die Maintenance	3.00	3.00	3.00	3.00
Ship		SEE ASSY.		
	$ 121.00	$ 71.50	$ 54.25	39.10
Plus .... %	–	–	–	–
TOTAL	$ – per M	$ – per M	$ – per M	–

Del. Tooling .... Production ....

speaking, the whole press, die and material is synchronized as stated in our previous articles.

The Screw Machine Estimate

Estimating a screw machine part depends primarily upon part geometry, material, machine tool, and the machining process. For the purpose of this estimate we will use a shaft (Fig. 5) which requires drilling, turning, forming, and tapping. There are a variety of ways of producing this part, but rather than go into the pros and cons of different techniques we will isolate each operation and estimate individually.

The material is given as B1113 cold rolled steel which has a standard commercial tolerance of +0.000,-0.002-inch on 0.500-inch diameter cold finished rounds. The machine tool will not be a specific screw machine but a machine with a hexagonal turret and front and rear cross slides. The process used will be the following:

1. Feed stock to turret stop (turret position No. 1).

2. Spot drill (turret position No. 2).

3. Drill (turret position No. 3).

4. Tap (turret position No. 4).

5. Turn 0.312-inch diameter (turret position No. 5).

6. Form (front cross slide).

7. Cutoff (rear cross slide).

An important feature to realize in the machining estimate is that it is possible to overlap operations and thereby reduce cycle time. Combination tools may be used to spot drill and turn. Form tools can also cutoff in some instances. These combinations go on and on so we won't elaborate further other than to note their possibility.

Having determined the process, we will now select a spindle speed based on the machinability of B1113. Again, recommended surface speeds in feet per minute vary, but we will use a speed of 200 sfm. The formula for determining the spindle rpm given a surface speed is

$$\text{rpm} = \frac{\text{Surface speed}}{0.262 \times \text{Material dia.}} \text{ or,}$$

$$\text{rpm} = \frac{200}{0.262 \times 0.500} = 1526$$

We will assume the machine has a high-speed/low-speed change and we will use 1500 rpm as the high-speed and 500 rpm as the low speed. The low speed will be used on the tapping operation; all others will be at the high speed.

To determine a cycle time we will convert all the operations into the number of spindle revolutions required to perform them. Using a standard reference table to obtain feed rates:

1. Spot drill, 0.003 ipr.

2. Drill, 0.003 ipr.

3. Turn, 0.006 ipr with a 1/16-inch depth of cut.

4. Form, 0.0015 ipr with a 3/8-inch wide form tool.

5. Cutoff, 0.0015 ipr with a 0.100-inch wide cutoff tool.

From these feed rates we can determine the number of revolutions required by dividing the distance being machined by the feed rate:

Estimating

1. 90-degree spot drill, 0.125/0.003 = 42 revolutions. (Note: When breaking the corner 1/32 inch, the spot drill has a diameter of about 0.250 inch. With a 90-degree point, this corresponds to a 0.125-inch depth.)

2. Drill, 0.750/0.003 = 250 revolutions.

3. Turning, 1.000/0.006 = 167 revolutions.

4. Forming, 0.154/0.0015 = 103 revolutions (stock size minus formed diameter divided by 2 equals 0.154).

5. Cutoff (half stock size) 0.250/0.0015 = 167 revolutions.

In addition to actual chip-cutting time, we must include time for turret indexing, tool approach, and tool dwell which is necessary to hold tolerances. These various noncutting times should be minimized as much as possible by operation overlap and imaginative processing. For this quote, however, we will simply add 5 percent to the number of revolutions.

To determine the number of revolutions for tapping we multiply the threads per inch by the length of tap. The tapping cycle will be performed using the low spindle speed of 500 rpm both to tap in and tap out. Calculate: 32 pitch x 1/2-inch deep = 16 revolutions to tap in, and 16 to tap out. But, remember that the 32 tapping revolutions are at 500 rpm and must be converted into the equivalent figure at the higher spindle speed. Thus 32 x 1500/500 = 96 revolutions at 1500 rpm.

Again we will allow 5 percent to include the revolutions required for approach and indexing time. Another point to be considered is that "tapping out" can be done at the high spindle speed which would reduce the number of revolutions.

Now, let's look at the total revolutions of the various operations:

1. Spot drill. 42
2. Drill .250
3. Tap. 96
4. Turn. 167
5. Form 103
6. Cutoff. 167
                           825

7. 5 percent. 41

Total revolutions. 866

A cycle per part of 866 revolutions, when divided by 1500 rpm, indicates a total operating time of 0.577 minutes, or about 35 seconds. In addition to the gross machine cycle time of 0.577 minutes or 104 pieces per hour, al-

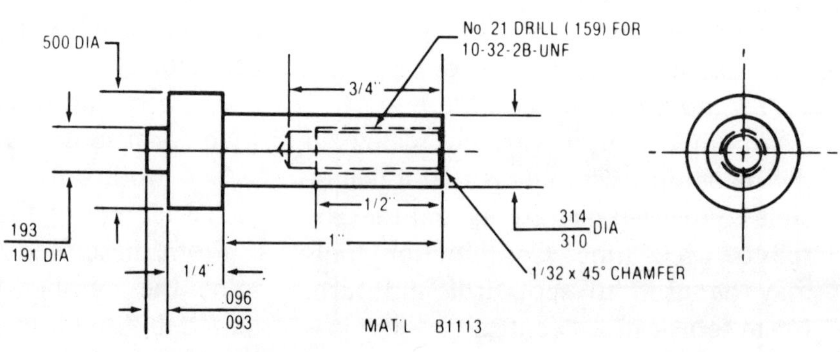

Fig. 5. *Step shaft of B1113 CRS must be estimated at two points: (1) as a screw machine manufactured part and, (2) as part of a subassembly operation. Screw machine manufacturing costs are broken out on accompanying estimate form.*

QUANTITY	2000	5000	10000	20000			
SET-UP	26.25	13.12	6.56	3.28			
RUN	192.00	192.00	192.00	192.00			
MAT'L.	32.00	32.00	32.00	32.00			
H.T.							
PLATE							
TOTAL	250.25	237.12	230.12	227.28			

PART NO. 12345

COMPANY XYZ

DIVISION

BUYER

REPRESENTATIVE

COMMISSION ☐ YES ☐ NO

X-L ENGINEERING CORP.

DATE

TOOLING $50.00 Form Tool

OPER. NO.	OPERATION		MACHINE-TOOLING		MINS. Per Pc.	SET-UP (Hours)	PIECES Per Hr.
10	Purchase Material (100 Pieces/Bar)		.500 Diameter B-1113				
		Revolutions	Automatic Screw Machine				
20	Feed Stock						
	Spot Drill	42					
	Drill	250					
	Tap	96					
	Turn	167					
	Form	103					
	Cutoff	167					
		825					
	Plus 5%	41					
	866 Revolutions @ 1500 RPM			Gross	.577	3.5	104
				Net	.769	3.5	78
30	Tumble to deburr						
40	Final Inspection - Pack - Ship						
				Total	.769	3.5	

STAMPING QUOTATION

CUSTOMER VIII   PART NO.   SUB-ASSEMBLY   DATE 11/1/76

NOTES:

1. Prog......... Blank........ Comp......... Die $ SEE PREVIOUS QUOTES FOR TOOLING

2. 1st Form Die

3. 2nd Form Die

4. Other (Pierce) Die

5. Mill/Drill Fixt.

6. Other

Total $.

PRODUCTION MATERIAL

QUANTITY LEVEL		2 M	5 M	10 M	20 M
Material Costs	$	$	$	$	
Inspection, Check, Set-up					
Blank STAMPING		121.00	71.50	54.25	39.10
Deburr SCREW MACH. SHAFT PART		250.25	237.12	230.12	227.28
1st Form					
2nd Form					
Other ASSEMBLY) STAKING)		45.00	40.00	35.00	35.00
Mill.					
Grind.					
Drill/Ct'sink FINISH BARREL)					
Deburr ZINC CHROMATE))		15.00	14.00	13.00	13.00
Tapping					
Plating					
Inspection		12.00	5.00	3.00	2.00
Die Maintenance					
Ship		5.00	3.00	2.00	2.00
	$	448.25	$ 370.62	$ 337.37	318.38
Plus .... %		44.83	37.06	33.74	31.84
TOTAL		$ 493.08 per M	$ 407.68 per M	$ 371.11 per M	350.22

Del. Tooling   ROUND OFF TO $493.00   Production   ROUND OFF TO $371.00

F.O.B. Our Plant

Estimating

lowance must be made for tool sharpening, inspection, chip removal, and sorting. We will approximate net production at 75 percent of gross production. Therefore the net cycle time will be 0.577/0.75 = 0.769, or approximately 78 pieces per hour.

For setting up the screw machine allow 1/2 hour to get prints, cams, and other tooling and 1/2 hour to set each position; a total of about 3 1/2 hours. Naturally, different machine tools will require more or less time. If quantities warrant a multiple-spindle machine, setup time will increase but cycle time will decrease.

Using an hourly rate of $15.00 (25 cents per minute) we have the following:

1. Setup = 3.5 hours x $15.00/hr = $52.50

2. Cycle time = 0.769 minutes x 0.25/minute = 0.192 each or $192.00/M

3. Material @ 40 cents per pound
   Piece part wt. = 0.08 lb
   0.40 x 0.08 = 0.32 ea. or $32.00/M

4. Form tool = $50.00

The form tool is separated from the rest of the quote since it will be a one-time charge.

From the accompanying estimate sheet, the final quote is seen to be as follows:

    2M $250.25/M
    5M $237.12/M
    10M $230.12/M
    20M $227.28/M
    Tooling: $50.00 form tool

The Subassembly Estimate

The illustrated subassembly estimate, briefly, is a combination of the costs for tooling, stamping, screw machine parts and assembly. Other additions such as plating and shipping are the final cost additions. The final figures are as follows:

Tooling—Design and Build

Progressive die $5,988.00
Screw machine form tool   50.00

    Total $6,038.00

Production—Complete Subassembly

    2M . . . . $493.00 per M
    5M . . . . $407.68 per M
    10M . . . . $371.00 per M
    20M . . . . $350.22 per M

This concludes our estimating series. It is hoped that, through this primer, some of the estimating mystique has been removed. Success in estimating comes about through daily association and practice; cross checking the actual cost of the jobs you have estimated and keeping abreast of the ever-changing technology in the metalworking field.   ● ● ●

CHAPTER 3

DIE DESIGN

Reprinted from Modern Machine Shop, March 1971

Facts Every Die Designer Should Know

By Lawrence A. Wacker
President
Sterling Tool & Manufacturing Company
Milwaukee, Wisconsin

Quantity production of stamped parts at lowest cost requires a study of the various methods of manufacture. This study is the art of process planning, and ramifications such as the ultimate cost of the finished product, accuracy requirements and comparative ease of tool maintenance must be considered. The facts which follow will help you achieve greater production at lower cost.

PLAN AHEAD

1. Study the part itself before designing a die. Will the part be comparatively easy to make or will some simple redesign make it easier and cheaper to produce? Can a lighter die material be used by taking advantage of ribs or beads? Possible stress points must be evaluated and strengthening beads added where necessary.

2. Determine the elongation of material from forming, drawing, beading, and so on. Generally, elongation of steel should be no more than 12 percent in two inches.

3. Consider the secondary operations to keep handling costs to a minimum.

4. Costs of assembly are usually high; therefore, ease of assembly is a large factor in redesigning a part for high production.

5. Consider finishing operations such as polishing, plating or painting so the appearance won't be impaired.

6. When the best finished product design is finalized, make a further study of the part for the best method of manufacture. This type of study may seem elementary, but, when the various types of dies are considered, it could become complex. Just think of some of the types of dies - beading, bending, bevel-

ing, blanking, bulging, burring, cam, closing, coining, cold heading, crimping, cupping, combination, compound, curling, drawing, dimpling, dinking, double-action, embossing, expanding, forming, hemming, horn, joggle, lancing, multiple, perforating, piercing, progressive, riveting, sectional, shaving, shimmy, side-action, single-action, swaging, trimming, two-step, two-stage, wedge, extruding, flaring, flattening, fluting, indenting, indexing, ironing, necking, nosing, parting, planishing, redrawing, reducing, reeding, restriking, seaming, shearing, sizing, slitting, staking, stretching, striking, tapering, wiring, pinch-trim, reverse-draw, reverse-trim and so on.

7. Determine tooling requirements. High production tooling may consist of any number of the above dies or operations built into a single component. That is why it is not easy to design such a tool. The size or shape of the workpiece may indicate that single-operation dies answer the cost problem although progressive dies, transfer dies or multiple operation single-stage dies using Die Draulic principles may be the most economical in the final analysis.

8. Study the workpiece for stock grain - especially if bending, forming or drawing is involved.

9. Make a stock-strip layout to show the sequence of operations, keeping in mind the following:

Will dimensional requirements be easily achieved and maintained as the strip progresses through the die?

Will the carrier strip be of sufficient strength for continuous operation of the die? (A rule of thumb is that the minimum carrier strip should be equal to the stock thickness or 1/16 inch, whichever is greater)

Will the burrs on sheared, blanked or pierced parts be located where the least difficulty will be encountered in future handling and where costly burr removal will be minimized?

Will the press selected for the job be adequate; that is, can the die be located properly in the press to achieve balanced tonnage?

DESIGN THE DIE

10. Use formulas for piercing, punching, shearing, forming and drawing pressures to establish the center of force and work from this center as closely as possible.

11. Establish the locations of relief areas, notch and pierce punches, pilots and so on.

12. Don't forget the strength required for punch and die sections - make them as heavy as possible and "heel" sections where necessary. Correct designing of sectional joints for ease of manufacturing and maintenance is of prime importance.

13. Consider the necessary station stops, automatic stops, die safety, workpiece ejection and scrap removal when designing a progressive die.

14. It is rather difficult to keep certain areas flat in progressive dies. Restriking may be necessary; remember, pressure alone seldom does the job because spring-back continues to occur. The overbending principle - using a center point of force and two high points of pressure - will usually solve the problem. Where angular areas spring back, it is best to overbend to the opposite angle.

15. Use positive-return type cams in progressive dies to accomplish side forming, lancing, piercing, and so on; damage usually occurs when a spring-return type fails to retract to its retarded position.

DIE STEELS

16. Use oil and air hardening steels when production will not exceed 500,000 pieces and if the operations involved are not too strenuous. When production is greater or when high wear points are detected, the use of high carbon, high chrome steels are in order; these are usually known in the trade as the high type and the low type. The high type, D3, contains 2¼ percent carbon, whereas the low type, D2, contains 1½ percent carbon. Other additions such as molybdenum and cobalt increase the wear resistance of these steels.

17. When production runs into the millions or when materials are abrasive, carbide dies may be required for satisfactory die life.

18. In recent years, steels such as Ferro-Tic - a hardenable tool steel impregnated with carbide - have taken over in many wear applications where only carbide would work previously. While this material costs about $30 a pound, it is not as expensive as carbide. Its machinability is excellent-similar to a heat treated alloy machine steel. Distortion in heat treating is so minute that grinding can usually be eliminated. For many applications, its wearability is so close to that of solid carbide that very little difference can be detected.

DRAWING OPERATIONS

19. The cold forming of flat blanks into hollow shells or sections of shells is commonly referred to as the drawing of metal. A metal blank is held between the die and the pressure ring (which holds the material flat) while the punch begins to form the metal into a shape.

20. When drawing round shell, the metal on the sides takes on a different shape and the action of the pressure ring is instrumental in lengthening the sides of the shell being drawn. A shell two inches in diameter and two inches high requires a blank of only a little over four inches in diameter.

21. The old rule of thumb stipulated that, in one draw, the diameter of the shell and the height of the shell should be equal; for instance, a cup two inches in diameter could be drawn two inches deep. In practice, however, stay slightly below the maximum when possible. Try to limit the diameter of the first draw to 45 percent reduction of the diameter of the blank. For instance, a five-inch-diameter shell six inches deep requires a 12.040-inch-diameter blank. A 45 percent reduction would result in a 6.620-inch-diameter shell after the first draw.

22. In subsequent draws, limit the reduction to 30 percent. For example, the shell cited above can be reduced to 4.634 inches in diameter; therefore, there will be no difficulty in drawing a five-inch-diameter shell six inches deep in two draws.

23. Some materials, such as stainless steel, work harden rapidly and must be annealed between subsequent drawing operations.

24. Zinc is one of the easiest

metals to draw because it work
hardens very slowly. In exper-
imenting with zinc quite a few
years ago, we drew a 36-inch-
long cup without an anneal between the myriad operations required to make this length.

25. When drawing square or
rectangular shells, a round or
an elliptical blank, respectively, can often be used for
the first draw. Subsequent
draws are made on a sleeve the
shape of the first draw - which
can often be a round draw in
the case of a finished square
shape or an elliptical draw in
the case of a rectangular box.

26. The drawing of rectangular
or large square boxes almost
invariably requires stretching
the sides of the box after the
last draw to eliminate "oil
canning." This stretching can
be accomplished during the last
draw by increasing the holding
pressure between the pad and
the die ring and stretching
from ¼ to 3/4 inch. If the radius is to be made smaller, a
reversible draw ring - with a
smaller radius on the opposite
side - can be used at the same
time.

27. When using progressive
dies, the metal strip must usually be lanced in the first
stage or two to permit free
movement of the metal in the
carrier strip. Usually, an
idle station follows the last
lancing station to facilitate
the greatest possible freedom
of movement with the least a-
mount of distortion of the carrier strip.

28. A series of draw reductions can be made without the
use of draw bushings provided
the reduction is approximately
10 percent per stage. It is
imperative that the exact area
of metal drawn in the first

station will be maintained for
all subsequent draws. Normally,
the metal will not withstand
crowding or stretching in this
type of progressive drawing
operation.

PROGRESSIVE DIE FEEDS

29. Various feed innovations
have been developed in recent
years and one of the most interesting is the endless belt feeding of parts through a progressive die. Due to the difficulty
of carrying some parts through
a die of this type, it may be
easier to stamp out blanks, ori-
ent them through various means
such as a vibratory feeding hop-
per, then use an endless belt
of spring steel to carry the
parts through a progressive die
to perform further operations
such as shaving, piercing, flattening, forming and so on. The
endless belt contains many
holes, which have the same configuration as the part to be
progressed through the die. A
standard type of hitch feed ad-
vances the part to the approximate location. The endless belt
is notched with a register and a
pilot to guide the belt to its
exact operating location. Safety
units, such as micro switches,
can be installed in this type of
die to eliminate problems of
misfeed or accidental improper
orientation. Production of 4000
to 6000 parts per hour are common with this type of feed.

30. Another method of moving
parts through various press
operations is by the use of
transfer feeds. This method is
becoming more popular each year.
The carrier strip is eliminated;
stock is fed into a blanking
station and the blank then carried through the series of operations by the use of a shuttle
device actuated by means of a
rack and gear, a cam, or an air

cylinder. The part is gripped
by a finger bar located on each
side of the die. The finger
bar is the reverse shape of the
part at its particular station.
Actually, each station is a
separate punch and die. For
this reason, it is comparatively easy to make repairs should
damage or wear cause a shut-
down. Spare dies can be maintained for those stations that
may require frequent maintenance. Because the operations
are broken up into simple stations, this type of die normally requires very little maintenance.

31. The motion of each station
is coordinated with the transfer mechanism so that on the
upstroke a dwell occurs during
which time the contact of each
part is transferred from the
punch to the fingers of the
transfer mechanism. The fingers close in at a constantly
decreasing speed and practically come to a standstill when
grasping the part.

32. Recently, a new idea in
transfer has appeared in multistation presses wherein each
station is individually cam
operated - completely independent of each other. This permits raising the speed of drawing; for example, carbon steel
parts can be increased from 80
feet per minute to 200 feet per
minute. Other materials can be
similarly advanced in speed.
This type of transfer press had
been limited to small parts,
but recently the same principles have been applied successfully to steel blanks 16 inches
in diameter and 0.100 inch
thick. A 7-inch draw on a
16-inch blank at 33 strokes per
minute and smaller jobs running
at speeds up to 250 strokes per
minute are common.

33. While speeds of 30 strokes
a minute have been considered
tops for large transfer mechanisms, the use of air transfer
feeds has raised this figure.
Occasionally, devices have to
be designed to restrain the part
because of the speed of transfer
and the weight of the part being
moved.

MULTIOPERATION, SINGLE-STATION
DIES

34. For this operation it is
necessary to have variable die
pad pressures. With hydraulic
pistons built into the die itself, it is possible to blank,
draw, form, pierce, trim, coax
and curl - all in one single
stroke of the press. Hydraulic
cylinders, controlled by valv-
ing, take over the necessary
timing of the various die parts
in both the punch and the die
sections. Multistage relief
valves release the pressures
from certain die sections when
they are no longer needed.

35. Hydraulic principles are
also useful for knockouts where
maximum spring pressure would
be inadequate; thus, a job previously impossible is made a
simple task. Delayed forming
and area preforming are also
easily accomplished through the
use of hydraulics.

PRECISION BLANKING

36. It may not be practical to
produce some parts in a conventional high production die because of costly edge finishing
or hole shaving or reaming operations. Parts such as small
cams, gear segments, instrument
components and so on can be produced through a method called
precision blanking or fine
blanking.

37. Triple-action presses are
used in this method of blanking.

Pressures are required to hold the strip from which the part is to be blanked to produce the blank and to eject the part. In a precision blanking press, the cam is driven with a toggle action which slows the punch speed to 1/2 to 2/3 the normal rate at the actual moment of blanking and then has a rapid return stroke.

38. Part of the success of the operation is the scoring of a deep groove around the blank by the upper pressure pad, which has a raised V or wedge ring machined on its lower face close to the cutting line.

39. Zero clearance between punch and die would normally force the metal outward; however, the annular ring and the high pressure between plates cause transverse compression which forces the material to flow toward the punch before and during blanking. In fact, when cutting thin material, the punch doesn't actually enter the die.

40. Extremely rigid die sets are required and dies have to be four to five times as heavy as standard types because of the lack of clearance between the punch and die.

41. Operation is fast with blanking speeds of 900 to 2100 strokes per hour being quite common when using coil stock and automatic feeds. A greater amount of scrap is produced because of the required impingement ring around the blank.

42. High pressures are required for fine blanking because of the forces needed to press in this wedge ring and to eject the part which, because of zero clearance, is practically a press fit in the die block. A 200-ton press, for instance, is required to pro-duce a blank which would nor-mally require a 120-ton press.

Triple-action presses normally required for fine blanking are very costly; therefore, the number of companies doing this type of work is limited to a comparative few. Recently, a new development has minimized the investment for this process to approximately 15 percent of the cost of a fine blanking press. Sterling Tool & Manufacturing Co. engineered and developed a system which does the work of fine blanking by adding pressure pads to a standard press - either mechanical, hydraulic or knuckle joint. The only requirement is that the press or presses have sufficient shut height over and above the die height to accept the pressure pads. With the added factor of interchangeability between different presses it brings the possibility of fine blanking within the financial scope of many organizations which normally could not afford the cost of fine blanking presses.

DIE LONGEVITY

43. Heat treatment and grinding greatly affect the longevity of a die.

44. It is most important to heat treat tool and die steels correctly; that is, according to the recommendations of the tool steel company. Some metal treaters have a tendency to lump similar steels into lots for heat treatment, and this can cause estimable damage. When improper heats are used, the heat treater may be lucky and not have the material crack, but the useful life of the tool will be curtailed and the part will often go out of shape to a much greater degree than it would have if the proper range of heat had been held.

45. Heat treatment to minimize
scaling and decarburization
should be followed and the latest equipment and practices
should be used. The small a-
mount of additional time and
labor in proper heat treatment
is reflected hundreds of times
in the improved life of the
finished die.

46. Grinding contributes substantially to short die life.
Grinding after heat treatment
should be kept to a minimum.
Metal should be removed at high
work speeds with light cuts so
as to spread out the heat generated to as large an area as
possible. Proper wheels are of
great importance as a wheel
that glazes the steel will damage it - sometimes beyond repair. Some steels are more
sensitive to grinding than
others, and extreme care must
be exercised in grinding the
high carbon high chrome types
of steels, Relatively soft
grades of wheels must be used.
Wheels that will break down and
continue to present a porous
cutting surface usually are
best for these types of steels.

47. Direction of grinding is
always important. Die steel
should be ground in the same
direction as the travel of
stock through the die to lessen
the stresses set up in tool
steels through friction. Grinding also sets up stresses, and
in the case of extremely intricate tooling it is advisable to
stress relieve the die parts at
about the same temperature as
that used in the original heat
treatment.

48. Regrind a dull die before
the edges get badly rounded;
greater die life will be obtained by sharpening the die
more frequently.

49. When finish grinding a

die block or punch always remove the burr with a fine
stone. A burr left on a die
part causes the corner to break
on the first hit of the press;
consequently, the die is slightly dulled before production actually begins.

50. Jig grinding the holes in
high production dies removes
any decarburization that may
exist after heat treatment and
usually increases the burrfree
life of the hole by 30 to 35
percent. This operation will
increase the life of the punch
by about the same percentage.

51. Sometimes, due to the
smoothness of a jig-ground hole,
the slug will stick to the punch
and pull out with it. Ejector
type punches are not the real
answer to the problem on high
production dies due to the relatively short life of the spring
and plunger. The proper way to
eliminate this problem is to
jig grind the holes with a
½-degree reverse taper for at
least the thickness of stock
being punched. This almost invariably solves the problem and
does not interfere with die
life.

Principles for Blanking and Perforating

By Karl A. Keyes
General Manager
International Fineblanking Corporation

Since conception, metal stamping dies have basically looked alike. The early improvement which introduced the die set resulted in all dies appearing very similar in size, shape and basic construction. For this reason, it is difficult for many to believe there has been any change in dies over the years. We find from close examination, however, that some very radical changes have taken place. It might be said that these changes in the design and construction of metal stamping dies have occurred through a slow moving evolutionary process; and, it actually did take several years to span the gaps between major breakthroughs.

Most of the early breakthroughs resulted in reducing the burden of construction rather than increasing the production efficiency of the die. One example of such a change took place in the mid-40's when the chase type construction, used to retain shape blade punches, was replaced by the

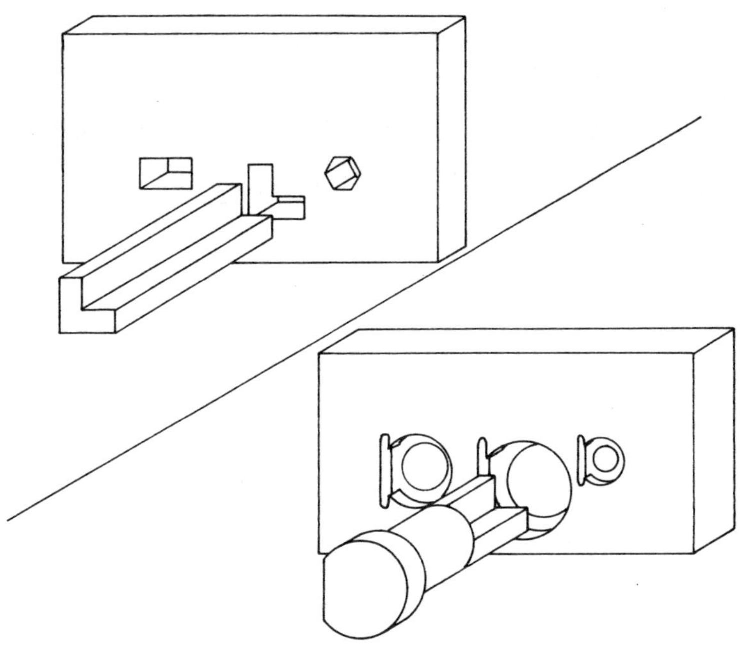

Figure #1
Shaped Chase vs Rd. Componentry

shaped point round shank perforator. Some improvement in
production efficiency was experienced. But the primary reason for the change was to eliminate the tedious task of
machining and filing necessary to fit the shaped blade punches
into the chase.

After the switch was made to the round shank perforator,
adding shapes in a die became as easy as machining a round
hole at the right location. The only precautions necessary
were maintaining the diameter and squareness of the hole in
the retainer plate. After the machining was completed, the
punch was inserted into the hole, oriented to position the
shape and then keyed in place.

Almost instantly, die construction changed across the

United States. Now more time was available to concentrate on areas in dies which offer the greatest challenges in order to meet the requirements of part drawings.

An interesting point to consider is that the shaped internal diameter matrix button was introduced at approximately the same time as was the shaped point round shank perforator. Yet, the shaped ID matrix button did not receive the same degree of acceptance as the perforator. A close examination of the benefits from incorporating shaped matrix buttons into a die revealed that not only a savings in die construction could be expected but also longer production runs with a reduction in die maintenance problems. Further, the acceptance of the shaped matrix button could have made the dream of a totally renewable, high production metal stamping die a reality. But today, shaped perforators are still used 4 to 1 in relation to shaped ID matrix buttons.

Other important benefits did come from the use of round shank perforators in die construction. One of these was the placing of the prime responsibility for die construction in the hands of the die designer. Although this statement is true, it took a few years before the industry accepted the fact. Another benefit, coordinate dimensioning of dies was now not only possible but practical. Another few years would again pass before the average die designer spend much time in coordinate dimensioning dies.

The fabulous 60's put other demands on dies besides the ability to produce good parts. Now the die had to operate more efficiently over longer periods of time producing the same stampings still within a degree of accuracy acceptable by product engineering. With this new demand, further examination was made of the die construction in hopes of isolating the areas in the die which contributed to the greatest frequency of die failure. The findings pointed out that the areas which were the most difficult for the die maker to make were not the areas which most frequently broke down causing the die to be removed from the press.

Determination was made that the prime source attributing to press down time was the cutting sections. More dies have been taken out of production because of excessive burrs on the piece part than because of any other type of failure. Of the various cutting punches, the major contributor to die failure was the perforators. To make matters worse, certain perforators would outlast others, so the production life of a die in some instances could be controlled by the wearability of one single perforator.

Perforators possessing acute angles with sharp corners will be the first ones to break down. Sharp corners formed by right angles are next to go, then radius corners, then oblongs and finally round hole perforators. When a die is built with only round punches, the smallest round punch will usually fail

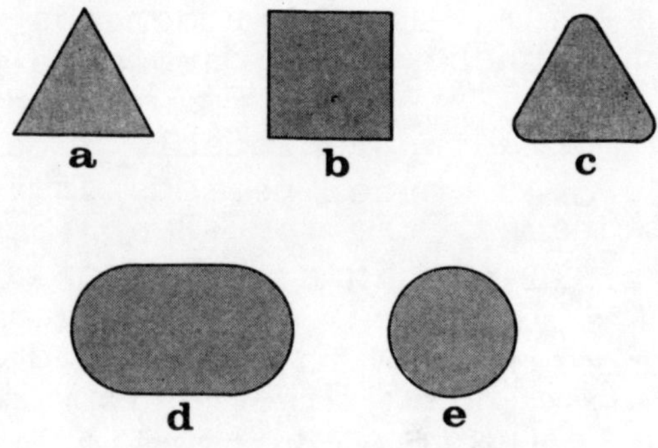

Figure #2
Shapes of Punch Points

first. Once it was determined that the perforating sections
played such an important part in the production life of a die,
steps were taken to prolong the life of perforators.

There are six die principles governing cutting sections
which must be followed in order to extend die life for any
prolonged period of time.

These die principles deal with the considerations
necessary to properly control Alignment, Stock, Shock, Slugs,
Wear and Loads. Conditioning the perforating sections of a
die by these fundamentals could virtually lead to the construction of the ultimate die.

DIE PRINCIPLES

ALIGNMENT. Alignment is by far the most important
of all the die principles. Without proper align-

ment, an unbalanced load exists around the periphery of the punch point, causing a premature uneven wear of an unpredictable nature. As this condition worsens, the resultant load will force the punch off location, eventually oversizing the shank retention hole. The result can be as minor as an excessive burr appearing on the stamping or as extreme as the breaking of the punch.

Establishing and maintaining proper alignment in the die is a result of a combination of skills on the part of both the machinist and the die maker; as well as the condition of the press in which the die will be run. A good die design can also aid in assuring better alignment.

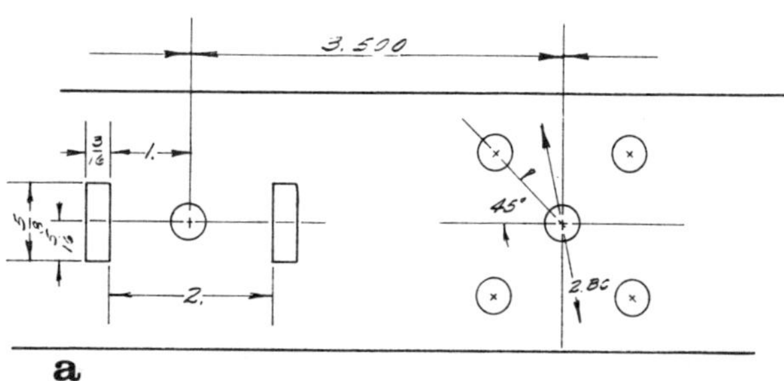

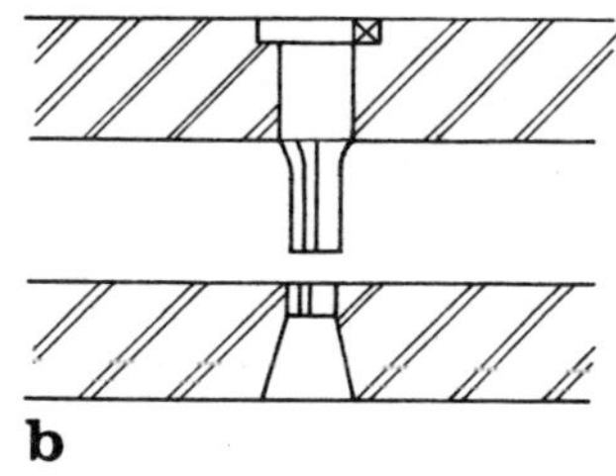

Figure #3
Alignment Problems

Coordinate dimensioning from one zero datum point
should be used to obtain the needed relationship
of all the tooling and pilot locations throughout
the die. Coordinate dimensioning will place most
of the responsibility for proper alignment in the
machine tool, thus reducing the chance for human
error.

Matched Set Tooling (punch shank and matrix button
of the same diameter) incorporated into the die
design will allow a relatively easy machining operation to eliminate the difficulty involved in
aligning a shaped point round shank perforator with

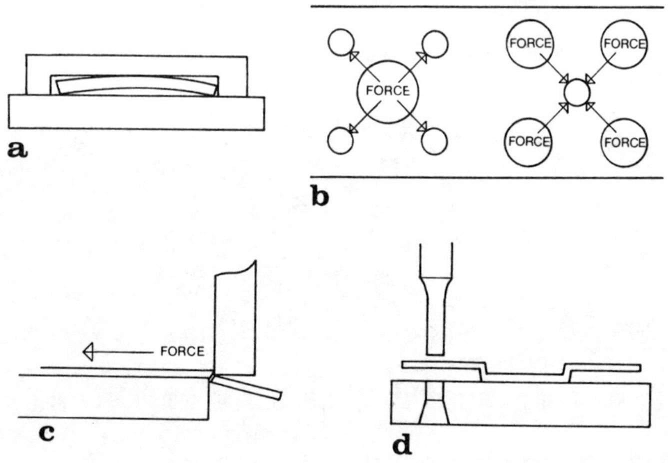

Figure #4
Stock Control Problems

its mating hole in the matrix block.

<u>STOCK CONTROL</u>. Stock control deals with the ability to maintain position and support of the stock during the entire stamping process. This principle considers stress and strain as well as restrainment of the stock.

Proper restraint of the stock can be accomplished by:

> (1) a spring loaded stripper with leveling pads,
>
> (2) pilots to accurately position the stock
>
> (3) anchor pins to support against lateral movement created by notching, severing and bending operations.

Small perforators must be positioned so that the stress influence of larger perforators will not promote premature wear.

Adequate support must be placed under the stock to eliminate buckling or other deflection which could result in the reduction of punch life.

<u>SHOCK CONTROL</u>. Shock control has to do with adequately designing the punch to withstand the shock involved when the punch penetrates the stock, as well as when fracture occurs. The punch must be designed in such a way as to allow the shock to

pass from the punch point into the punch shank
and finally out into the mass of the die. Any
restriction to these tremendous forces could
literally cause the punch to explode.

To properly stabilize perforators against shock
it was determined that the area of the perforator
shank should be two times the area of the point
with the head 4 times the point area.

The length of the punch point must also be considered. When using unguided punches of tool steel,
the point length should not exceed two and a half
times the point diameter. This ratio may be in-

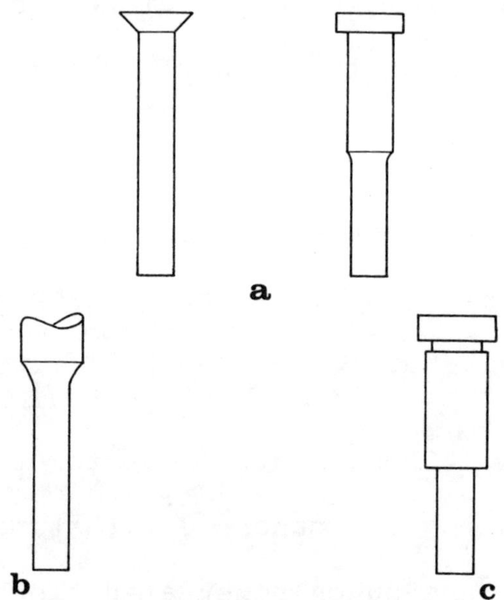

Figure #5
Shock Control Problems

creased to four to one when the perforators are
made from high speed steel.

Supporting the punch to withstand shock is not
enough. The punch must also be designed to control shock. This is accomplished by placing radius
blends at the transition point between the point
and shank and also where the shank meets the head.
The proper radius assures a smooth transition of
the shock wave from the point of the punch through
the punch plate into the mass of the die.

<u>SLUG CONTROL</u>. Slug control pertains to the elimi-
nation of slug problems which is considered by many
to be the number one problem in dies. It might be

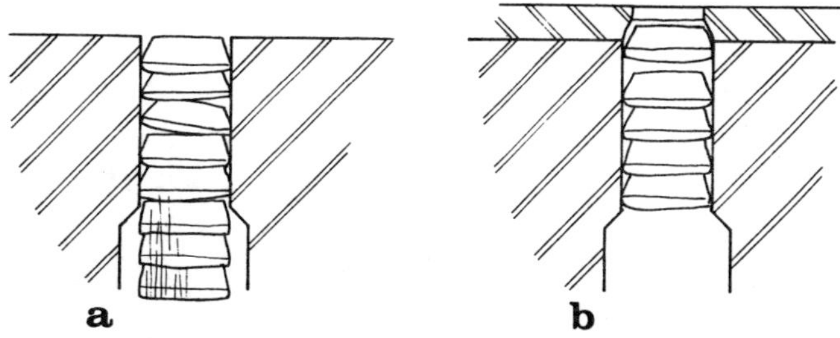

Figure #6
Slug Problems

interesting to note, in passing, that the original
reason why clearances were selected was to offset
the difficulties that occurred from slug pulling
problems. Pulling is, however, only one of the
slug problems; the other is jamming. Slug problems
are difficult to solve because solutions used to
overcome slug pulling may result in promoting slug
jamming or vice versa.

When slugs jam into the die cavity, restricting the
flow of additional slugs through the die, the punch
is subjected to an extremely high compressive load
which will cause breakage of the punch point. The
tendency for slugs to jam can be reduced by following these recommendations;

    (1) the die land should contain no more than
        six slugs

    (2) a tapered relief is needed below the
        straight land

    (3) a clearance hole of the proper size must
        be put through the lower die shoe.

(Note: With stock thicknesses less than .020 the
clearance hole should be tapered to match the relief hole to assure the slugs maintain proper
orientation as they leave the die.)

Slug pulling takes place when the slug is lifted
from the die cavity and partially trapped in the

stock strip. Thus, the strip is restricted from
further travel. Slug pulling can be overcome by:

    (1) using Jektole punches (a slug ejector
type punch designed for ease of maintenance)

    (2) emitting air through the punch

    (3) installing vacuum devices in or under
the die

    (4) attaching resilient materials to the
punch to hold the slug in the matrix
cavity as the punch is withdrawn.

(There are other ways to reduce slug pulling, but
they tend to reduce die life, and should be left
to the die expert for use only as a last effort!)

WEAR CONTROL. Wear control considers the selection
die steel as well as engineering the correct punch
to matrix clearance.

It has been determined that 2/3 of the perforator
wear occurs as the punch is withdrawn from the
stock. Since clearance plays an important part in
development of the hole size, selecting the correct
clearance is essential to reduce the amount of hole
closure on the punch at withdrawal. There is no
single right answer for proper clearance since
various materials will react differently. It is
suggested that the correct clearance be determined

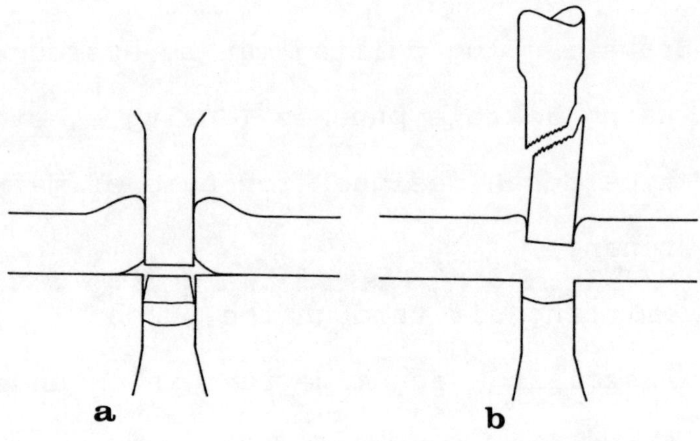

Figure #7
Wear Problems

through experimentation. Recent research has told
us that doubling the presently accepted clearance
will do much to increase tool life with little or
no effect on the stamping.

Further promotion of better wear control requires
use of the right tool steel. Tool steel must be
coordinated with the part material and the operation to be performed. It cannot just be selected
for the ease of machining and heat treating or
ready availability. The greatest concern in tool
steel selection is the shape of the cut to be
made. (See Figure #1)

A die should not be restricted to one tool steel,
or a single tool steel be selected just because
one perforator within the die requires that particular material. Selection of more than one type of

tool steel may be necessary to create balanced wear
throughout the die.

The following chart offers some tool steel recommendations as they relate to the shock and abrasive
characteristics of piece part material.

PART MATERIAL CHARACTERISTICS		TOOL STEEL
Shock	Abrasive	AISI
med	med	A-2
low	med-high	D-2
med	med-high	M-2
med	high	M-2/Nitride
high	high	Ferrotic CM
low	ultra-high	Carbide

LOAD CONTROL. Load control relates to the proper
placement of the resistance load in the die so that
this load is located very near the center of the
action force of the press, side to side and front to
back.

In cases where it is impossible to properly center
the resistance load in the die, it may be necessary
to add load leveler blocks to minimize the amount of
deflection from an unbalanced load condition.

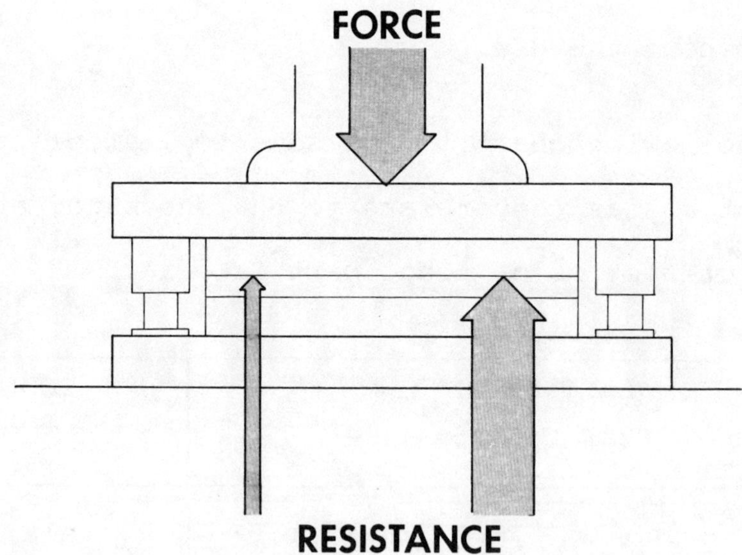

Figure #8
Load Control Problem

Stripper springs as well as shedder knockouts must
be placed in a central position related to the stripping of the stock with the least amount of cocking
action.

These examples illustrate some of the ways the die
principles affect die design.

COMPUTERIZED DIE DESIGN

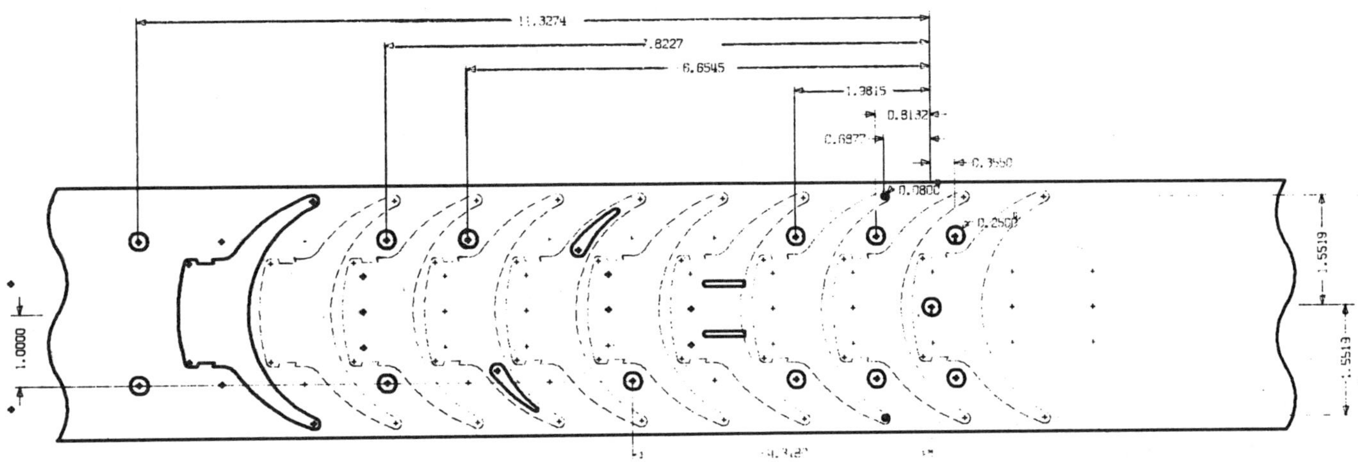

Computer-created strip layout orients part for minimum scrap, allows for stock size and grain direction requirements, and gives sequence of operations according to best accepted design practice. If, for some reason, layout is not acceptable, it can be readily redesigned.

Progressive dies are completely designed in less than a day by computer. Input is a product drawing with material specifications. Automatic output is a computer-generated strip layout, die and punch layouts, and fully dimensioned shop drawings giving die-part details. Complete bill of materials and management die and piece cost information are simultaneously developed. NC tapes for die machining can be a by-product.

H. BREDIN[1]

Now there's a computer solution to the die designer shortage that is expected to cut the lead time for new-product introduction to a mere fraction. Literally, the equivalent of hundreds of years of die designing experience have been encoded into the electronic memory of a fully computerized engineering system developed to speed the design of progressive dies. From its customer's dimensioned product drawing, stock description, and quantity requirements, DieComp, Inc.,

South Plainfield, N. J., claims its system will produce a complete progressive tool design—including detailed shop drawings—all in a day's time. Within a year, with the aid of faster plotters, the company expects to reduce its designing time to a matter of hours. In comparison, similar dies designed by traditional methods may take several months.

Management Benefits a Bonus

Almost as a by-product, the PDDC (Progressive Die Design by Computer) system simultaneously develops much critical and essential management information, often unobtainable in common practice. The system will calculate the cost of the tooling-up process, die-building time, and the cost per piece at various production quantities. Since the computer has generated the design of the tooling, it also knows how much machining and material are involved. This information is converted automatically into die-building time and cost values. The data can also be helpful in scheduling die-component machining and manufacturing control. An automatic output of the system is a complete bill of materials for the die in a form useful in inventory control and in simplifying purchasing procedures.

The basic system, for which patents are now pending, has been five years in development. DieComp, Inc. was formed in 1969 and just recently produced its first

---

[1] Associate Editor, MECHANICAL ENGINEERING.

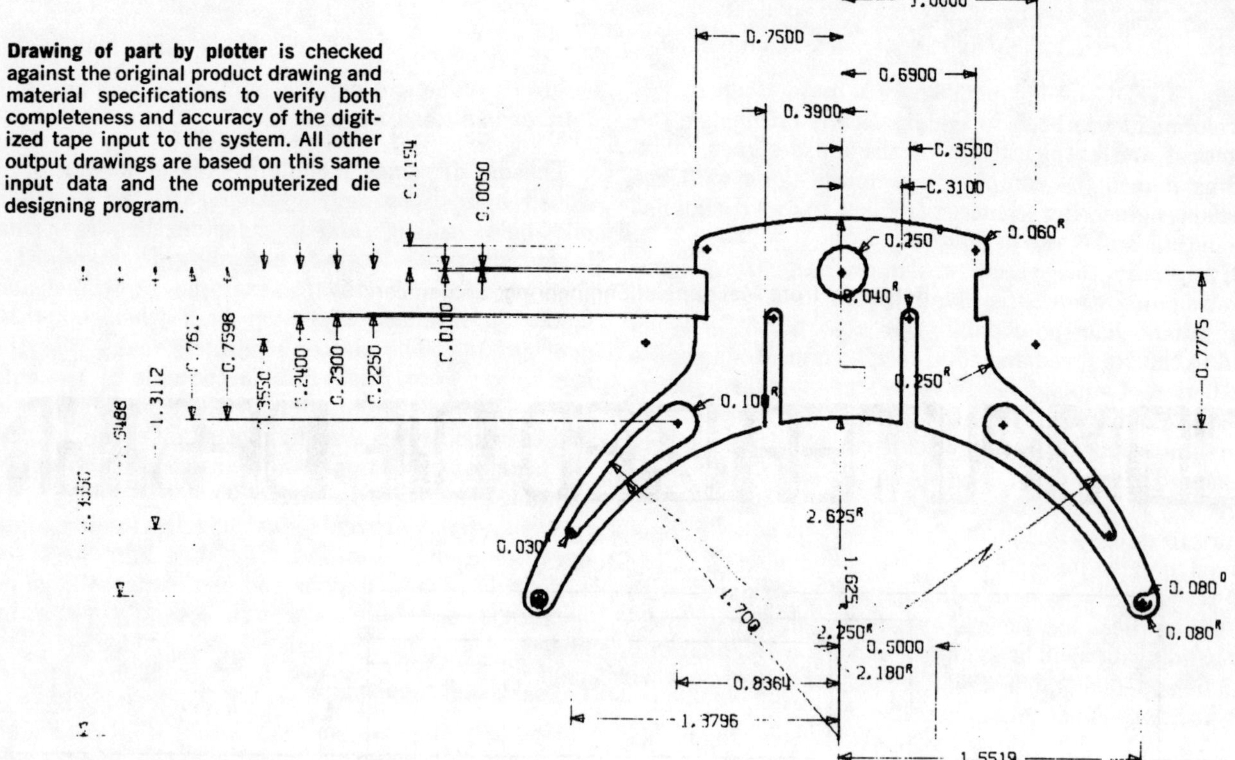

Drawing of part by plotter is checked against the original product drawing and material specifications to verify both completeness and accuracy of the digitized tape input to the system. All other output drawings are based on this same input data and the computerized die designing program.

Die-component drawings, fully dimensioned for shop production are automatically produced as part of the computerized die-design package. Enlarged insert drawings, assembly drawings, a full bill of materials with listings for standardized parts, their manufacturer, and catalog number are additional outputs.

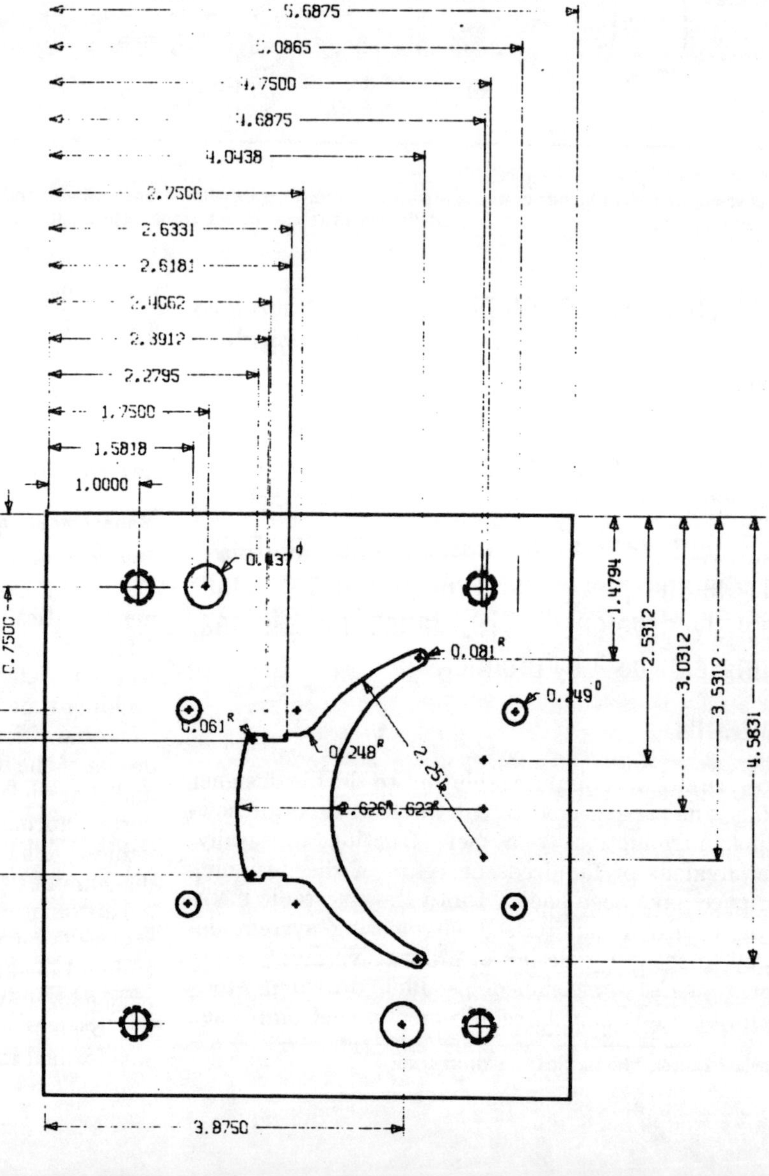

salable PDDC design package. A major step in the development has been to systematically rationalize the practical and empirical art of the die designer. Encoding it into the computer's memory along with the various engineering sciences pertinent to tool design has taken time and is still in progress.

Eventually the system will have many times more die-designing knowledge available at its keytips than any human has the capability to absorb in a lifetime. Add to this its speed and total recall, and the company's prediction of cutting design times from months to hours seems reasonable. Actually, even under present system limitations, the computer time requirements are measured in minutes. The big bottleneck in the system currently is in the plotters, which now require several hours to automatically draft a complete set of dimensioned die drawings.

Presently the system's full capability is directed at stamping dies. In the very near future full capability for bending dies will be available and draw-die capability will be added shortly thereafter. Even now some work is being done experimentally on bending dies.

How the System Works

The customer provides a hand drawing or sketch of the new product complete with all dimensions, tolerances, and material thickness and specifications. Die-Comp first digitizes the drawing, punches the information on tape, and feeds it into the computer. A plotter immediately produces a dimensioned check drawing which the customer then is asked to verify as correct. This is the only time in the entire process that a print needs to be checked. Every other output drawing is guaranteed as correct with all dimensions being derived from the original input data. Output drawings can be automatically dimensioned to different reference points for inspectors, assemblers, or for the machine shop.

Next, the system produces a layout with the part oriented on the strip material for production with minimum scrap. The minimum web and the geometry of the strip layout are automatically calculated and decided by the computer according to standard shop practices. By incorporating input information on any restrictions due to the customer's metal grain direction specifications or stock strip widths, the system will accordingly select the most acceptable strip layout. The computer also calculates the stock utilization factor, the actual weight of the stock and scrap, and the exact periphery being pierced or cut. This length of cut, along with the thickness of the material and its shear stress, is used by the computer in calculating the minimum tonnage of the press and, going one step further, determines the number of stations required.

After the sequence of operations is determined by the computer, a strip layout is drawn by the plotter, which can at this point be accepted, or rejected by the customer for redesign. Once accepted, the system produces assembly die drawings along with complete detailed shop drawings of the component parts for the die shop and a full bill of materials. Standardized parts which can be purchased will be listed by description along with manufacturer's name and catalog number as well as similar data for an alternate supplier.

The die drawings include details of inserts dimensioned in inch or metric values with all clearances, pilot-hole shrinkage, and other factors that are normally taken into consideration according to standard die shop practice. Angles, for example, are dimensioned across an inserted inspection ball, whenever this is appropriate. The plotter accurately makes the drawings to any appropriate scale acceptable by the equipment. Enlarged computer-produced detail drawings of die segments can even be plotted on mylar for direct use in an optical comparator for inspection or for optical grinding of a critical die surface.

The system can create stamping dies for any contour part that can be digitized for coding into the system. Parts with artistic designs and cam surfaces have been handled successfully, although coding took a little longer.

Design Costs Comparable

Basically the company is offering what an in-house die-design department would produce under usual conditions but much faster, more accurately, and with extra benefits to management.

For the future, DieComp is considering leasing a special-purpose computer which will be kept up to date in programmed design capability. The user will need only to code in a part drawing and the system will automatically create the progressive tooling layout and detail drawings of the die parts.

NC Tapes for Die Parts

Another output of the system that will be made available at slight extra cost are NC tapes for the production of specific die parts. Customers that are equipped with NC machines can add this to system benefits. In creating the die-part design, it generates it in a form easily converted to an NC tape. The low cost of the tapes could make NC economical for die production even when the machining is all one-of-a-kind.

An interesting and typical sidelight is that while developing the system for draw-die design, the company found it necessary to develop a more accurate and usable metal springback formula in order to eliminate any need for trial and error. The equations are based on the exact physical properties of the actual metal to be drawn and not on its nominal specification properties. Thus the computerized design system is amenable to handling die design variations required by lot-to-lot or age-induced changes in metal properties. The PDDC system can design interchangeable die pieces to handle wide variations of stock properties. The pieces can be interchanged on the production line without the days or weeks of grinding and adjustment now often necessary. Conversely, the system can predict whether or not a high-precision part can be produced within tolerance limits with a single die, knowing the variation in available stock properties.

Selecting Cold Work Die Steels

Reprinted from Manufacturing Engineering, May 1976

OPTIMUM GRADE SELECTION of cold work die steels is determined by individual tool design, part requirements, metal thickness and condition, production requirements, and manufacturers' know-how. A general understanding of the metallurgical characteristics of such steels is also necessary for proper grade selection. To provide this understanding, Charles N. Younkin, metallurgical engineer with the International Div. of Latrobe Steel Co., discussed the chemistry, metallurgical principles, and properties of these steels, and stressed the importance of a continuing evaluation of their performance in a recent SME Tech. Paper, TE76-918.

Steel Chemistry. Compositions of the most commonly used cold work steels are listed in the accompanying table. To achieve the high-working hardnesses required, practically all these steels have high carbon contents — ranging up to more than 2%. Total alloy content, however, can range up to more than 20%. For this reason, the steels have been classified by the AISI into four general groups.

One group, termed the water hardening steels, is identified by the letter *W*. The second group is identified by the letter *O*, which denotes the low alloy types with an oil hardening capability. The third group bears the letter *A*, denoting a medium alloy content and the ability to air harden in relatively large sections. The fourth group, commonly called the high carbon-high chromium steels, is identified by *D*, and may be either air or oil hardened.

At the extreme ends of the chemical composition range in any of the four groups, a relatively wide variation in metallurgical properties can be expected. Between individual grades within any one of the four groups, however, only a slight variation in chemical composition may have been formulated to meet a specific need. For example, both D1 and D2 steels have no real difference in chemical composition with the exception of a 0.50% higher carbon content for the D2 steel. Metallurgically, this increase in carbon content is significant in that D1 steel is tougher than D2, and D2 has greater wear resistance.

Metallurgical Principles. The more important crystal structure phases in hardenable tool steels are called austenite — face centered cubic (FCC), ferrite — body centered cubic (BCC), and martensite (BCC). As molten iron freezes (crystallizes) from the liquid state, iron atoms form a BCC lattice (delta iron). As the crystals cool, the atoms reassemble into an FCC lattice, and on cooling still further they reassemble into a BCC lattice. In the annealed (as supplied) condition, the basic crystal structure of cold work steels is BCC, and is called ferrite. In the fully heat treated (hardened and tempered) condition, the crystal structure is also BCC, but is now referred to as martensite.

Upon heating from room temperature, the BCC structure (ferrite) transforms to FCC (austenite). Basically, this austenitizing treatment controls the exact hardness which will be obtained, determines eventual size change, develops the red hardness, and influences other desirable and undesirable metallurgical characteristics, as determined by the chemical composition. In controlled cooling, which allows the steel to harden, the FCC structure transforms to a body centered tetragonal lattice. During the final phase of heat treating, which is tempering for the primary purpose of toughening the steel, this tetragonal structure rearranges to the BCC or beta martensite phase.

Higher alloyed die steels, particularly the high carbon-high chromium types, are considerably more sluggish than carbon and low alloy tool steels in dissolving carbides. As a result, much longer soaking times at the austenitizing temperature are necessary to develop full hardness. Also, the minimum cooling rate that is necessary to develop full hardness, which is determined basically by the chemical analysis, is quite rapid for low alloy steels, but much slower for steels containing appreciable amounts of alloys such as molybdenum, nickel, chromium, and manganese. Steels are therefore classified as to the required cooling (quench-ing) rate — water, oil, or air.

Comparison of isothermal transformation diagrams (temperature vs. time) show the variations in transformation times as alloy content increases. For example, the required quenching rate for W1 steel is about 168 times faster than for A2 steel to completely transform to martensite. At a temperature of 1000 F (538 C), only 2 1/2 seconds are required for the beginning of the austenite to ferrite transformation. Therefore, water cooling is necessary for W1 steel to successfully achieve a partial transformation to martensite, which begins to occur at about 300 F (149 C). In comparison, the shortest time period for the austenite to ferrite transformation of A2 steel is 7 minutes, and occurs at 1300 F (704 C). Therefore, air cooling from the austenitizing temperature will provide a sufficiently fast rate to allow austenite to martensite transformation.

Properties Influencing Selection. In general, an increase in alloy content improves properties, but some factors can be mutually exclusive. For example, if an application requires high wear resistance, some sacrifice in toughness and workability will be necessary. *Figure* 1 illustrates comparative wear resistance of commonly used cold work die steels at their usual hardness levels. Such comparisons are relative, and the test conditions used should be the same for all grades tested. For the data shown, a Type T1 high-speed steel was used as a basis for comparison, with its wear factor under a specific set of test conditions given an arbitrary rating of 100. Under the same test conditions, the wear factor for O1 steel was 180, that is its wear factor was 80% poorer than the T1 sample.

Increasing the carbide formers in a steel is more important than increased alloy content for improved wear resistance. For example, Type O1 steel has over double the alloy content of W1, but only slightly improved wear resistance. This is because manganese, which is the major alloying agent in O1, is not a carbide former. The positive effect of the strong carbide former chromium is evident in the high carbon-high chromium types, group *D*. The effect of vanadium, tool steel's best carbide former, is seen in the increase in wear resistance of D7 over D3.

The relative toughness characteristics of common cold work die steels are shown in *Figure* 2. Alloy content influence is reversed with respect to this property — as alloy content increases, toughness decreases. Higher carbon

COMPOSITIONS OF COMMONLY USED COLD WORK STEELS

AISI Type	Composition (%)							
	C	Si	Mn	W	Cr	V	Mo	Co
W1	1.00	0.20	0.20		0.10			
O1	0.95	0.30	1.20	0.50	0.50			
A8	0.55	1.00	0.30	1.20	5.00		1.20	
A2	1.00	0.30	0.70		5.25	0.25	1.10	
D2	1.50	0.30	0.30		12.00	1.00	0.75	
D5	1.50	0.50	0.25		12.25		0.85	3.10
D3	2.10	0.50	0.50		13.00			
D7	2.30	0.40	0.40		12.50	4.00	1.10	

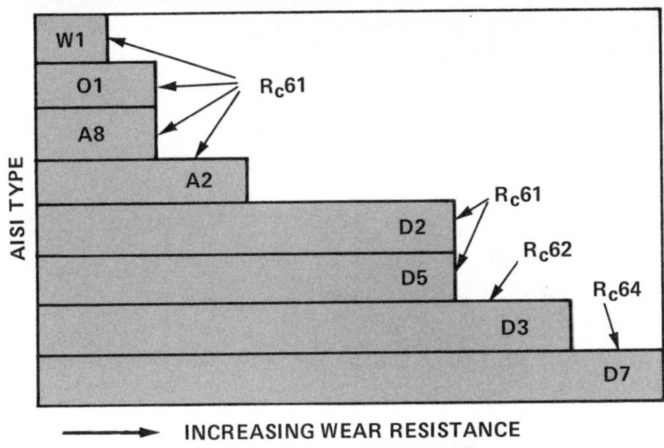

1. COMPARATIVE WEAR

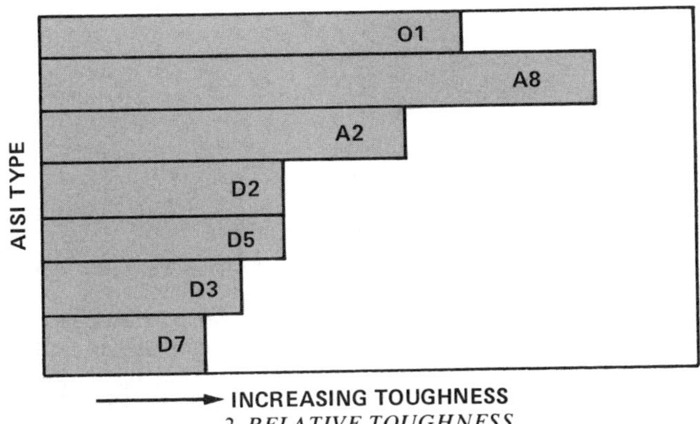

2. RELATIVE TOUGHNESS

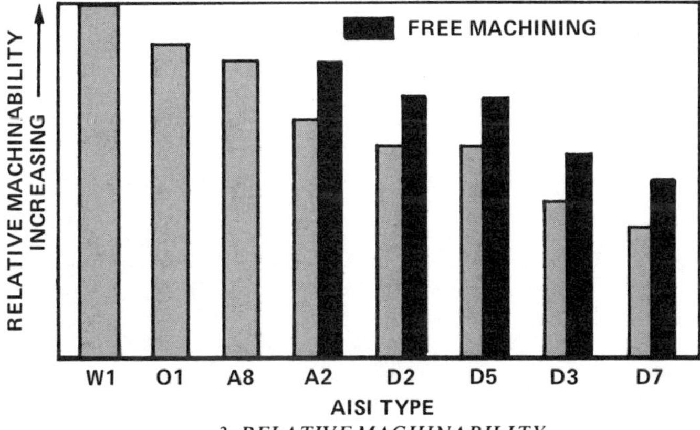

3. RELATIVE MACHINABILITY

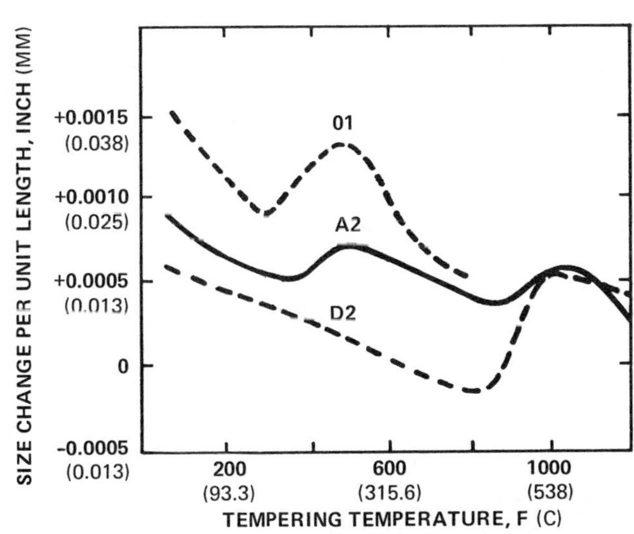

4. SIZE CHANGE AVERAGES

levels also contribute to lowering toughness as a result of the formation of the hard, brittle carbide phase.

As seen in *Figure* 3, relative machinability of these steels in the annealed condition decreases from Type W1 through D7. The relationships are basically the same as the relative wear characteristics — as alloy content increases, machinability decreases. A number of the higher alloyed steels, however, can be obtained as free machining grades containing a fine dispersion of sulfide particles which improve machinability 20 to 25%, and make it practical to produce relatively intricate shapes.

Size and shape changes can occur during heat-treatment as a result of shrinkage during phase transformation, and during tempering due to a continuing crystallographic rearrangement of the atoms. This deviation, which decreases gradually with increasing alloy content, is illustrated for Types O1, A2, and D2 in *Figure* 4. Figures shown refer to an average size change, and are not uniform for all three dimensions.

A better word for such shape changes is distortion, since it is a warping or bowing caused by stresses, rather than a dimensional increase or decrease. Sources for these internal stresses include machining, mechanical flexing and bending, peening, uneven heating, and poor support during heating and quenching. Water hardening grades are the most susceptible to quenching, and consequently exhibit the poorest resistance to distortion of the cold work die steels. Oil hardening grades would be next, and air hardening grades are the least affected. Corrective methods to minimize distortion include keeping parts as uniform and symmetrical as possible, stress relieving before finish machining, and slow heating and uniform quenching.

Selection and Continuing Evaluation. While lower alloy, water hardening grades have the poorest resistance to distortion in heat-treatment, they provide adequate service life for many applications. Also, they develop a hard case-tough core relationship when heat treated, which affords less susceptibility to weld cracking. The increased alloy content of the oil hardening grades provides a slight increase in wear resistance, less possibility of distortion problems, and permits through hardening in sizes up to the 2 to 3-inch (51 to 76-mm) range. The high alloy, air hardening grades provide a minimum in size and shape changes, a maximum in wear resistance, and are through hardening in sections up to 6 inches (152 mm). They are, however, relatively less tough, and more difficult to machine and weld, but their excellent service performance often outweighs these disadvantages.

To take full advantage of the individual characteristics of any steel, and prevent money losses from poor utilization, a continuing evaluation of tool and die performance is essential. Tool steel manufacturers offer many grades from which a selection can be made to best suit a particular application, or to replace those not giving optimum results. Performance data should again be accumulated on replacement steels. ∎

The Productivity Aspects of Die Design

BY

CORWIN J. BRAY, Sales Engineer, Production Engineering Company
Elmhurst, Illinois

ABSTRACT. Productivity is synonomous with profit; and the greatest opportunity to lower cost comes through better methods. A very complex process begins when we take a piece part drawing, convert it into a set of mechanical tool drawings, have the tool built, put the tool into production, and hopefully, make a profit from an estimate that was made previously. For some companies, this seems to be extremely easy. For other companies, it is extremely difficult.

Consider the cost ramifications of the stamping operation, and the various facets involved, such as labor, materials, facilities, equipment, and tools. What are investments today as compared to, say, five years ago, or think about today's costs as compared to what they will be five years hence. Are any of today's jobs potential runners for the next five years? Let's carefully analyze what our position will be five years from now when we are attempting to run that same stamping for nearly the same cost. Will we compete? Will we retain the business? These are questions that relate to productivity and consequently, to the methods of die design and engineering used within the operation.

Think about the stamping when it comes in the door for cost estimating. There are normal routine considerations to be made, but is the stamping really evaluated? Most times, an engineer designing a mechanical part for a particular product only understands his product needs. Consideration is seldom given to the problems he may be designing into that particular piece part. We should not try to change their designs or their concepts, but it is our obligation to inform designers when they have gone to extremes. This assistance will also promote stamping sales. The problems that we see at a glance could require hours of study on the part of a non-stamping oriented design engineer.

A plus or minus .001 tolerance on a form, an extruded hole, or on a lanced leg, may cost the customer thousands of extra dollars. Most certainly, it is going to cost many headaches. Very possibly, the same dimension could just as well be plus or minus .005.

How about materials? Can we help the customer there? Does he need to use cold rolled special killed steel? Perhaps, because of experiences

	LABOR	MATERIAL	FACILITIES	EQUIPMENT	TOOLS
5 YEARS AGO 1975	TOOLMAKER 8.00 TOOLDESIGNER 8.00 OPERATOR 5.00	CRCQ .16#	17.50 SQ. FT.	20.00/HR.	20.00 PER HR.
TODAY 1980	TOOLMAKER 11.50 TOOLDESIGNER 12.50 OPERATOR 6.50	CRCQ .42#	20.00 SQ. FT.	25.00/HR.	25.00 PER HR.
5 YEARS HENCE 1985	TOOLMAKER 17.50 TOOLDESIGNER 19.00 OPERATOR 10.00	CRCQ .70#	31.00 SQ. FT.	38.00/HR.	38.00 PER HR.

and problems he has had in the past, he has standardized. Between $1,200
and $1,500 per year extra may be spent, just to specify a special killed
steel, when commercial quality would do just as well. We can identify
these things quickly by utilizing our internal expertise.

The quality-related characteristics of the stamping depend upon many
different things. Particularly, the capability that we have within our
organization. Of course, we don't all stamp the same quality parts. We
most certainly stamp them to conform to the piece part drawing, and consequently, we satisfy the quality control of our customer. But the quality
of parts that one stamper makes, as compared to that of another, will vary
with the type of product dealt with. A stamper running bumper brackets
from 3/16" hot rolled steel certainly has quality problems, as does the
stamper running .008 thick nickel silver to make integrated circuitry lead
frames. However, the part quality and their equipment are going to be considerably different. The approaches to their stamping problems are radically
different. Therefore stamping quality is related back to our own people.
Other factors to consider are: what are we accustomed to running? How have

we have trained? What type of stamping equipment do we have? What type of
tools do we have? And lastly, what are our overall facilities? All these
questions relate to the evaluation that must be made before even taking on
a new stamping.

Tool design analysis starts with the piece part drawing which we have
already discussed in some detail. How do we lay out the strip? What is the
best method to run that particular part? Let's assume that we are going to
run a progressive die. With today's hardware and software that is available
to the tooling industry, we have a tremendous opportunity to capitalize on

the technology that has been developed from processes such as numerical control and computerization. Tool designs can now be made by computer and computer controlled plotters that will actually draw up the design and detail
each section. That is the business of some specialty tool design companies.
But the productivity aspect of tool design still has to be designed into the
die. It must be programmed even on a computer. Now, through the use of
computers, a layout of the developed blank can be made, fed into the computer, and printed out with the most economical progression and the best
angularity to minimize scrap loss. To optimize the bending lines of a
formed leg to give it maximum strength, merely ask the computer to give
another five degrees bias on the progression angle, and it can replot the
entire strip layout. An entirely new strip layout with the proper adjustments for progression and all the die section locations can be achieved.
Utilizing optimum strip layout may save five to ten percent in material.
On a medium sized part, this saving could easily represent eight to ten
thousand dollars per year. Even though productivity ideas may not be
generated by the computer, it can free our designers to apply their expertise to that end. This approach could also save hours and hours of manual
layout. Computerized die design is available today!

Eliminating Press Stoppages

Press stoppages must be eliminated. Every time that line is shut down, it
costs money. We do not produce on time, our customers are upset, and most
of all, we are not making a profit. The press stoppage requires the attention of the operator, supervisory people, die setup men, staff engineers,
tool makers, and tool room supervisors. Very easily, as many as five or six
people may be involved with the press stoppage. We have all seen it happen
just because of a so-called ordinary press stoppage. Why did we get this

stoppage? Let's examine some of the basics. . .The things we should think
about, but do we? No matter how hard we work to improve productivity, only
by using intelligence, imagination, and probably a little more capital, are
we going to resolve the majority of press line stoppage problems.

We want our tool designers to be productivity conscious. But we have
the same tool designers that we have had for the last ten years. Are we
going to fire all tool designers and hire all new ones with all new ideas?
It must be an attitude change throughout the entire organization. Be willing to look at the other side of the coin. Attention must be given to ideas
that have not been considered before. Perhaps we have tried ideas and
failed. A lot of new thinking must be drawn out of our tool designers.
Maybe it is going to require getting them off the beaten track and taking a
fresh approach to figure out new ideas and concepts. Above all, everyone
throughout the organization must be willing to support and implement new
programs.

Tool designers must know sectioning, thickness for perforator plates,
punch length, and all the normal die standards. These are the basics, but
let us consider the subtle areas that are always there.

<u>How to Improve Die Set-Up</u>

A major area that always seems to create a lot of problems and can be
reconciled by the tool design people is die set-up. In most cases, tool
designers do not even consider die set-up. It is just a matter of sliding
the die onto the bolster plate, grabbing a couple of straps and bolts, and
clamping the upper show. Some other clamps and bolts lock it to the bolster plate. Now it is all set to go. It is not realized that many times,
this in itself, has caused the press stoppage. Walk into the shop tommorow

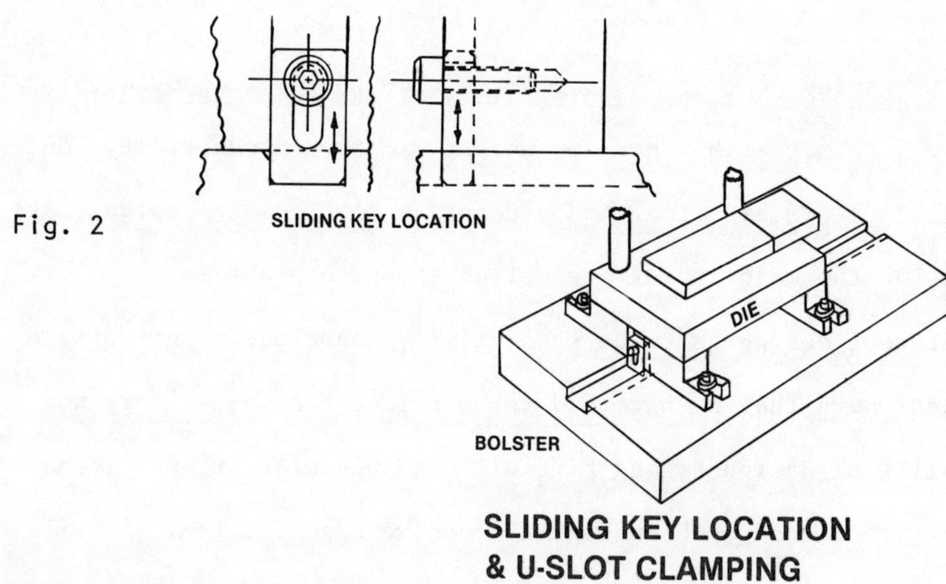

Fig. 2

SLIDING KEY LOCATION
& U-SLOT CLAMPING

morning and check every set-up. How well is it located? How well is it clamped into the press? How well is it aligned for material flow? At least one half of the tools will be inaccurately set. Is it the fault of the die setter? At the time of tool design, these considerations should be made to improve the die setter's job. Die set-ups could be faster with less press stoppage with proper tool design.

Locate the tool with a centerline keyway in the bolster plate and a sliding key at each end of the die. The sliding key can be retracted upward after the alignment and clamping is completed. During maintenance in the tool room, sliding keys will not cause unbalance at the base of the die. This is a very easy way of locating the tool on the centerline of the press and on the centerline of the feed. It also gives feed to die perpendicularity needed to insure smooth flow of the material through the stock guidance of the die.

Location can also be made with shot pins. And again, these are things the tool designer could put into the tool when it is being built. Jig boring a pattern of holes in the bolster plate, which are compatible to the family of dies run in the press, permits a fast accurate set-up. This will achieve centerline alignment, perpendicularity to the feed, and right to left location balance for off-center tonnage loading.

Commercial, hardened bushings, and shot pins can be used to avert wear over the long term.

Backstop locators are probably least popular due to the variation of die sizes. It is unusual to have similar size die shoes to utilize backstop locators. Die location, whether it comes through keyways, shot pinning, or backstops should be done by the die designer at the outset.

Clamping is another subject. It is easy in some cases to grab a set of clamps and find a few blocks of cold rolled steel, seven nice slugs, stack them all up, and make a nice clamp block assembly. That is not the way it should be approached. Commercially, there are many clamps available that do an excellent job. Preferably, have the tool builder put the holes directly through the die shoe. With tee slot and U-notches in the edge of the die show, bolt through the shoe into the tee nut. Tee slotted bolster plates and rams offer very efficient clamping of tools. Clamping blocks or building blocks are not required with tee slots. The same can be applied to the upper half of the die shoe. This is probably the best method to use. There are other ways that are well applied. One would be die clamps that insert into holes directly drilled into the edge of the die show. Predetermined location of holes above the lower face, or below the upper face of the shoe, are required. These clamps require no blocking or shimming because the dimensions from the shoe face to the hole centerlines are standardized to suit the commercial clamp. There are other types of commercial clamps available, but they require clamping blocks. Clamps with clamp blocks require more set-up time and normally are prone

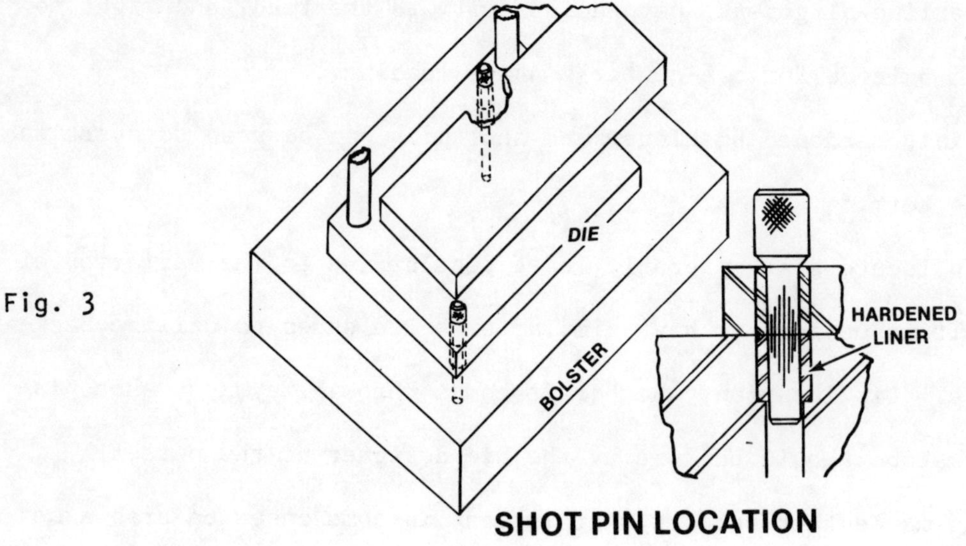

Fig. 3

SHOT PIN LOCATION

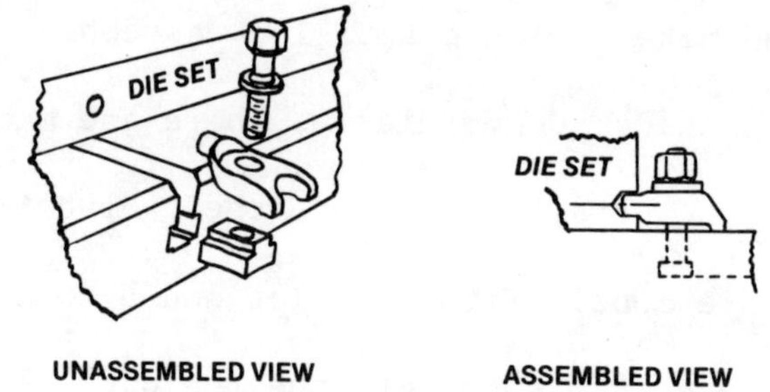

Fig. 4

UNASSEMBLED VIEW

ASSEMBLED VIEW

DIE SET CLAMP

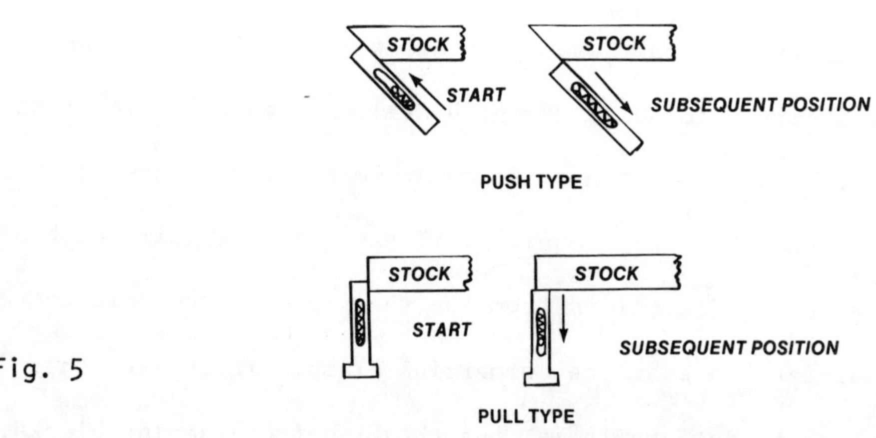

Fig. 5

PRIMARY STOPS

to being lost. For this reason, I would emphasize the use of a pre-engineered type of clamping system.

A few years ago a hydraulic clamping system was devised and designed to accommodate the average tool within the average stamping facility. The concept utilized a positive-locking hydraulic clamp system, gibbed upper and lower die adapter plates, backstops, and a roller platform die cart. By merely selecting adapter plates large enough for the family of tools to be used in a press, all dies are drilled and tapped to accept the common adapter. This provides common die location, easy die handling, and fast, positive clamping. Combined with the powered shut height adjustment and digital readout, die setting could be accomplished in a very short time. This system applied to a 60 ton OBI press and a family of twenty dies would cost approximately ten to twelve thousand dollars. Averaging two die changes per day, the entire system would be justified on a one-year payback.

Improved Stock Threading

A great deal of die damage occurs at the start of a new coil, or at the tail end of an old coil. We must provide the operator with a pre-trimming station, or appropriate staging stops in the die. On the newer equipment with extremely accurate feeds, after the first hit is made, the strip can walk through without any problem. This will override the need for having manual die stops. But what about that leading edge that leaves a one-quarter moon section on the die? These sections cannot be removed with a bridge or solid stripper on the tool. Blowing loose sections out of the tool probably will not work. Why not make a pre-trim station for the tool. An arbor press fixture with a pre-trim die seems to be the

most popular method. These design considerations do not cost a great
deal of money and they will eliminate press stoppages.

With less accurate feeds, one must design in primary and secondary
strip stops. The number of secondary stops depends upon the die layout,
but one too few can cause major miss-hits in the tool. Analyze the tool
to determine exactly what is needed.

French stops are very effective when hand feeding coil stock. However, they are the "nemeses" of today's modern roll feeds. The theory of
use to the designer is great, but in actual practice they end up destroying new, accurate feeds. The reason lies in the fact that die setters
rarely take the trouble to back off the feed once it hits home on the
French stop. The result is slipping rolls, broken gears, and worn out
clutches. Consequently, dies are broken because of badly mangled, inaccurate feeds. All this because of an innocent French stop. The real classic
case comes when one utilizes a French stop, a solid stripper, and a form
in the die. The result is a strip locked in the tool. It cannot be
removed forward or reverse. This really gives the die setter something
to do with his spare time. Elimination of French stops also saves material.
A five to seven percent reduction in material usage can be achieved on a
2 1/2" x 3" x .06" thick piece-part. Converted to dollars, this would
represent five hundred to one thousand dollars per year savings, based
upon a quantity of one hundred thousand to two hundred thousand parts.
Multiply this sum by a dozen parts and it represents a tidy little savings
to put into the profit column. To the die setter and press operator,
French stops, damaged equipment, and downtime are anonyms.

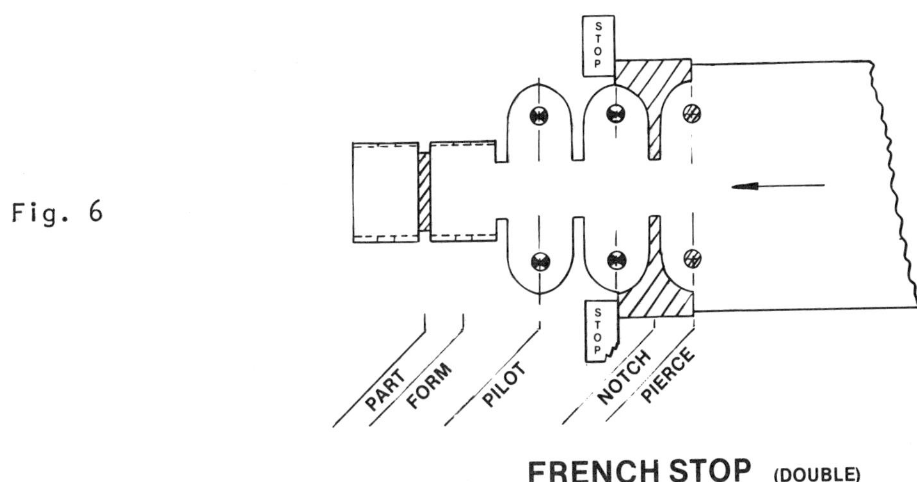

FRENCH STOP (DOUBLE)

Fig. 6

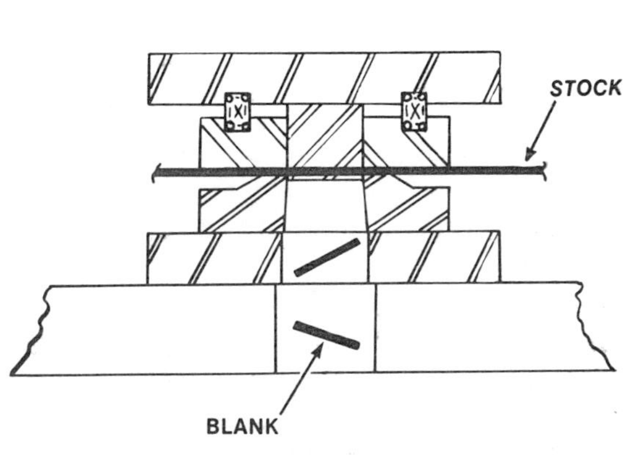

DROP-THRU BLANKING DIE

Fig. 7

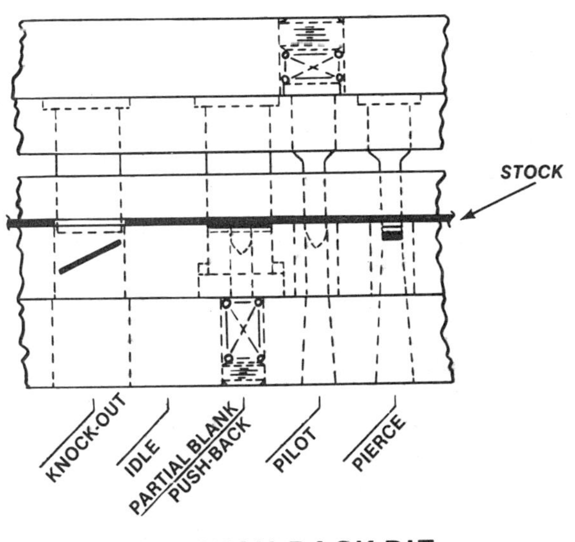

PUSH-BACK DIE

Fig. 8

<u>Part and Scrap Ejection</u>

This is the area that separates the men from the boys in die design.
Whenever possible, blank the part through or cut off the part. Do not
use blow offs, shedders, or knock outs. If the part will tolerate blank-
through ejection, it is far superior to any other method. Once this is
designed into the tool, it is fixed. Now the tool is capable of running
at much higher speeds. The press speed is not limited by the piece-part
ejection from the die. The same thing is true with scrap. When possible,
always push the scrap through. Do not leave the little whiskers, the
little skeleton sections, or the little slug on top of the die. This only
creates grief and downtime. The law of gravity guarantees that everything
is free and clear. Other methods of getting that piece-part to scrap out
the tool does not need to be engineered or set up. The part or scrap is
clear to fall directly into conveyors, or into containers, to be carried
away.

Many times parts must be extremely flat, and a blank through die can
cause dishing. When this is the case, consider a pushback die and a knock-
out station. This gives the ability to punch out flat parts, push them
back into the scrap skeleton, advance the progression, and simply knock
them out of the strip through the bottom. Again, we are striving to eliminate auxiliary methods for ejection. When a new die is being constructed,
the additional cost of a pushback station may be in the area of an
additional ten percent of the total die cost. Shearing the die one time
due to non-ejection can easily pay for this additional investment.

A compound die station is normally used to obtain flat parts. Slow
press speed is a resulting problem. A push-back die station will permit
the press to run three or four times as fast. Again, considering

productivity added profits, we would typically spend one thousand dollars
more on the die to gain two thousand five hundred dollars, based on one
million parts per year. Net savings would be one thousand five hundred
dollars, not including the major factor of reduced tool repair.

The alternate methods for ejection are cyclic air blow-off, mechanical
shedders, or knockouts. How well can we apply knockouts? The press
closes and punches out the piece part. The upstroke of the press is wasted because ejection can be accomplished only at the top of the stroke.
When the part is finally ejected, the race is on. Now, during the down
stroke, the part must move faster than the press ram in order to eject
clear, prior to die closure. Failure to clear in time results in a double
in the die. Press speeds are normally reduced to prevent this very occur-
ence.

If ejection must be mechanical, use a cam style knockout. It permits
ejection early during press upstroke and allows seventy to eighty percent
more time to clear the die. This extra ejection time will permit the press
to be run fifty to one hundred percent faster.

Better Stock Guiding

Extra attention must be given to the flow of material in the tool. When
press speeds are high or feed lengths are long, consider the effects of
stock guiding.

Coil materials that have a low ratio of thickness to width tend to
present major problems because of camber. High ratios present columnar
strength problems and must be pulled through the guide. These conditions,
coupled with slitting burr, utilizing raw strip edges, and coil curvature,
create a situation requiring exact decisions by the die designer. A part

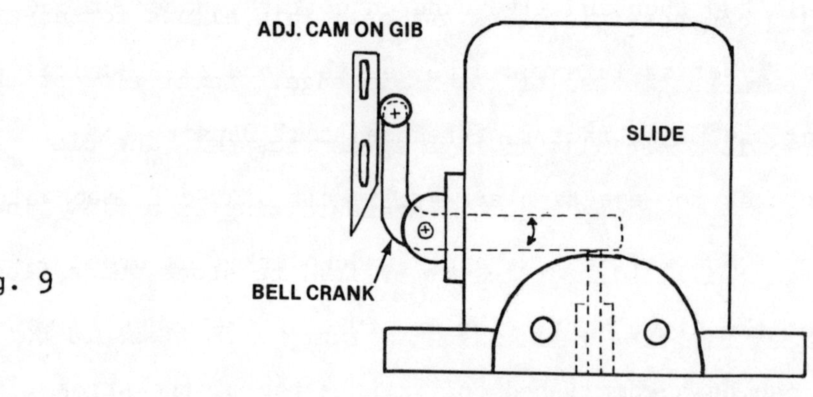

Fig. 9

CAM TYPE KNOCK-OUT
(SLIDE SHOWN AT TOP)

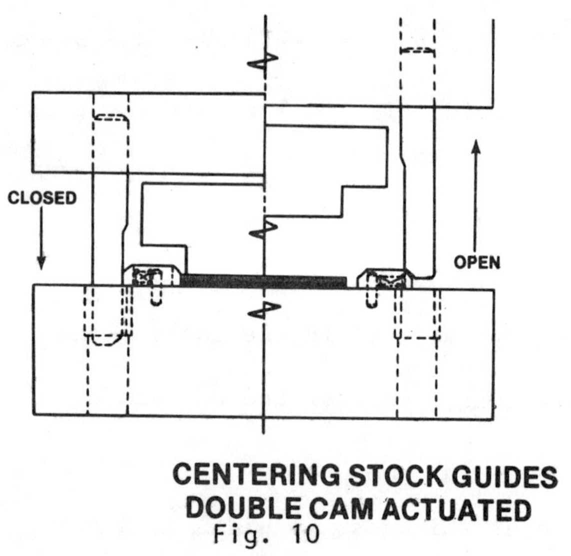

**CENTERING STOCK GUIDES
DOUBLE CAM ACTUATED**
Fig. 10

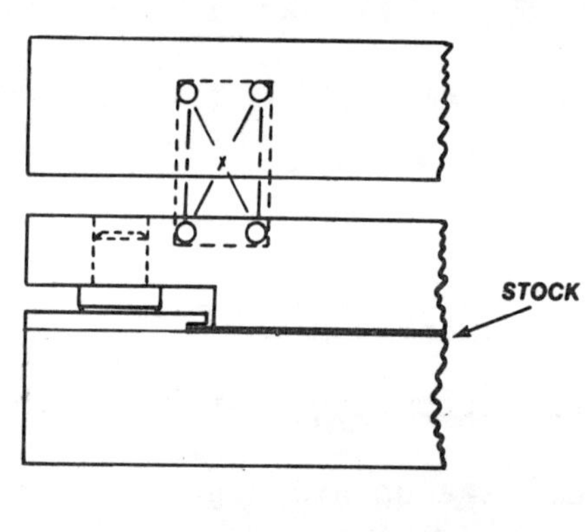

**SPRING STRIPPER
W/BALANCE PADS**

Fig. 11

prone to camber, but necessitating the use of a raw edge, would be a prime
candidate for centering cam actuated stock guides. This allows for adverse
flow conditions, yet permits the use of raw strip edge. Level flow of the
strip is always essential, especially in high speed presses. The die
designer can improve the set-up man's job by standardizing on a flow line
height. Too many times height misadjustments of feed to stock guide flow
line result in friction, creating faulty feeds. Every effort should be
made to design flow heights to a standard.

Die Stripper Functions

Solid strippers are economical. They fall into the class "You get what
you pay for." They do not control materials during punch penetration and
withdrawal. Therefore, with the strip floating, there will be an excessive
amount of broken perforators, pilots, and distorted material. It will also
limit access into the tool for clean-up and periodic visual inspection.
Solid strippers can be well utilized to guide punches. This seems to be
the most appropriate application outside of initial cost.

Spring strippers give you the best of everything in controlling the
strip, but they are expensive. The initial cost will be saved many times
over through less die breakage and downtime. One area of caution is stripper balance. Do not permit cocking when starting a new strip. This is
especially important with materials 1/16" thick and greater. Cocking will
be prevented with the use of balancing pads. These pads should be of
equal thickness to the material.

Preventing Pilot Breakage

Pilot breakage can be more of a problem than punch breakage. Materials
with a low thickness to pilot diameter ratio will present more pilot

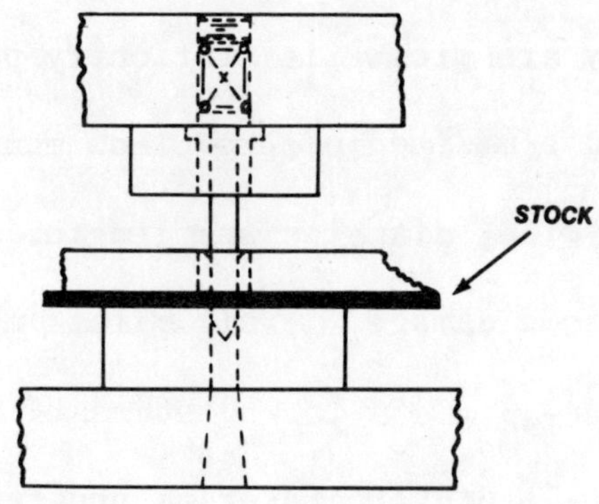

SPRING LOADED PILOT

Fig. 12

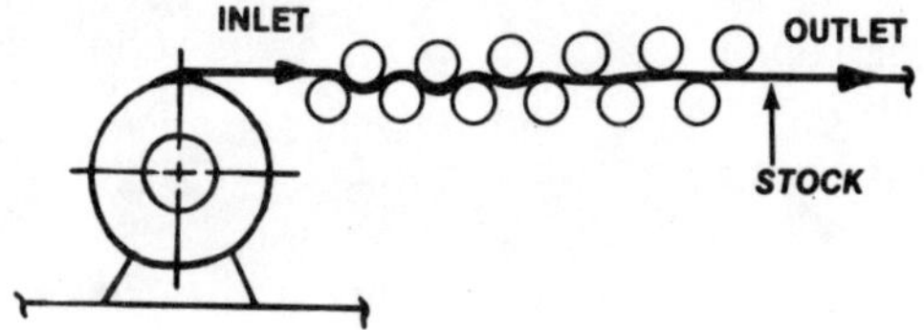

FLEX ROLLING

Fig. 13

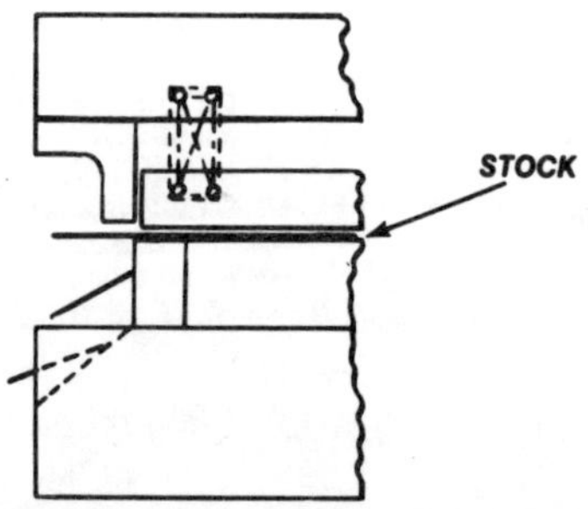

**DIE MOUNTED
SCRAP CHOPPER**

Fig. 14

breakage problems. If they are press fit stationary pilots, they must penetrate the material when a misfeed occurs. How many misfeeds they can withstand depends upon the pilot diameter and length. Rest assured that the first time will cause some damage. Frail pilots must be supported with guides to prevent deflection.

Spring loaded pilots will prevent material penetration in most cases. No pilot penetration; no breakage. Many designers say accuracy will be lost if slip fit clearances are used. Honestly, how many jobs are really that critical? Spend a few dollars on spring loaded pilots and keep those presses running.

Cures for Slug Retention

Books could be written about slug retention. When a slug pulls, do we set the die deeper to drive the slug on through? Do we put shear on the punch, notch the punch, dull the punch, put ejector pins in the punch, notch the die, or angle the die? Let's examine material stress equalization. Consider the coiled stock that is being fed through the press. There are stresses in that coil due to rolling, slitting, and rewinding. As we start gutting, perforating, notching, lancing, forming, and drawing, we are setting up more stress in the material. For example, let us consider a part from fifteen thousandths thick phos bronze alloy that has 1/8" diameter holes. Punch out the slugs and they pop right back up. The strip jams, causes a misfeed, and shears the tool. Now we have a one hundred dollar regrind job, excluding lost press time. Why does it happen? Could it be resolved? Precision stock straighteners are available. They operate in the realm between conventional stock straighteners, which simply remove coil set, and the roller leveler, which can remove

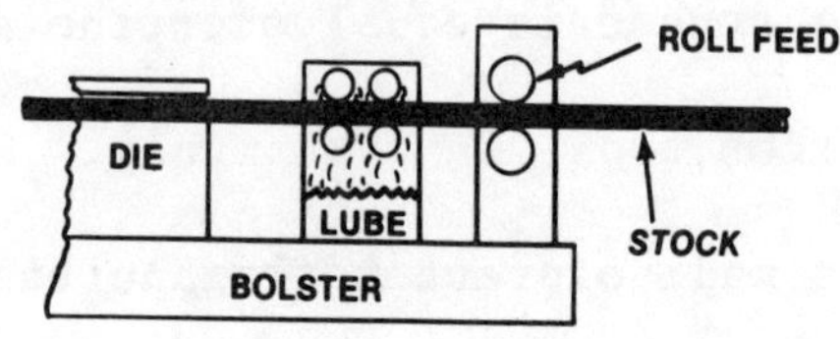

**RECIRCULATING
STRIP LUBRICATOR**

Fig. 15

Fig. 16

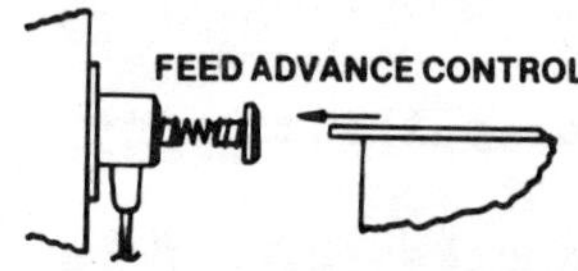

**ELECTRONIC SENSORS FOR
DIE MALFUNCTION DETECTION**

residual stress in most material. A precision stock straightener flex
rolls the stock enough to equalize the stress across the strip. Those
1/8" diameter slugs no longer pop out of the die bushing. Why? Because
the stock is flat and the stress is equal over the entire area of the slug.
Therefore, when grinding the tool, we save money. Special shapes and
angles need not be maintained to prevent slug pulling. Less punch and die
will be ground away because penetration is shallower. And this is possible
only if we prevent slug pulling.

Today's carbide lamination dies are worth at least two hundred dollars
per one thousandths inch or die life. Now we can fathom the extreme importance minimizing die grinding.

Choosing Scrap Choppers

The best scrap chopper is the die mounted type. The main reason is maintenance. Grinding the die grinds the scrap chopper and it is easily timed
with the tool. All other scrap choppers rarely are maintained. They
usually run until they form the material instead of chopping it. Then and
only then, does someone look at its cutting condition. Most scrap choppers are also inaccurately timed.

They may be cutting or pulling when the feed is released for piloting.
Net results are misfeeds, broken pilots, and sheared discs.

Facts About Die Lubrication

A recirculating die lubrication system is highly desirable on most operations. It applies proper amounts on either side of the strip, conserves
lubricant, provides better overall housekeeping and can be installed between the feed and the die. The location is important because if mounted

before the feed, the rolls or gripper jaws could iron off most of the lubricant. This could also cause the feed to slip and feed inaccurately.

Should designers consider die lubrication? Designing pockets in the tool that could become filled with die lubricants will cause hydraulic action, thus marking or even distorting the piece part. What lubricant should be used in the tool? The press operator normally has this decision delegated to him. If running a carbide die, a sulphur base lubricant should never be used. This would cause the carbide binders to erode, and the tool will fall prematurely. Lubricant specifications should be part of the routing process. Many times a tool will make beautiful samples. Then in the production press, after fifty pieces, it is galled. The part is welded to the draw punch. Everything is designed properly. But what has happened? Correct lubrication specifications were not established at the outset. Processing and operating people must be advised as to the importance of proper lubrication selection.

Die Malfunction Detection

Damage to tools or presses can be prevented by electronic die sensing. Consider the piece part that you have not been able to blank through. Design in a sensor to check feed advance, stock buckle, end of material, misfeed pilot, and high strippers.

If a misfeed pilot is used, it is easy to install a commercial limit switch and achieve misfeed detection by actuating the press stop circuit. In order to detect a misfeed with a pilot, the press and die must be almost closed.

Unless the press is extremely slow, or has an oversized brake with a fast reaction time, a miss-hit is inevitable. This probably will shear

the tool, or at least make the location pilot punch through the strip.
During a normal run, there may be several misfeeds. Multiply the times
a misfeed occurs by the number of times the pilots punch through the stock
and there will most certainly be a broken or bent location pilot. This
will cause serious damage to the tool. Misfeed pilots must be extra long
to provide ample press stop time. This extra length will dictate the feed
cycle. Therefore, the feed cycle will be limited by the extraction and
entrance of the misfeed pilot.

A more logical approach would be to sense the end of the strip for
feed advance. The feed cycle can then be timed early to enhance ejection
or stop time. Consult a reputable press room equipment supplier for
particulars, as this requires intimate knowledge of die design and malfunction detection.

Checking for stock buckle between the feed and the die is another
method to eliminate pilot probes. Most failures to advance the stock
accurately comes from slugs pulling, or poor material conditions. Either
case would cause the feed to buckle the strip when resistance occurs in
the tool. Again, we can utilize our feed cycle to our advantage and permit more stopping time.

The cost of one miss-hit cannot be precisely defined. Outwardly, the
operator may see little or no change in die performance. But quite possibly the number of hits will be reduced between regrinds. A die that
should run one hundred thousand hits now only gets eight thousand hits
before excessive burr appears. The one miss-hit at the beginning has
been forgotten, but it is the true reason for getting 20 percent less
from the die.

These figures are representative of cost involved with only one miss-hit. The product of the total cost for miss-hits would stagger the mind if it could be accurately determined.

Do, by all means, take advantage of the electronic sensing equipment that is available today. And do it in the tool design stages. Do not make the tool room retrofit those tools. It is much easier for the tool designer to do it on paper than later, when working with hardened tool steel.

Once the basic die design is completed with the appropriate considerations for productivity, the tool builder must be selected. The state-of-the-art in the tool builders' industry is changing radically. A tool builder must stay abreast with the latest developments of wire E.D.M., numerically controlled machines, duplicating equipment, and computerization of machine tools. To the leaders in the tool building industry, the byword now is numerical control. This is going to be a way of life. The most recent, most revolutionary new approach is the wire E.D.M. Within five years, anyone running or building dies will be required to use this method. It will be a must to remain in a competitive position.

We have talked about new tools. What about the existing tools? Can we upgrade what we now have to utilize some of these new concepts or new ideas? It very well may be economical to do so. One viewpoint would be extending of die life. In some cases a very minor change is required to accomplish this. We may double, triple, or quadruple, the life of a die simply by making a minor modification at the cutoff end of the tool. Let the piece part fall free and clear, instead of trying to blow it out of the tool. If there are fewer stoppages, it indicates there are fewer miss-hits, sheared and broken dies.

At the time of tool construction, the cost to go one station longer
will be approximately 10 percent more on the average. Due to part configuration and processing, the cost to retrofit in this productivity feature on existing dies will range form 10 percent of the initial cost to
complete replacement of the die. Generally speaking, dies are not designed
with extra room nor are they conveniently compatible to retrofitting ideas.
For this reason, much serious consideration must be given to the tool
design job at hand, in order to achieve optimum productivity features.

Equipment selection and configuration is crucial to productivity of
tooling. Standardize, where possible, in the areas that will enhance die
setting. You will gain greater flexibility in machine utilization. A
few examples would be: (1) Pre-engineered bolsters and rams. This permits easier die mounting and alignment. (2) Pre-engineered electrics.
This provides extra contacts for use of cyclic air, die malfunction and
pre-determined part counters. (3) Powerized machine adjustments. Included
would be speed controls, shut height setting, and feed length setting.
(4) Unitized control points. This includes installing manifolded air
outlets, eye level rotary limit switches, and auxiliary hand tool outlets.

On existing equipment, the application of new techniques and methods
by retrofitting can be accomplished in most cases. Utilize outside expertise to obtain new ideas. The latest state-of-the-art can be obtained from
competent metal stamping suppliers. Infusion of new outside thought, in
most cases, can be very productive in itself.

The implementation of new ideas that will enhance productivity is not
extremely simple. It involves a system to report press stoppage. Once
the report is made, the reason for stoppage must be identified, and

corrective action taken. For this reason, the entire organization must
take part in any effort to upgrade and improve productivity.

The execution of all the detailed items we have discussed now becomes
paramount. Operational procedures must be passed down to the line supervisors and operators. Without aids and guidelines, productivity improvements will be of little value.

Every feature of the press, tool, feed, coil support equipment, safety
devices, material handling, and automation devices must be right. Any
shortcuts, at this point, will cause a weak link in the program, resulting
in operational inability to conform to the new standards.

Our final task to achieve the "productivity aspects of die design"
lies within the heart of the company. People must make the job happen.

Management's firm stand to attain more productivity is mandatory.
Management must give direction and guidance to the overall approach.
Establishment of communications, up and down the organizational ladder
is essential. Dedication to properly trained personnel, at all levels,
is imperative. Limitations of personnel and overall facilities should be
recognized and placed into proper perspective. Manufacturing management
at all levels must also learn the economics of productivity. The portion
of burden costs applicable to inadequate tool design and execution must be
isolated. This total figure will then become the "bogie" or goal towards
which to work. Because management's major role is devoted to the P and L
statement, they are the ones who must put the final stamp of approval on
the entire productivity program. Information must be converted to motiva-
tion.

Staff personnel must be willing to consider new ideas and diligently
work toward their success. Personal upgrading, through additional schooling,

may be needed. When the communication lines are established, they are to be used, not abused. Information put into these channels must be precise and accurate. Training is an everyday occurrence for staff and line. They must constantly talk to outside sources and evaluate new information. This filtered and adapted information can then be used to improve their never-ending program toward achieving more "productivity." Failure to carry on continuing training at all levels will result in the eventual collapse of the productivity program.

Line personnel are the end of the organizational sequence. <u>They</u> must put into operation all programs and improvements that management and staff have implemented. Line attitudes must constantly be reviewed to insure that they are able to carry out new assignments. The real load is on them to produce efficiently, and to do it according to the new procedure.

Communication and training is a daily requirement to make old habits disappear and new productive habits routine.

Productivity aspects of die design are rarely attained for three major reasons: (1) Management fails to recognize the economics of productivity. (2) Staff fails to dedicate their time and efforts toward the implementation of productivity. (3) Line personnel fail to run presses at specified rates because making parts to print is more compelling than achieving operational efficiency.

"Optimum productivity" has never been achieved. This would be utopia and it will never happen. A better and more profitable venture is always just around the corner. Some companies can produce rings around their competition. Why? Because, the best organizations invariably do these things.

No stamping organization could go far wrong by taking a good look at
its operations in terms of the factors discussed here or by challenging
its suppliers to come up with new ideas.

TOOL STEELS —
TRADE NAMES AND SUPPLIERS

Here is a helpful guide that will save you time in finding sources for the many tool steels available. Steels are arranged in groups generally conforming with AISI types. Under each type, trade names of the tool steels are listed alphabetically, followed by the supplier. A list of producers, suppliers,and their addresses is provided at the end of this guide

WATER HARDENING TOOL STEELS

Type W1
Extra Carbon Tool Steel
Annealed . . . . *Achorn*
Blue Label . . . . *Ackerlind*
Pompton, Pompton Special,
Pompton Extra . . . . *AL Tech*
Standard Steel, Atsco Tool Steel,
Atsco Extra Tool Steel . . . .*Atlantic*
X-10, X-12, Alpha, XX-95 . . . . *Atlas*
X, XCL, XX . . . . *Bethlehem*
Extra . . . . *Braeburn*
11 Special, Comet, Titian, H-9
Double Header, Reading Tap,
Green Label Drill Rod . . . . *Cartech*
Special, Extra, Extra
Headerdie, Standard . . . .*Columbia*
Sanderson Extra, Labelle Cold
Header, Black Diamond . . . . .*Crucible*
W-1 . . . . *Fagersta*
Lion Extra, Lion . . . . *Jessop*
Pure-Ore Extra . . . . *Kloster*
Carbon . . . . *Latrobe*
Apex Drill Rods, Conqueror,
Conqueror Drill Steel,
Utility . . . . *Lehigh*
Ajax, Tru Cor . . . . . . *Precision-Kidd*
Red Label . . . .*Simonds*
Dukane, Reliance . . . . *Teledyne*
Colonial No. 14 . . . . . . *Teledyne Vasco*
US Extra Tough . . . .*U.N. Alloy*
UHB 20 . . . . *Uddeholm*
Zivco . . . .*Ziv*

Type W2
V-85 Tool Steel Annealed . . . . . *Achorn*
White Label . . . . *Ackerlind*
Python . . . . *AL Tech*
ASA-10 . . . . *Atlas*
Superior, Best . . . . *Bethlehem*
Coldie . . . . *Braeburn*
Nitro, Special Vanadium . . . . *Cartech*
Econo #1-V . . . . *Champion*
Vanadium Extra, Vanadium
Standard . . . .*Columbia*
Alva Extra . . . .*Crucible*
Lion Vanadium . . . . *Jessop*
Renown . . . .*Latrobe*

Blue Label, SF-2 Blue Label
Ground . . . . *Peninsular*
Colonial No. 7 . . . . *Teledyne Vasco*
Extra S . . . .*U.N. Alloy*
UHB 19 VA . . . . *Uddeholm*
Zivco Vanadium . . . .*Ziv*

Type W5
Q . . . . *Atlas*
U.D.R. *Cartech*
Waterdie Extra, Waterdie
Standard . . . .*Columbia*
CFS . . . .*Latrobe*

Other Water Hardening Tool Steels
V-35 Double Header . . . . *Cartech*
Carbon Cold Header,
Hedervan . . . .*Latrobe*

SHOCK RESISTING TOOL STEELS

Type S1
U B C Steel Annealed . . . . *Achorn*
A. S. #7 (oil) . . . . *Ackerlind*
Seminole, Seminole Hard . . . . *AL Tech*
Tuncro Tool Steel . . . .*Atlantic*
Falcon-6 . . . . *Atlas*
67 Chisel . . . . *Bethlehem*
Vibro . . . . *Braeburn*
Excello . . . . *Cartech*
Buster Alloy . . . .*Columbia*
Atha Pneu . . . .*Crucible*
Top Notch . . . . *Jessop*
Pure-Ore Super-Alloy, Super-Alloy,
Chiz-Alloy . . . . *Kloster*
XL Chisel . . . .*Latrobe*
Brown Label . . . . *Peninsular*
Commando . . . .*Simonds*
Par Exc . . . . *Teledyne Vasco*
KLD . . . .*U.N. Alloy*
UHB Regin . . . . *Uddeholm*
Cyclops S1 . . . . *Universal-Cyclops*
Maxtuff . . . .*Ziv*

Type S2
Triton . . . . *Braeburn*
Solar . . . . *Cartech*
R.T.S. . . . . *Jessop*
Venago Special . . . . *Universal-Cyclops*

Type S3
Case Die . . . .*Simonds*
Wizard . . . .*Ziv*

Type S4
SM Punch . . . . *Cartech*
CEC Impact . . . .*Columbia*
Damascus . . . .*Latrobe*
Silman . . . . *Teledyne Vasco*

Type S5
U S I Steel Annealed . . . . *Achorn*
A.S. #5 . . . . *Ackerlind*
AL 609 . . . . *AL Tech*
Monarch-2 . . . . *Atlas*
Omega . . . . *Bethlehem*
Alloy 10 . . . . *Braeburn*
481 . . . . *Cartech*
255 . . . . *Champion*
Silico Alloy . . . .*Columbia*
La Belle Silicon #2 . . . .*Crucible*
No. 259 Shock Steel . . . .*Jessop*
Pure-Ore V-76, V-76 . . . . *Kloster*
Lanark . . . .*Latrobe*
Rocket . . . . *Lehigh*
Silver Label . . . . *Peninsular*
Orleans . . . .*Simonds*
Mosil . . . . *Teledyne Vasco*
SVM . . . .*U.N. Alloy*
UHB Tirfing 41 . . . . *Uddeholm*
Cyclops S5 . . . . *Universal-Cyclops*
Plancher . . . .*Ziv*

Type S6
Columbia S6 . . . .*Columbia*
La Belle HT . . . .*Crucible*

Type S7
Airhardening Shock Steel . . . . . *Achorn*
AL-7 . . . . *AL Tech*
Bearcat . . . . *Bethlehem*
Carpenter S-7 . . . . *Cartech*
Shock-Die . . . .*Columbia*
S-7 . . . .*Crucible*
Pure-Ore Chiz-Alair,
Chiz-Alair . . . . *Kloster*
Air Shock . . . . *Peninsular*
Simoch . . . . *Teledyne Vasco*

TOOL STEELS — TRADE NAMES AND SUPPLIERS *(continued)*

UHB Prega ....Uddeholm
Toughdie ....Ziv

Type S9
Shoe-Die ....Columbia

Other Shock Resisting Types
408 Punch, Hi Carbon Solar,
Hi Shock 60 ....Cartech
Staminal ....Latrobe
EZ Carb ....Simonds

COLD WORK TOOL STEELS OIL HARDENING TYPES

Type O1
Superior Oilhardening Tool Steel
Annealed ....Achorn
Green Label ....Ackerlind
Saratoga.....................AL Tech
Atlan Oil Hardening
Die Steel ....Atlantic
Keewatin ....Atlas
BTR ....Bethlehem
Kiski ....Braeburn
Carpenter O-1 ....Cartech
Choyce 77 ....Champion
Exl-Die ....Columbia
Ketos ....Crucible
Oil Hardening AISI-O1 ....DoALL
O-1 ....Fagersta
Truform ....Jessop
Swed-Oil ....Kloster
Badger ....Latrobe
Torpedo ....Lehigh
Oilcrat ....Marshall
Yellow Label ....Peninsular
Keystone, Presco .....Precision-Kidd
Teenax ....Simonds
496 Oil Hardening Precision
Ground Flat Stock ....Starrett
Warplis.....................Teledyne
Colonial No. 6 ....Teledyne Vasco
AMUTIT-S ................U.N. Alloy
UHB Arne ....Uddeholm
Wando ....Universal-Cyclops
Hargus ....Ziv

Type O2
Deward ....AL Tech
Stentor ....Cartech
MST .......................U.N. Alloy

Type O3
Oildie Smoothcut (O3S) ....Columbia

Type O6
A.S. Graphitic ....Ackerlind
Oilgraph ....AL Tech
O6 Graphitic ....Bethlehem
Col-Graph ....Columbia
Halgraph ....Crucible
Gray Diamond ....Latrobe
Graph Mo ....Peninsular
Graph-Mo ....Timken
UHB Grane ....Uddeholm

Type O7
Utica ....AL Tech
Tapdie ....Columbia
Red Star Tungsten ...Teledyne Vasco

Other Oil Hardening, Cold Work Tool Steels
O.H. ....DoALL

COLD WORK TOOL STEELS: MEDIUM ALLOY, AIR HARDENING TYPES

Type A2
CVM Airhardening Tool Steel
Annealed ....Achorn
A.S. #5 ....Ackerlind
Sagamore.....................AL Tech
Hardnair Die Steel ....Atlantic
Cromoloy ....Atlas
A-H5 ....Bethlehem
Airque ....Braeburn
484,484 FM ....Cartech
Econo #5 ....Champion
E-Z-Die, E-Z-Die
Smoothcut ....Columbia
Airkool ....Crucible
Air Hardening AISI-A2 ....DoALL
Windsor ....Jessop
Pure-Ore Air-Chrom,
Air-Chrom ....Kloster
Select B FM ....Latrobe
Airtem ....Lehigh
Aircrat ....Marshall
Pen Air #5 ....Peninsular
Air True ....Precision-Kidd
Airtrue ....Simonds
497 Air Hardening Precision
Ground Die Stock ....Starrett
Pittsburgh ....Teledyne
Airhard ....Teledyne Vasco
Special K-5 ..............U.N. Alloy
UHB Rigor ....Uddeholm
Sparta ....Universal-Cyclops
Dumore ....Ziv

Type A3
Airque V ....Braeburn

Type A4
Air-4 ....Bethlehem
A 4 ....Braeburn
Tempair ....Latrobe

Type A6
CM Airhardening Tool Steel
Annealed ....Achorn
Ack Low ....Ackerlind
Apache ....AL Tech
Nutherm ....Atlas
A-6 ....Bethlehem
Vega, Vega FM ....Cartech
Uni-Die ....Columbia
CSM 6 ....Crucible
Air Hardening AISI-A6 ....DoALL
Jessair ....Jessop

Diecrat ....Marshall
UHB 1550 ....Uddeholm
Lo-Air ....Universal-Cyclops
Lo-Air-Hard ....Ziv

Type A7
Sagamore V ....AL Tech
E-Z-Die V ....Columbia
BX3 ....Jessop
BR-2 (3), Br-3 ....Latrobe
A7W ....Simonds
Chromewear ....Teledyne Vasco

Type A8
AL-158 ....AL Tech
Cromo W-55 ....Bethlehem
Pressurdie 16 ....Braeburn
Carpenter A-8 ....Cartech
A8 ....Columbia
MGR ....Latrobe
Airtrue LC ....Simonds
Hotform No. 3 ......Teledyne Vasco

Type A9
Formdie ....Columbia
Thermold J ......Universal-Cyclops

Type A10
Graph Air ....Peninsular
Graph-Air ....Timken

Other Air Hardening, Cold Work Tool Steels
A-HT ....Bethlehem
Cr Ni Rotor FM,
Cr Ni Rotor ....Cartech
Chromewear 300, Sistal,
Vasco Die, Vasco Tuf,
Vasco Wear ....Teledyne Vasco
AHT ....Ziv

COLD WORK TOOL STEELS: HIGH CARBON, HIGH CHROMIUM TYPES

Type D1
AL 38 ....AL Tech
Superior 4 ....Braeburn
Chipper Knife ....Latrobe

Type D2
High Production Die Steel
Annealed ....Achorn
Tri Ack ....Ackerlind
Ontario ....AL Tech
Atlan HCC Steel ....Atlantic
FNS ....Atlas
Lehigh H ....Bethlehem
Superior 3 ....Braeburn
610, 610 FM ....Cartech
Trudie ....Champion
Atmodie, Atmodie
Smoothcut (D2S) ....Columbia
Airdi 150 ....Crucible
CNS1 ....Jessop
Pure-Ore Hi-Run, Hi-Run .....Kloster

Olympic FM ....*Latrobe*
Hyco .... *Lehigh*
White Label .... *Peninsular*
CCM ....*Simonds*
Beaver .... *Teledyne*
Ohio Die .... *Teledyne Vasco*
Special KMV ...........*U.N. Alloy*
UHB Sverker 21 .... *Uddeholm*
Ultradie 3 .... *Universal-Cyclops*
Darwin #1 ....*Ziv*

Type D3
Huron.................... *AL Tech*
Superior 1 .... *Braeburn*
Hampden .... *Cartech*
Superdie ....*Columbia*
CNS2 ....*Jessop*
GSN ....*Latrobe*
Special K ...............*U.N. Alloy*
UHB Sverker 1 ....*Uddeholm*
Neor ....*Ziv*

Type D4
V. Chrome ....*Ackerlind*
AL 124 ....*AL Tech*
NN ....*Atlas*
At 2 .... *Braeburn*
Atmodie 4 ....*Columbia*
HYCC ....*Crucible*
CNS3 ....*Jessop*
GSN + Mo .... *Latrobe*
Crocar .... *Teledyne Vasco*

Type D5
5 .... *AL Tech*
Superior 2 .... *Braeburn*
Truedie Special .... *Champion*
Atmodie 5 ....*Columbia*
Cobalt Chrome FM ....*Latrobe*
Special KCO .............*U.N. Alloy*
PRK-33 ....*Ziv*

Type D6
VI Chrome W ....*Ackerlind*
UHB Sverker 3 ....*Uddeholm*

Type D7
BR-4 FM ....*Latrobe*
ARS ....*Simonds*

Other High Carbon, High Chromium Cold Work Tool Steels
Versasteel ....*Crucible*
Cyclops SCK ..... *Universal-Cyclops*

HOT WORK TOOL STEELS: CHROMIUM TYPES

Type H10
173 .... *AL Tech*
Pressurdie 6 .... *Braeburn*
Peerless 56 ....*Crucible*
Thermotem 10 ....*Heppenstall*
Dart ....*Latrobe*
WMD ...................*U.N. Alloy*

Type H11
#33A Hot Work Steel
Annealed .... *Achorn*
Potomac A .... *AL Tech*
H-11 .... *Atlas*
Cromo-V .... *Bethlehem*
Pressurdie 3L .... *Braeburn*
882, 882 FM .... *Cartech*
HW 1 .... *Champion*
Firedie ....*Columbia*
Nu-Die, Halcomb 218 ......*Crucible*
Shelldie .... *Finkl*
Thermotem 11 ....*Heppenstall*
Dica B Mod ....*Jessop*
Dycast No. 1 ....*Latrobe*
Howord A ....*Simonds*
Hotform No. 2 ...... *Teledyne Vasco*
US ULTRA ...............*U.N. Alloy*
Thermold H11 ..... *Universal-Cyclops*
Lo-Van ....*Ziv*

Type H12
#33 Hot Work Steel
Annealed .... *Achorn*
Potomac .... *AL Tech*
Crodi .... *Atlas*
Cromo-W, WV .... *Bethlehem*
Pressurdie 2 .... *Braeburn*
345, 345 FM .... *Cartech*
HW 2 .... *Champion*
Alcodie ....*Columbia*
Chro-Mow ....*Crucible*
Thermotem 12 ....*Heppenstall*
Dica B ....*Jessop*
Pure Ore D-C-33, D-C-33 ..... *Kloster*
LPD ....*Latrobe*
Ferno .... *Lehigh*
Howord B ....*Simonds*
Hotform No. 1 ...... *Teledyne Vasco*
US ULTRA 4 .............*U.N. Alloy*
Thermold H13 ..... *Universal-Cyclops*
HPD ....*Ziv*

Type H13
#33M Hot Work Steel
Annealed .... *Achorn*
A.S. H13 Microdized ....*Ackerlind*
Potomac M .... *AL Tech*
Dievac ....*Atlas*
Cromo-High V ....*Bethlehem*
Pressurdie 3 .... *Braeburn*
883, 883 FM .... *Cartech*
HW 3 .... *Champion*
Firedie 13 ....*Columbia*
Nu-Die V ....*Crucible*
Thermotem 13 ....*Heppenstall*
Dica B Van ....*Jessop*
D-C-33-V .... *Kloster*
VDC ....*Latrobe*
Penco Hi-Van .... *Peninsular*
Hotform V .... *Teledyne Vasco*
Modified ISO DISC
Number 1 ................*U.N. Alloy*
UHB Orvar 2 Microdized .. *Uddeholm*
Thermold H13 ..... *Universal-Cyclops*
Maximold ....*Ziv*

Type H14
Red Indian .... *Atlas*
Pressurdie 1 .... *Braeburn*
Halcomb 425 ....*Crucible*
Lumdie ....*Latrobe*

Type H16
Chrome Tungsten ....*Latrobe*

Type H19
B-47 .... *AL Tech*
H-19 .... *Atlas*
Pressurdie C .... *Braeburn*
Halcomb 425 ....*Crucible*
WCC .... *Teledyne Vasco*
WCO ...................*U.N. Alloy*

Other Chromium Type, Hot Work Tool Steels
Cromo-N ....*Bethlehem*
Hardtem, C Annealed, Pyrotem, Pyroneal, Super Hardtem II, Special C ....*Heppenstall*
Cromo-N .... *Peninsular*
Die Flex ....*Simonds*
UHB Orvar 2 Microdized .. *Uddeholm*
WMD Extra .............*U.N. Alloy*

HOT WORK TOOL STEELS: TUNGSTEN TYPES

Type H20
Hotpress .... *Teledyne Vasco*

Type H21
Atlas A .... *AL Tech*
Seneca .... *Atlas*
57 HW ....*Bethlehem*
T-Alloy A .... *Braeburn*
TK .... *Cartech*
Formite 21 ....*Columbia*
Peerless A .... *Crucible*
Thermotem 21 ....*Heppenstall*
2B-LC ....*Jessop*
CLW ....*Latrobe*
Marvel .... *Teledyne Vasco*
WKZ ...................*U.N. Alloy*
Thermold H21 ..... *Universal-Cyclops*
Lotung ....*Ziv*

Type H22
Atlas B .... *AL Tech*
T-Alloy .... *Braeburn*
TK Modified .... *Cartech*
Peerless LCT 2 ....*Crucible*
2B-MC ....*Jessop*

Type H23
HCA .... *Braeburn*
Kalkos ....*Latrobe*
W.W. Hotwork ...... *Teledyne Vasco*

Type H24
Mohawk.................... *AL Tech*
T-Alloy B .... *Braeburn*
Formite 24 ....*Columbia*

CHW . . . .*Latrobe*
S.C. Special . . . . *Teledyne Vasco*

Type H25
T-Alloy C . . . . *Braeburn*
EHW No. 1 . . . .*Latrobe*
Forge Die . . . . *Teledyne Vasco*

Type H26
LXX (Low Carbon) . . . . *AL Tech*
Special HS 55 . . . . *Atlas*
Vinco Hot Work . . . . *Braeburn*
H-26 . . . . *Cartech*
Clartie HW26 . . . .*Columbia*
REX AA PX . . . .*Crucible*
Electrite No. 5 . . . .*Latrobe*
Red Cut Superior J . . . *Teledyne Vasco*

**Other Tungsten Type,
Hot Work Tool Steels**
Pyrotool A, Pyrotool EX,
Pyrotool V, Pyrotool M,
Pyrotool W, Pyrotool 7,
Pyrotool 9, Pyrotool 16 . . . . . . *Cartech*
Lesco HW114 . . . .*Latrobe*

**HOT WORK TOOL STEELS:
MOLYBDENUM TYPES**

Type H41
Low Carbon Tatmo . . . .*Latrobe*

Type H42
DBL-2 (Low Carbon) . . . . *AL Tech*
Braemow Special . . . . *Braeburn*
Mustang LC . . . . *Jessop*
Electrite No. 7 . . . .—*Latrobe*

Type H43
HW 8 . . . .*Bethlehem*
Montemp R.S.P. . . . . *Braeburn*
Low Carbon TNW . . . .*Latrobe*

**Other Molybdenum Type,
Hot Work Tool Steels**
Prestem, Presneal . . . .*Heppenstall*
MCH, MCL . . . .*Latrobe*
Thermold 75 . . . . . . *Universal-Cyclops*

**HIGH-SPEED STEELS:
TUNGSTEN TYPES**

Type T1
18-4-1 High Speed Annealed . . . *Achorn*
LXX . . . . *AL Tech*
H.S. Steel . . . .*Atlantic*
Spartan-7 . . . .*Atlas*
T-1 . . . . *Bethlehem*
Vinco . . . . *Braeburn*
Star Zenith . . . . *Cartech*
Clarite . . . .*Columbia*
REX AA . . . . —*Crucible*
T-1 . . . . *DoALL*
T-1 . . . . *Fagersta*
Supremus . . . . *Jessop*
Pure-Ore Clipper, Clipper . . . . . *Kloster*

Electrite No. 1 XL . . . . *Latrobe*
Volcano . . . . *Lehigh*
Red Streak . . . .*Simonds*
Red Cut Superior . . . . *Teledyne Vasco*
UN T-1 . . . .*U.N. Alloy*
Super High Speed . . . .*Ziv*

Type T2
M-L . . . . AL Tech
Twinvan . . . . *Braeburn*
Vanite . . . .*Columbia*
Supremus Extra . . . . *Jessop*
E. No. 19 . . . .*Latrobe*
Lehigh XXX, Lehigh XXX
Tool Bits . . . . *Lehigh*
Lock Port Special . . . .*Simonds*
E.V.M . . . . *Teledyne Vasco*

Type T3
E. Vanadium . . . .*Latrobe*

Type T4
Panther Special . . . . *AL Tech*
Cobalt . . . . *Braeburn*
REX AAA . . . .*Crucible*
Purple Label . . . . *Jessop*
E. Cobalt . . . .*Latrobe*
Red Cut Cobalt . . . . . . *Teledyne Vasco*
UN T-4 . . . .*U.N. Alloy*

Type T5
Super Panther . . . . *AL Tech*
ACT Carbide H.S. Steel . . . . . .*Atlantic*
Nipigon . . . . *Atlas*
Bonded Carbide Jr. . . . . *Braeburn*
Cobite . . . . *Columbia*
T5 . . . . *DoALL*
Purple Label Extra . . . . *Jessop*
E. Super Cobalt . . . .*Latrobe*
Circle C . . . . *Teledyne Vasco*
UN T-5 . . . .*U.N. Alloy*

Type T6
Bonded Carbide . . . . *Braeburn*
King Cobalt . . . . *Jessop*
E. Ultra Cobalt . . . .*Latrobe*
Speedcut Cushnd Tool Bits,
Lehigh S.S. . . . . *Lehigh*

Type T8
Maxite . . . .*Columbia*
REX 95 . . . .*Crucible*
T-8 . . . . *DoALL*
Jessop T-8 . . . . *Jessop*
Marshall . . . .*Marshall*

Type T15
Panther V . . . . *AL Tech*
Sabre . . . .*Atlas*
T15 . . . . *Braeburn*
Maxite 15 . . . .*Columbia*
CPM REX T-15 . . . .*Crucible*
T-15 . . . . *DoALL*
E. Dyna-Van XL . . . .*Latrobe*
Vasco Supreme . . . . . . *Teledyne Vasco*
UN T-15 . . . .*U.N. Alloy*
Ziv's T-15 . . . .*Ziv*

Other High-speed, Tungsten Types
CPM REX 76 . . . .*Crucible*

**HIGH-SPEED STEELS:
MOLYBDENUM TYPES**

Type M1
M-1 Type High Speed
Annealed . . . . *Achorn*
LMW . . . . *AL Tech*
Amotun H.S. Steel . . . .*Atlantic*
Mohican-8 . . . . *Atlas*
M-1 . . . .*Bethlehem*
Mocut . . . . *Braeburn*
Star Max . . . . *Cartech*
Molite 1 . . . .*Columbia*
REX TMO . . . .*Crucible*
M-1 . . . .*Fagersta*
M-1 . . . . *Hi-Alloys*
Mogul . . . .*Jessop*
E. Tatmo . . . .*Latrobe*
Hi-Moly . . . . *Peninsular*
STM . . . .*Simonds*
8-N-2 . . . . *Teledyne Vasco*
UN M-1 . . . .*U.N. Alloy*
Motung, Motung
P&D . . . . *Universal-Cyclops*

Type M2
M-2 Type High Speed Steel
Annealed . . . . *Achorn*
A.S. #66 . . . . *Ackerlind*
DBL-2 . . . . *AL Tech*
Sixix . . . .*Atlas*
M-2 . . . .*Bethlehem*
Braemow Mocarb . . . . *Braeburn*
Speed Star,
Hi Carbon M-2 FM . . . . *Cartech*
Molite 2 Class 1, Molite 2
Smoothcut Class 1 . . . .*Columbia*
REX M-2 . . . .*Crucible*
M-2 . . . .*Fagersta*
M2 . . . . *Hi-Alloys*
Mustang . . . . *Jessop*
Pure-Ore Moly 6-6, Moly 6-6 . . *Kloster*
E. (Double Six) M2 XL,
Lesco HS29 XL
(High Carbon M2) . . . .*Latrobe*
Marshall . . . .*Marshall*
Penco 66 . . . . *Peninsular*
Molva T . . . .*Simonds*
Vasco M-2 . . . . *Teledyne Vasco*
UN M-2 . . . .*U.N. Alloy*
Motung 652 . . . . *Universal-Cyclops*
Red Shadow . . . .*Ziv*

Type M3-1
M-3 Type High Speed Steel
Annealed . . . . *Achorn*
DBL-2 1/2 . . . . *AL Tech*
M-3 . . . .*Atlas*
Braevan . . . . *Braeburn*
Molite 3 Class 1, Molite 3
Smoothcut Class 1 . . . .*Columbia*
REX M-3-1 . . . .*Crucible*
M-3-1 . . . .*Fagersta*

E. Corsair XL . . . . *Latrobe*
Van Cut Type 1 . . . . . . *Teledyne Vasco*
UN M-3 . . . . *U.N. Alloy*
Unicut . . . . *Universal-Cyclops*

Type M3-2
M-3 Type High Speed Steel
Annealed . . . . *Achorn*
DBL-3 . . . . *AL Tech*
Braevan-2 . . . . *Braeburn*
Molite 3 Class 2 . . . . *Columbia*
REX M-3-2 . . . . *Crucible*
M-3-2 . . . . *Fagersta*
E. Crusader XL . . . . *Latrobe*
Van Cut Type 2 . . . . . . *Teledyne Vasco*
Unicut 2 . . . . *Universal-Cyclops*
M-3 Type 2 . . . . *Ziv*

Type M4
DBL-4 . . . . *AL Tech*
M-4 . . . . *Atlas*
Braefour . . . . *Braeburn*
Four Star . . . . *Cartech*
Molite 4 . . . . *Columbia*
CPM REX M4 . . . . *Crucible*
E. Stark . . . . *Latrobe*
Neatro . . . . *Teledyne Vasco*
Cyclops M4 . . . . *Universal-Cyclops*

Type M6
Congo . . . . *Braeburn*

Type M7
LMW-V . . . . *AL Tech*
M-7 . . . . *Atlas*
M-7 . . . . *Bethlehem*
Motuf . . . . *Braeburn*
Molite 7 . . . . *Columbia*
REX M-7 . . . . *Crucible*
M-7 . . . . *Fagersta*
M7 . . . . *Hi-Alloys*
E. Tatmo V . . . . *Latrobe*
Molva C . . . . *Simonds*
Vasco M-7 . . . . *Teledyne Vasco*
UN M-7 . . . . *U.N. Alloy*
Motung CV . . . . *Universal-Cyclops*

Type M10
VLM . . . . *AL Tech*
M-10 . . . . *Atlas*
M-10 . . . . *Bethlehem*
Motemp . . . . *Braeburn*
Ten Star . . . . *Cartech*
REX VM . . . . *Crucible*
M-10 . . . . *Fagersta*
E. TNW . . . . *Latrobe*
Van Lom . . . . *Teledyne Vasco*
UN M-10 . . . . *U.N. Alloys*

Type M15
E. Ultra-Van . . . . *Latrobe*

Type M30
Super LMW . . . . *AL Tech*
Como . . . . *Braeburn*
E. Lacomo . . . . *Latrobe*

8-N-2 Cobalt . . . . *Teledyne Vasco*
Super Motung . . . . . *Universal-Cyclops*

Type M33
Super LMW Extra . . . . *AL Tech*
M33 . . . . *Braeburn*
REX M-33 . . . . *Crucible*
E. Kelvan . . . . *Latrobe*
STMCO . . . . *Simonds*
8-N-2 Cobalt 8 . . . . *Teledyne Vasco*
Super Motung
33 . . . . *Universal-Cyclops*

Type M34
Super LMW Special . . . . *AL Tech*
M-34 . . . . *Atlas*
E. Tatmo Cobalt . . . . *Latrobe*

Type M35
Braeco . . . . *Braeburn*
CPM REX M-35 . . . . *Crucible*
M-35 . . . . *Fagersta*
UN M-35 . . . . *U.N. Alloy*

Type M36
Super DBL . . . . *AL Tech*
E. CO-6 . . . . *Latrobe*
Victory Cobalt . . . . . . *Teledyne Vasco*

Type M41
Molite 41 . . . . *Columbia*
REX 49 . . . . *Crucible*
M41 . . . . *DoALL*
M-41 . . . . *Fagersta*
RC70 . . . . *Jessop*
Marshall . . . . *Marshall*

Type M42
Exocut . . . . *AL Tech*
M-42 . . . . *Atlas*
Braemax . . . . *Braeburn*
Super Star M-42 . . . . *Cartech*
Molite 42 . . . . *Columbia*
REX M-42 . . . . *Crucible*
M-42 . . . . *Fagersta*
M42 . . . . *Hi-Alloys*
E. Dynamax . . . . *Latrobe*
Vasco Hypercut . . . . . *Teledyne Vasco*
UN M-42 . . . . *U.N. Alloy*
Cyclops M42 . . . . . . *Universal-Cyclops*
Ziv's M-42 . . . . *Ziv*

Type M43
E. Dynacut . . . . *Latrobe*

Type M44
Braecut, Braetuf . . . . *Braeburn*

Type M46
AL-46 . . . . *AL Tech*
REX M-46 . . . . *Crucible*

Type M47
Exohard . . . . *AL Tech*

Type M50
HTB-2 . . . . *AL Tech*

REX M-50 . . . . *Crucible*
E. MV-1 . . . . *Latrobe*

Type M52
REX M-52 . . . . *Crucible*
M-52 . . . . *Fagersta*
E. MV-2 . . . . *Latrobe*

Other Molybdenum Type, High-speed Steels
FB-52 (D-952), WKE-4,
WKE-45, D-950 . . . . *Fagersta*

SPECIAL PURPOSE TOOL STEELS: CARBON TUNGSTEN TYPES

Type F1
Berkshire . . . . *Cartech*
W. Tap . . . . *Latrobe*

Type F2
KW . . . . *Cartech*
Rapid Finish . . . . *Jessop*

Type F3
ESA . . . . *Latrobe*

Type F8
Silvanite . . . . *Columbia*

Other Carbon Tungsten, Special Purpose Tool Steels
Berkshire High Manganese . . *Cartech*

SPECIAL PURPOSE TOOL STEELS: LOW ALLOY TYPES

Type L1
33 Non-Tempering
Tool Steel . . . . *Atlantic*
Presto . . . . *Cartech*
M Chrome Regent . . . . *Latrobe*
Flexor . . . . *Pennsylvania*

Type L2
Caroga . . . . *AL Tech*
Tough M . . . . *Bethlehem*
Columbia L2 . . . . *Columbia*
Halvan . . . . *Crucible*
Crown (4) Superb . . . . *Latrobe*
Vanadium Type H . . . *Teledyne Vasco*
Cyclops L2 . . . . *Universal-Cyclops*
Zivan 45 . . . . *Ziv*

Type L3
JY Roll Steel . . . . *Cartech*

Type L4
V-Kut . . . . *Columbia*

Type L6
Tough 6 . . . . *Ackerlind*
Tioga . . . . *AL Tech*
Die Steel . . . . *Atlantic*
L-6 . . . . *Atlas*

TOOL STEELS — TRADE NAMES AND SUPPLIERS *(continued)*

Bethalloy . . . . *Bethlehem*
RDS . . . . *Cartech*
Econo #2 . . . . *Champion*
Nicrodie . . . . *Columbia*
Champaloy . . . . *Crucible*
NDS . . . . *Latrobe*
N.C. Alloy . . . . *Lehigh*
Nikro M . . . . *Teledyne Vasco*
Cyclops L6 . . . . *Universal-Cyclops*
Metalmold . . . .*Ziv*

Other Low Alloy, Special Purpose Tool Steels

Brake Die . . . . *Bethlehem*
Berkshire High Manganese . . *Cartech*
Machinery Steel #3 . . . .*Ziv*

MOLD STEELS

Type P1

Mirromold . . . . *Cartech*
Dura Mold C . . . . *Peninsular*

Type P2

Moldaloy Steel . . . . *Achorn*
Duramold B . . . . *Bethlehem*
OCS . . . . *Peninsular*
Hob-A-Die . . . .*Ziv*

Type P4

Duramold A . . . . *Bethlehem*
Super Samson . . . . *Cartech*
ACS . . . . *Peninsular*
UHB Premo . . . . *Uddeholm*

Type P5

Samson Extra . . . . *Cartech*
Vasco
Chromold VM . . . . *Teledyne Vasco*

Type P6

512 Steel Annealed . . . . *Achorn*
Super Impacto PQ . . . . *Atlas*
Duramold N . . . . *Bethlehem*
Carpenter 158 . . . . *Cartech*

Type P20

Heat Treated Mold Steel . . . . . . *Achorn*
A.S. P20 . . . .*Ackerlind*
Mold Special . . . . *Atlas*
P20 . . . . *Bethlehem*
CSM #2 . . . .*Crucible*
Mold Die . . . . *Finkl*
Indie ZP . . . .*Heppenstall*
Penco Mold . . . . *Peninsular*
UHB Impax . . . . *Uddeholm*
Newmax . . . .*Ziv*

Type P21

Cascade (3) . . . . *Latrobe*

Other Mold Steels

A.S. Stavax E.S.R. (Stainless),
A.S. Holder Block . . . . *Ackerlind*
Stainless Type 420 Mold . . . . . *Cartech*
CSM 414, CSM 420 . . . .*Crucible*
Flexor . . . . *Pennsylvania*
UHB Stavax ESR . . . . *Uddeholm*

TOOL STEEL PRODUCERS AND SUPPLIERS

Achorn Steel Co.
109 Smith Place
Cambridge, MA 02138

Ackerlind Steel Co.
45-15 Barnett Ave.
Long Island City, NY 11104

AL Tech Specialty Steel Corp.
P.O. Box 152
Dunkirk, NY 14048

Atlantic Steel Corp.
35-27 36th St.
Astoria, NY 11106

Atlas Steels Co.
50 Center St.
Welland, Ontario, Canada

Bethlehem Steel Corp.
Bethlehem, PA 18016

Braeburn Alloy Steel
Braeburn
Lower Burrell, PA 15068

Carpenter Technology Corp.
P.O. Box 662
Reading, PA 19603

Champion Steel Co.
P.O. Box 97
Penniman Rd.
Orwell, OH 44076

Columbia Tool Steel Co.
500 Lincoln Hwy.
Chicago Heights, IL 60411

Crucible Specialty Metals Div.
Colt Industries
P.O. Box 977
Syracuse, NY 13201

DoALL Co.
254 N. Laurel Ave.
Des Plaines, IL 60016

Fagersta Inc.
2 Henderson Dr.
West Caldwell, NJ 07006

A. Finkl & Sons
2011 Southport Ave.
Chicago, IL 60614

Heppenstall Co.
4620 Hatfield St.
Pittsburgh, PA 15201

Hi-Alloys Div.
Maryland Specialty Wire
100 Cockeysville, Rd.
Cockeysville, MD 21030

Jessop Steel Co.
Washington, PA 15301

Kloster Steel Corp.
224 N. Justine St.
Chicago, IL 60607

Latrobe Steel Co.
26265 Ligonier
Latrobe, PA 15650

Lehigh Steel Corp.
1775 Broadway
New York, NY 10019

Marshall Steel Co.
P.O. Box 340
LaGrange, IL 60526

Peninsular Steel Co.
P.O. Box 3853
Detroit, MI 48205

Pennsylvania Steel Corp.
12380 Beech-Daly Rd.
Detroit, MI 48239

Precision-Kidd Steel Co.
101 Erie Ave.
Aliquippa, PA 15001

Simonds Steel Div.
Wallace Murray Corp.
Lockport, NY 14094

L.S. Starrett Co.
Athol, MA 01331

Teledyne Pittsburgh Tool Steel
1535 Beaver Ave.
Monaca, PA 15061

Teledyne Vasco
P.O. Box 151
Latrobe, PA 15650

Timken Co.
Steel Div.
1835 Dueber Ave., S.W.
Canton, OH 44706

U.N. Alloy Steel
383 Dorchester Ave.
Boston, MA 02127

Uddeholm Steel Corp.
721 Union Blvd.
Totowa, NJ 07512

Universal Cyclops Specialty
Steel Div., Cyclops Corp.
650 Washington Rd.
Pittsburgh, PA 15228

Ziv Steel & Wire Co.
11952 Hubbard
Livonia, MI 48150

Stamping Dies — Tool Steel or Tungsten Carbide?

Production requirements, leadtime, and cost per unit stamped are major considerations in selecting a die material

GAIL P. McCLEARY
Carbidie Div.
Aiken Industries, Inc.
Irwin, PA

Reprinted from Manufacturing Engineering, February 1978

Decisions on whether to make punches and die sections of tool steel or tungsten carbide are important since they can mean the difference between profit and loss. By effecting manufacturing efficiencies they can also determine whether deliveries will be late or on time.

How do you go about making this important decision? Contributing factors which must be considered are:

1. The number of pieces to be stamped
2. Cost of the punch and die section, and the total cost of the die
3. Time required to obtain the component material, and total time to complete the die
4. Regrinding costs and press downtime
5. Finally, and most important, the die cost per unit stamped.

How Many Pieces? Considerations for the decision on which die material to use for a specific stamping operation should start with two simple questions — how many pieces must be made on the die for the current stamping run, and what are the prospects for future runs?

Trying to determine how many pieces of a particular stamping are required to justify a tungsten carbide die depends on so many factors that stating a specific number can be misleading. Die production can vary widely depending on workpiece material metallurgical characteristics, tolerances required, part consistency, and many other factors. For any given stamping, at some quantity there is a breakeven point where the cost is the same whether the die material is tool steel or tungsten carbide. Above this point, the probability of 8 to 12 times greater production per sharpening can more than justify the higher cost of the tungsten carbide die.

Although carbide is easy to justify for runs of 10 million or more pieces, it should also be considered for runs of 1 to 10 million pieces when handling materials that are difficult to stamp.

Die Material Procurement. How long does it take to obtain delivery of tool steel as compared to tungsten carbide? A time advantage can generally be obtained with tool steel since annealed material is usually stocked in rectangular or circular cross sections, and can be routinely machined into a punch or die section.

Making tungsten carbide punches and dies, particularly those of complicated configuration, is somewhat more involved. Tungsten carbide in the sintered state is extremely hard and can be efficiently formed only by grinding with diamond wheels or by EDM. However, in the soft (green) unsintered state, carbide can be formed by just about any machining method. Therefore, carbide is usually preformed in the soft presintered state, then sintered, and delivered as a hard punch or die blank to the diemaker who grinds, polishes and performs other finishing operations. A good tungsten carbide supplier who specializes in manufacturing die and wear parts can deliver carbide preforms in two or three weeks. In emergency situations, blanks can be delivered in less than one week.

How About Die Costs? Initial costs for finished punches and die section components in steel or tungsten carbide may vary from about the same to about five times more for carbide. For example, when making punches which are large and bulky, requiring a relatively small amount of finishing time, the cost is generally in favor of steel since much of the cost of the component is for the base material.

However, if the material volume of the punch is small and it is intricate in design, with very close tolerances requiring a large amount of diemaking time, the cost of the punch in carbide or steel may be the same or close to the same even though the steel blank may cost less than the carbide blank. The steel blank would have to be machined completely from square or round stock.

The same punch in tungsten carbide can be obtained in preformed shape from a carbide blank producer, and a relatively small amount of stock has to be removed in the finishing process.

Tool steel stock can be removed more quickly than carbide, but with carbide preforms there is much less stock to be removed. It is often easier to finish grind an intricate punch to a very close tolerance in carbide because tungsten carbide is three times as rigid as steel, and is much less flexible and heat responsive. Grinding error due to distortion is less likely.

Another advantage in finishing carbide punches is that time and money are not required to heat treat and harden the material before final grinding. Heat treating semifinished steel punches also presents a hazard since they may crack, causing the loss of the blank and the time and labor investment.

Die Construction. A tungsten carbide die requires stronger and more precisely built die sets and other accessories. The punch and die cavities must be more accurately aligned, and the components capable of holding this alignment. The punch press used with carbide dies must be well built, of adequate tonnage, and properly maintained.

Some stamping die setups which are not very stable may perform fairly well with steel components because the more flexible punches work under adverse conditions, without chipping, until worn out. A carbide punch may chip under the same conditions.

Another factor to be considered is maintenance. If a carbide die runs 10 times as long as a steel die before resharpening, other die components should be built well enough to work as long before maintenance is required. This all adds up to a higher initial cost for the total setup in carbide than it does for steel, but the resulting efficiencies can mean profit.

Regrinding. The number of regrinds

1. *TUNGSTEN CARBIDE*
billets being removed
from a cold isostatic press.

needed for steel punches and dies can be an important consideration in deciding to use carbide. Regrinding costs of steel and carbide range from about the same up to 25% more for the carbide. Steel can be ground more rapidly with higher stock removal rates, but a worn steel punch usually requires the removal of more stock than a carbide punch.

A steel die must be ground much more frequently. For example, in an operation stamping 0.012″ (0.30-mm) thick low carbon metal strips for paper fasteners, the pieces must be essentially burr free. The average number of pieces per grind with a steel die was 100,000. The first run on the same part with a carbide die produced over three million pieces before a 0.005″ (0.13 mm) chip on the punch made regrinding necessary. About 0.007″ (0.18 mm) of stock was removed during regrinding. This 30 to 1 improvement in die life is not uncommon.

Downtime required for regrinding can be very expensive. If the stamping run is only 100,000 pieces and the steel die runs that long between regrinds, downtime has no significance. However, if the run is three million pieces, one regrind probably would be needed for the carbide die and 30 regrinds for the steel die.

Another advantage is that a carbide punch will rarely score the die. At times, steel workpiece material will build on the edge of a steel punch and score the steel die cavity. This results in an expensive die rework or replacement.

After all costs have been considered and all other comparisons made between steel and tungsten carbide dies, the decision as to which material to use is generally based on one final comparison — what is the total cost per unit stamped?

Problem Solving With Carbide. The use of tungsten carbide in a stamping die provides more production per regrind, longer total life, and greater accuracy and repeatability of the stampings produced. Some very intricate parts can be made only with carbide dies because of their inherent accuracy and stability. Considering this, tungsten carbide is sometimes looked upon as a cure-all for everything in stamping die work. That's not the way it is.

If a steel die is running fairly well but has punch or die chipping problems, substituting tungsten carbide of any grade will rarely solve the problem. Since carbide is a relatively nonductile material and more prone to chipping, the problem will get worse rather than better. Improved die alignment, stronger die sets, hardened and accurate holding components, and other methods to stop chipping should be tried before changing the die material.

A carbide die must have the right grade of carbide for each station in the die to provide maximum performance. Optimum die life is obtained when the carbide in the punch and die is as hard as possible to give the best wear, but

not hard enough to chip. In areas of the die where heavy impact is encountered, a grade with a medium amount of cobalt and large grain size is generally used. Wear life of a punch used for stamping medium hard to soft material can be improved considerably by using the submicron grain carbides.

In order to design and make a tungsten carbide die for the best performance, an understanding of carbide composition, manufacture, properties and finishing is essential.

Die Carbide Composition. Tungsten carbide is the hardest, most wear resistant metal available for stamping dies and wear parts. It consists of extremely hard tungsten carbide grains held in a binder of cobalt, a malleable material. The grains are in the form of a tight skeletal structure bound together and strengthened by the cobalt.

Most tungsten carbide preform manufacturers make 15 to 20 die grades. These grades have been developed for specific applications, according to hardness and toughness levels, by varying the amount of cobalt binder. They vary from the relatively soft but very tough, 25% cobalt grade to the very hard, long wearing, 3% cobalt grade.

Hardness and wear properties of these grades are also varied, though to a more subtle extent, by varying the tungsten carbide grain size. Most carbide grades have a normal grain size varying from 1.1 to 3 microns. The submicron grades have a grain size of no more than 0.7 micron. Grades requiring high strength and impact resistance have larger grains, ranging up to 6 microns. Small grains tend to make the material harder and longer wearing. Large grains interlock better and make the material stronger with some reduction in wear resistance. Grade recommendations in the accompanying table are for a variety of tungsten carbide applications.

Carbide Properties. Hardnesses vary from R_A 92.8 for the 3% cobalt die

WC GRADES	
Applications	Grade (% Cobalt)
Stamping dies	
Normal	14-15
Lamination die, extra abrasive	11
High production, medium-hard steel	15 (submicron)
High production, medium-soft steel	10 (submicron)
Nonferrous	12
Powder-metal compacting dies	
Normal	6-9
Severe	10-12
Forming dies	
Normal	12-13
Severe	14-15
Draw dies	
Normal	6-9
Severe	10-13
Cold forming dies	
Severe wear applications	11 (Large grains)
No shock	3-5
Light shock	5-6
Impact applications	
Light	14-15
Medium	16-17
Heavy	20-25

Notes: Where two cobalt levels are shown, the more wear resistant grade is shown first. These recommendations are intended for general information only, and should not be used for any specific application without independent study and determination of specific requirements.

2. INDUCTION VACUUM FURNACES used for sintering carbide parts.

grades down to R_A81.5 for the softest 25% cobalt grade. R_C scale equivalents are from 81.5 down to 61. Tool steel hardnesses vary up to about R_C64 at their workable extremes. Tungsten carbide is a very heavy material, with density varying in die carbides from 15.3 g/cm³ for the hardest grades, down to 13.2 g/cm³. Most tool steels range around 8.7 g/cm³.

For ductile materials, such as steel, tensile strength data are usually provided by the manufacturer. Since tungsten carbide is a nonductile material it is difficult to get accurate tensile test data. As a result, the more accurate transverse rupture test is used. No comparison with steel for this property is valid since they are tested by different techniques.

Tungsten carbide will withstand higher compressive loads than any other die material. The compressive strength for die carbide varies from 700,000 down to about 450,000 psi (4827 to 3103 MPa). This compares to 550,000 psi (3792 MPa) for the hardest tool steels. The thermal expansion rate of carbide is less than half that of steel, which is important when making shrink fits.

Tungsten Carbide Manufacture. The tungsten carbide manufacturing process starts with carburization. Tungsten is blended with carbon in precise ratios by weight, mixed in a conical blender, and then heated in an induction furnace. During the carburizing process the powder changes chemically from tungsten and carbon to the tungsten carbide compound.

After carburization and quality testing, the tungsten carbide powder must be thoroughly smeared with cobalt. This is accomplished by ball mill techniques. Naptha is used as a milling agent forming a slurry. Melted paraffin is also added to provide green strength, and to allow the angular shaped tungsten carbide grains to flow easily and pack uniformly when the finished grade powder is pressed into machinable billets. The slurry is then thoroughly dried and separated into control lots.

The most commonly used method for compacting tungsten carbide die parts is the cold isostatic press process. Billets are pressed into rectangular shapes, solid rounds, or cylinders. A rectangular billet, for example, is formed in an open mold made of welded steel, with a series of holes to permit water to flow through easily. A rubber bag of the same rectangular shape is put into the mold, filled with grade powder, tamped down, and sealed water tight. A number of molds are placed in a cold isostatic press at one time, *Figure* 1. The hydrostatic press develops a pressure of 25,000 psi (172 375 kPa).

Tungsten carbide in the presintered state is an easily worked material. However, it is extremely abrasive and industrial diamonds are required for

3. CARBON ANALYSIS of tungsten carbide powders.

most machining operations. After the carbide parts have been machined into the required shapes, they are sintered in induction vacuum furnaces, *Figure* 2. The sintering operation shrinks the parts about 20 to 21.5% linearly, and 48% volumetrically. After sintering, the cemented carbide material is in a full hard condition, and cannot be annealed and hardened again.

Quality Control. The basic raw materials — tungsten, carbon and cobalt — are so costly that a rigid quality control system must be maintained. Control lots are maintained and samples checked metallurgically, *Figure* 3. Before the carbide blank is sent to the diemaker, several additional quality checks are made. First it is checked for hardness.

Material quality checks are made by microscopic inspection of a polished spot on the preform's surface. This can detect carbon excess or deficiency, and grain structure and distribution can be checked. Large pieces are ultrasonically tested for internal cracks and inclusions, and dye penetrant testing is utilized to detect any surface cracks. All surfaces are checked dimensionally.

Purchasing Carbide Preforms. When buying tungsten carbide blanks, the purchaser should give the carbide manufacturer finished part drawings, including tolerances. Unless otherwise instructed, the supplier will allow grinding stock on all surfaces. If there are areas, such as holes or counterbores, to be used as sintered with no further finishing, they should be indi-

cated on the drawing. A common error by those unfamiliar with ordering carbide preforms is to add grinding stock allowance without telling the carbide supplier. The result is that all surfaces end up with twice as much grind stock as needed.

The question of how detailed the carbide part should be preformed in the soft state, and how much should be removed during the finishing operation is often best left to the carbide supplier. Extensive preforming is best when a considerable amount of stock would have to be removed. However, in configurations such as small grooves, it is often more economical to omit a preforming operation and do all the locating and grinding in the hard state with only one setup cost. ∎

Improving flanging and forming with
Nitrogen cylinders in dies

Reprinted from Tooling & Production, October 1977

by **H. Arlan Heiser**
General Manager
Hyson Div
Teledyne Efficient Industries
Brecksville, OH

Pressworking of sheet metal often requires the flanging and forming of the material in a die. This generally requires a pressure pad to hold a part of the workpiece while the bending or forming action takes place.

In many cases, the force behind the pressure pad is a collection of springs that allow the pad to stop on the surface of the material while the ram of the press continues to close the die and complete the pressworking operation.

In many other cases the die designer or metal stamper concludes that springs just won't do the job, so he specifies nitrogen die cylinders in their place.

To be concise we can say that there are two basic reasons to put our Super Nitro-Dyne cylinders into a flange die: 1) To obtain more *force* than is practical with mechanical springs within the confined space available in dies. 2) To obtain more *travel* than is practical with mechanical springs within the

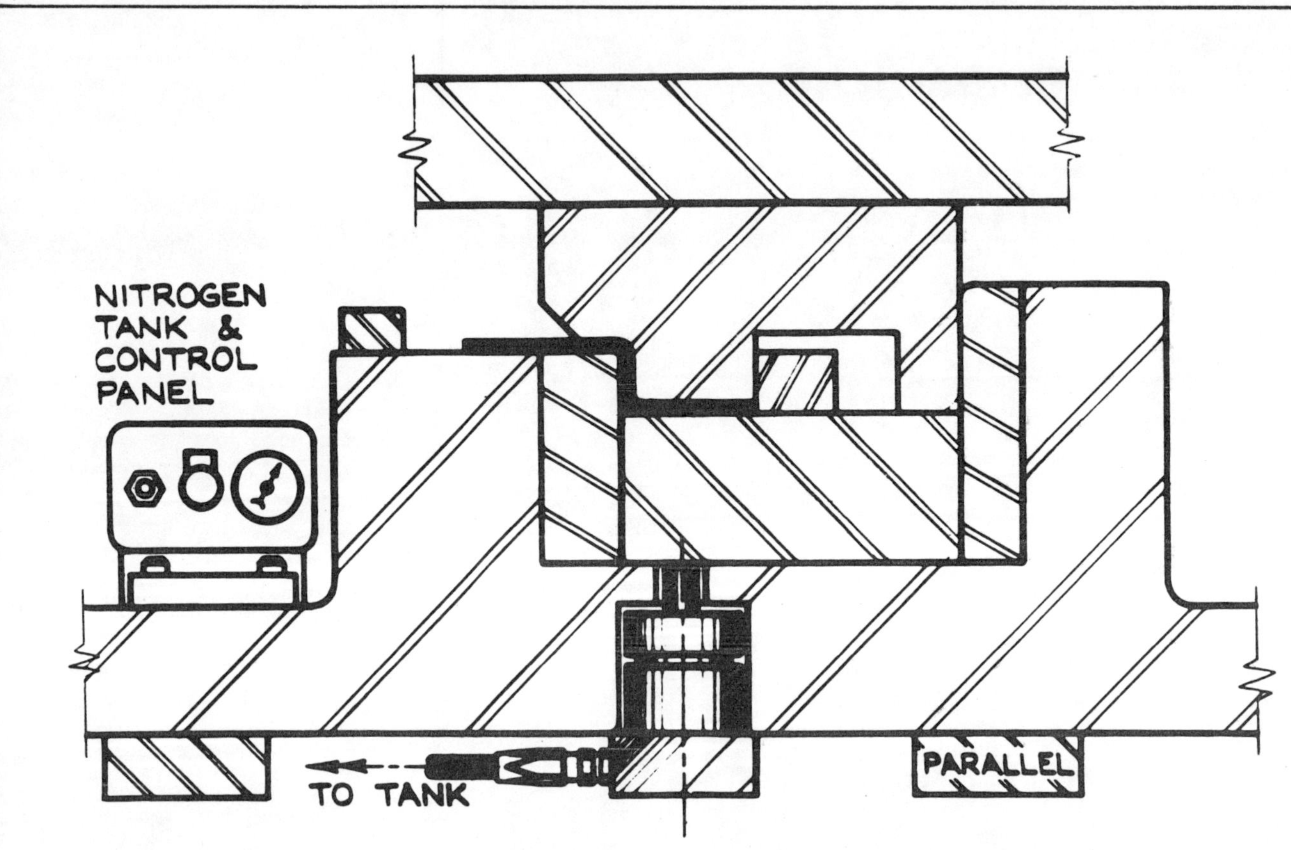

Forming in a step as shown does require much more holding force over the part than demonstrated in the Form Down and Form Up configurations at the bottom of these pages. For parts of this type, designers will try to increase the hold on the part by inserting hardened, serrated discs (cats' paws) in the pad to bite into the part. If the design permits, a pilot hole is punched in a previous operation and a locating pin put into the part to hold it firmly. These methods mark or alter the part. The designer may even add more metal to the part and thus increase the area that can be held. This excess material must of course be trimmed off in later operations and adds to part cost by increasing the scrap. Using a high pressure nitrogen system to hold the part more firmly offers a useful alternative.

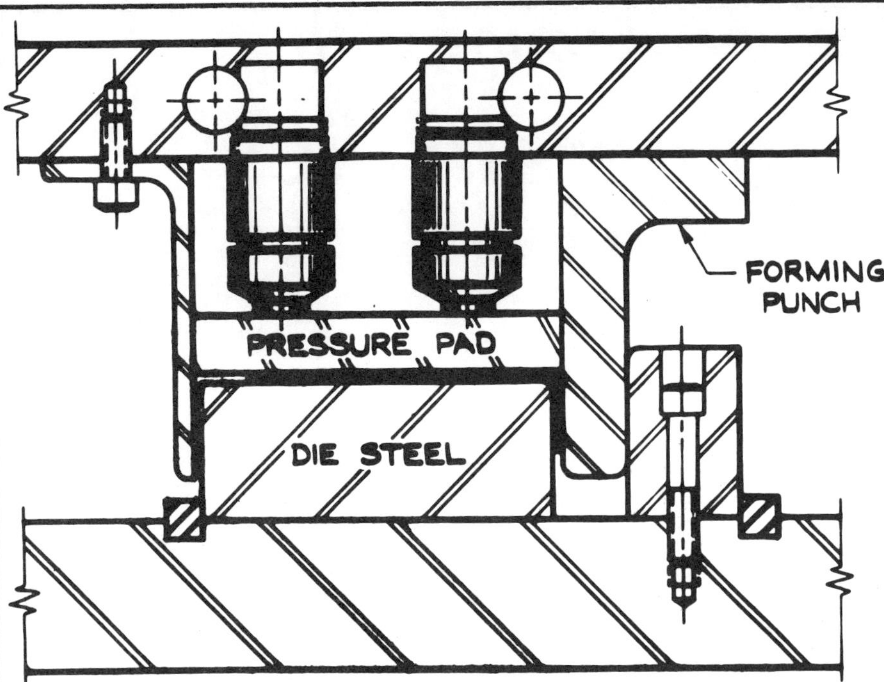

FORM DOWN

In the Form Down configuration, the forming is done with a solid die moving past the stationary part held in place with a die cylinder loading the pressure pad. Forming can be done upward by putting the pressure pad under the part and moving it past a solid die steel. A built-in press cushion can be used for Form Up. Springs normally do these jobs if the metal is thin and the area of the metal on the pad is large. Thicker metal and smaller pads need more force.

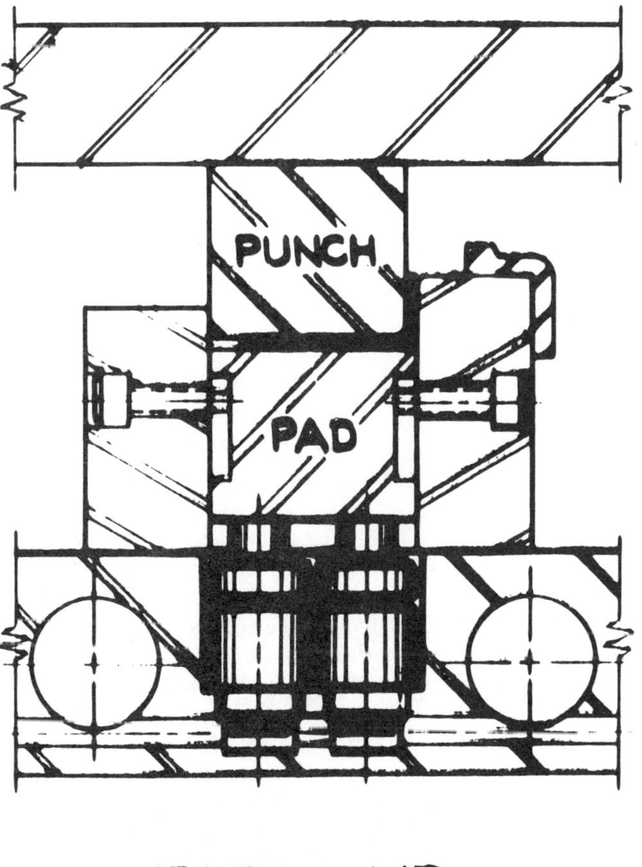

FORM UP

confined space available in dies.

Please note the inclusion of the word "practical" in the above statements. Mechanical springs may be very practical for a die making 4000 parts per year, and prove a costly maintenance item if the same tool must produce 400,000 parts per year.

Downtime due to broken, overworked springs can be an expensive nuisance. Pieces of broken springs also have a nasty way of getting between moving die sections and causing die crashes that truly compound maintenance costs.

Nitrogen die cylinders are made in a family of travel lengths to standardized cartridge dimensions, threaded, equipped with O rings and stocked for quick installation in standard SAE ports machined into the die plates.

The ports are connected to a manifold (which may be simply holes machined in the die plates, or an external cylinder tank) of sufficient volume that the complete displacement of the piston by an external force leaves the pressure behind the piston in the manifold virtually the same as it was.

The manifold system is charged with nitrogen from a commercial pressure bottle at the inlet fitting (like the check valve stem on a tire) on a control panel mounted on or near the die. The control panel includes an exhaust valve, a pressure gage and a rupture disc to protect the system from accidental overpressure. The cylinders are so well sealed that these systems generally do not need recharging during the run of a die. Nitrogen is used because it is readily available in pressure bottles and is safe should any of it leak from the system.

So the piston in the die cylinder acts like a spring in that it resists external forces to displace it. But it differs in that it resists with about the same force at any displacement, and that force can be adjusted without disassembly by changing the pressure in the manifold. Also, greater forces are generally obtainable in the same space and "bottoming" is less of a design problem.

Force and travel limitations are always considered in the light of space available to install springs or cylinders in a die. If a page of this magazine is a die pressure pad, a die designer could no doubt put 16 or 20 springs under it that are 2″ in diameter. He might even slip a smaller spring inside the big ones to pack in the force. Depending upon preload and work travel, he might get 9 or 10 tons of force under this pad.

If the stroke is very short and the springs permit a high preload, he can get a good deal more tonnage in a spot this

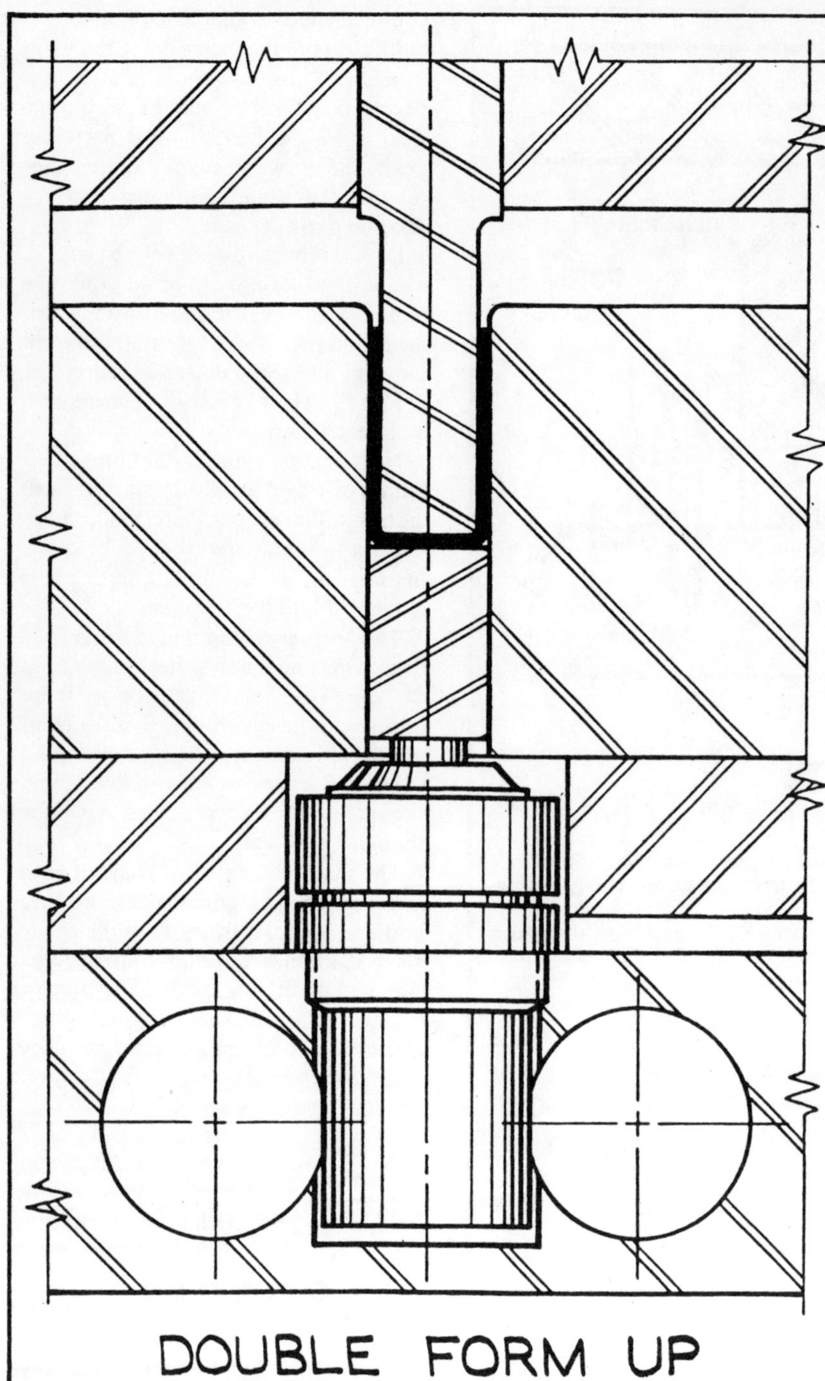

Producing a flat bottom can be troublesome on Double Form Up configurations. If the metal is thick or the part small, it can take a high pressure on the bottom pressure pad to prevent the bottom from forming into a bulged round bottom.

Some designs call for a "spanking die" operation to be sure the part is made to the proper shape.

The "spring-back" tendency of flanged parts to spread partially open after they are removed from the die is reduced or eliminated by holding the part firmly.

One installation of this type that is interesting to note is in the manufacture of familiar desk-top staplers by the Bostitch Div of Textron. For the staplers to work properly, the tolerances must be held closely. They make the channel shapes on some beautiful new Minster presses, automatically at speeds up to 115 spm. They commonly hold +0.001" on height and the same on hole locations 4" apart done in progressive dies.

The presses were purchased with air cushions; however, they are mostly idle and nitrogen cylinders built into the die sets are used. The cylinders are mounted directly under the pressure pad where their full force is directly under the spot it is needed.

Press cushions require some care in the distribution of load to insure that it is not unbalanced. Uneven loading is the cause of most cushion problems. Servicing a large press cushion is often a time-consuming job, while the seal kit to repack a Super Nitro-Dyne cylinder costs $18 and takes less than 10 minutes to install. Yet, in the five years Bostitch has been using nitrogen cylinders, they report zero maintenance. Generally they put a die into the press for a three week production run, which at 115 smp, 20 hours a day, comes to about two million parts without recharging.

size.

Nitrogen cylinders could deliver 30 tons in the space of this page and travel would not affect the usable force.

If the die designer determines that his die needs 16 tons in this area, he must make some evaluations:

1) If he preloads the springs to the maximum, he *may* make a part. He must decide how great a risk to take.

2) Will frequent downtime due to broken springs be a problem?

3) Can he compromise the design to increase the pressure pad size and thus have space to add additional springs? This may weaken other sections and certainly increase the overall die size.

4) Must he add another die operation to make the part? An expensive choice as it adds tooling costs at the beginning of the job and continues to cost in increased labor and press time.

5) He can put in a nitrogen cylinder system and easily get the 16 tons with only seven cylinders.

Now he is down to "What does it cost to put the nitrogen in?" Hyson has furnished systems of this size for around $2000. Allow another 10 percent to mount the system manifold plate to the die shoe.

Price comparisons must always be on an individual job basis. Generally, the higher the force or tonnage required or the longer the stroke the better a nitrogen cylinder system looks.

Illustrated here are a range of die applications from a single bend to compound bending. How far the idea of adding operations to a die can go is limited basically to the imagination of the die designer and the size of the presses available.

In the last three years we have noted a tremendous increase in the number of cylinder applications to compound dies. We see systems with two or more groups of cylinders, each having a different travel. In flange dies the cylinders may be in rows, and on draw dies they are often in concentric circles. Some factors

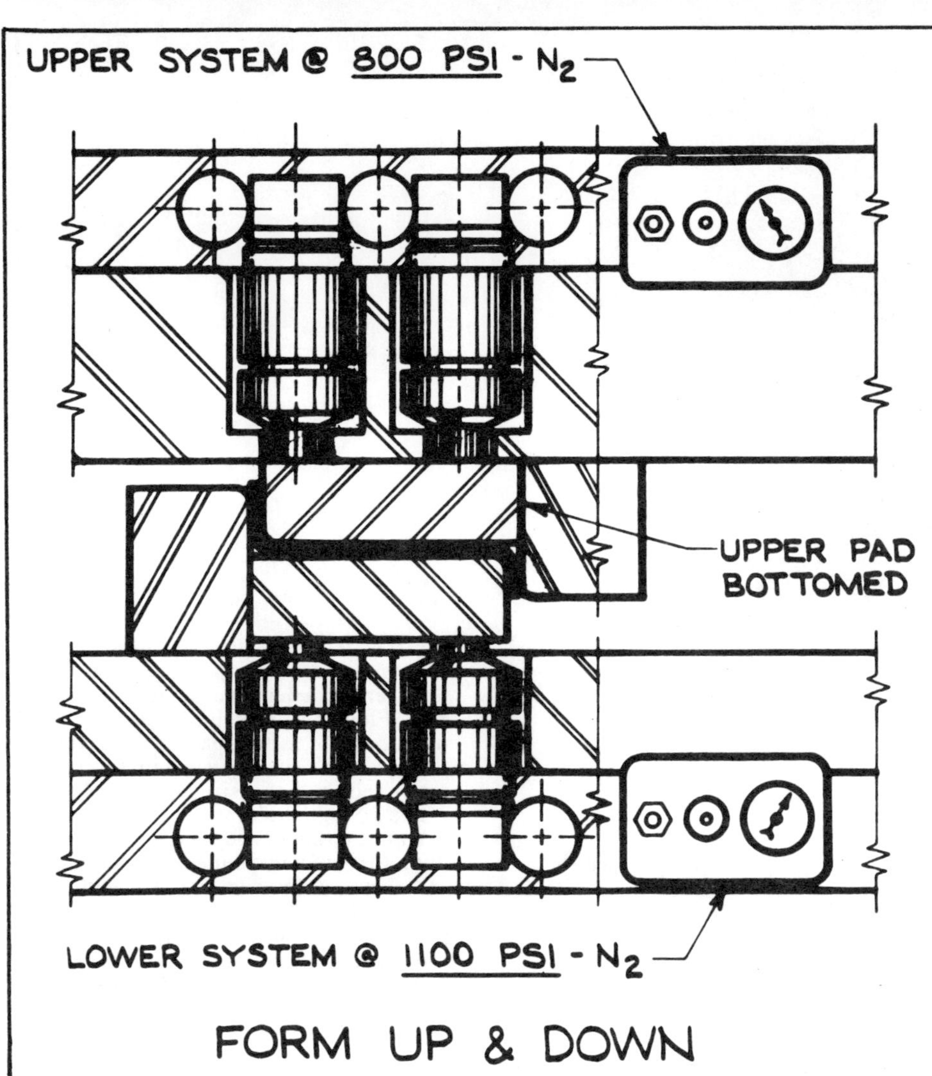

FORM UP & DOWN

By combining the actions of the Form Down and Form Up configurations, a part is made with the flange in both directions, provided space (press height) is available. Both die shoes contain a nitrogen cylinder of the same number and size. By charging the upper system to a lower nitrogen pressure than is used in the lower system, the part is formed down as the die first closed and the pressure of the lower system forces the upper system closed.

When the top pressure pad bottoms, the closing press action overcomes the lower system and the part is formed upwards.

This idea of forming up and down can be done with springs, press cushions and cylinders, or combinations of the three. The easy pressure adjustment and the accurate repeatability of pressure are major advantages of the nitrogen system. Note each system has a pressure gage and this makes very precise pressure setting a simple matter. With mechanical springs, setup is very much trial and error; as the springs age, the process often must be redone.

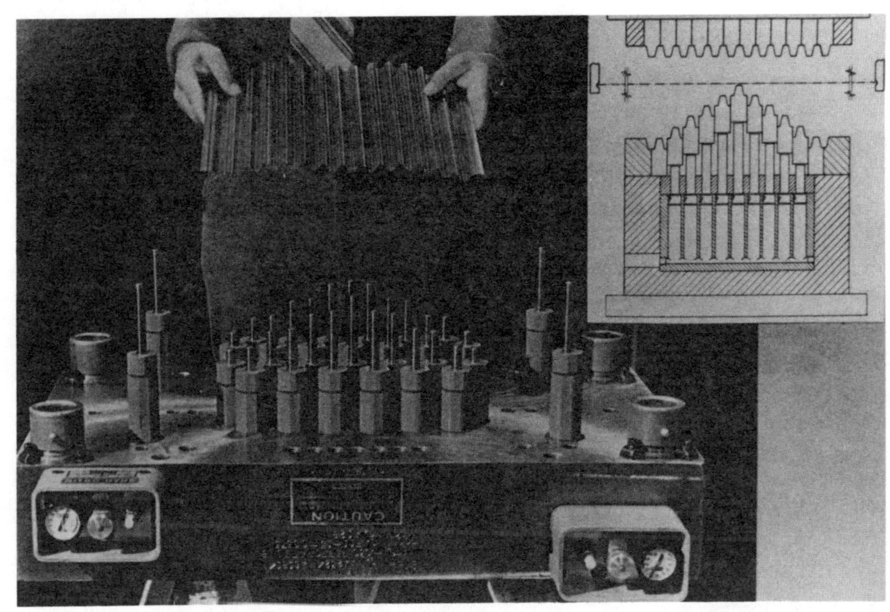

Corrugated parts are an excellent example of what can be done in imaginative use of nitrogen cylinders. Here the die section has solid forming steels on the upper die, while the lower forming steels are supported by a series of cylinders, with the longest stroke in the center. As the press closes, the metal is formed first in the center and then the steels on each side of center contact the metal and two more corrugations are made, continuing until the entire part is formed through what might be called a "vertical progression." Length of travel required of the center cylinder prohibits use of springs here.

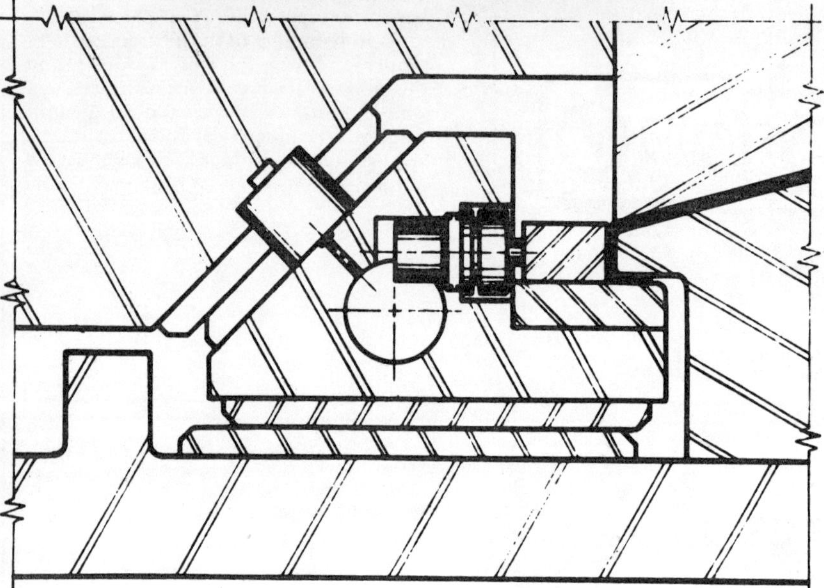

CAM FLANGE

Illustrated at the left is an installation from one of the big application areas for nitrogen cylinders, automotive flange dies. Flanging on major body parts is often difficult as the space available for the pressure units is frequently severely restricted, yet the size of the flange requires high holding forces. In this example, a cam flange die is used to produce a large flange along the lower edge of a new automobile bumper made of lightweight high-strength steel.

Each cam is tapped for 15 Hyson BC-2 two-ton capacity cylinders. When charged to maximum pressure of 1500 psi, the system can apply 30 tons of holding force on the part. The square cover plate seen on the end of the cam seals with an 0 ring a 3½″-dia hole which has been drilled through the length of the cam. This hole interconnects the cylinders and is the manifold for the nitrogen system. To the upper left of the cover plate is the control panel with inlet and exhaust valves and pressure gage.

It is quite common for users to have Hyson manufacture the manifold plate (in this case the cam) and pressure test it to insure a leak-free nitrogen system. The manifold then goes to the die builder for additional machining.

This is progressive die to make part of a truck instrument panel. Four nitrogen control panels are visible across the top shoe. There is a fifth in the lower die. Each station where high pressure is required has its own pressure system; this system permits individual force adjustment and travel. This simplifies die setup as there is no need to pull the die in and out of the press several times to make adjustments to mechanical springs to achieve pressure changes. While nitrogen cylinder systems are more common on the upper die members where space for mounting springs is so limited, systems are used in the lower die to supplement or substitute for the press air cushion. This eliminates cushion balancing and timing problems.

causing the increase are: 1) Combining operations means less dependency on labor, which is getting more scarce and expensive; 2) parts standardization programs reduce the number of dies and increase the production of the ones used, making more complex tools practical; 3) die designers have become more sophisticated; 4) availability and proven reliability of nitrogen die cylinders have given designers vast new freedom to design more complex tooling. ∎

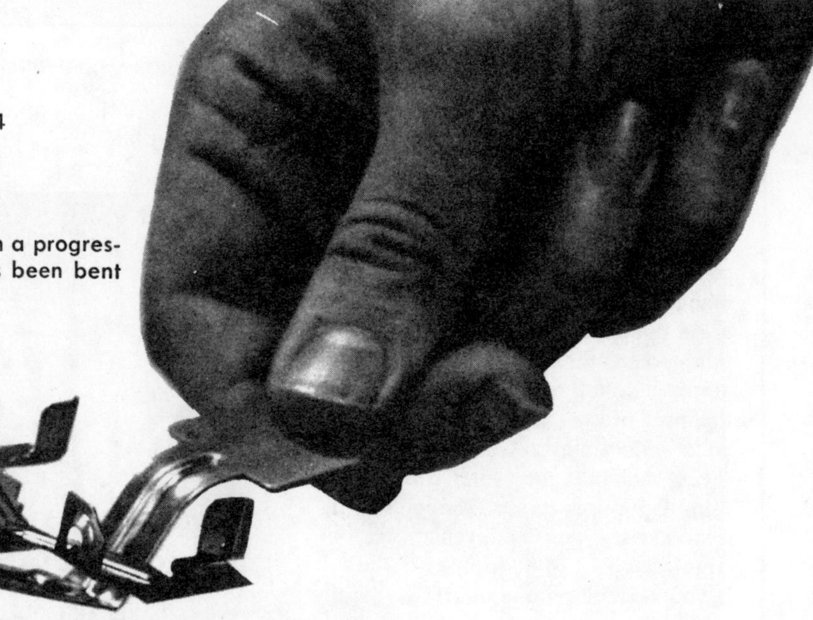

1. A complex part made in a progressive die. Note that it has been bent back against itself.

The hows and whys of progressive dies

By Nolan W. Rhea
Assistant Editor

THEY USED TO say, "You can't do it that way," but now they say, "you can't do it that way, can you?" this is the change in attitude toward progressive dies observed by Ed Stouten of Capitol Engineering Co., a firm specializing in die design.

In fact, the progressive die is an idea whose time has come. Instead of putting metal in, hitting it once, and taking it out, you can "walk" a strip through a series of tool stations and hit it repeatedly until you have complex finished parts. Many parts being run on a series of conventional presses can be made better on a single prog die press. Automobile ash trays, the wing nut that holds a car's jack in place, hose clamps, box car seals, and other complex parts are now being made this way, **Figure 1**. What is

more, this tool complies well with the OSHA attitude toward presses. You really can make a large number of parts without putting your hands in the die area.

The information in this story comes from the "Practical Design of Progressive Dies" clinic sponsored by the Society of Manufacturing Engineers.

Tips on design

There are some things your die designer should know when you order a progressive die set from him.

Stouten recommends that you specify the use of a "french notch." This notch is simply a piece knocked off one edge of the strip by a square punch. It becomes a stock stop and thus prevents overfeed. The thickest material Capitol has used in a progres-

sive die is ⅛″ and the thinnest in which it has successfully used the french notch is 0.010″. The notch is "cheap insurance" against half hits and restrikes which can produce scrap and damage the die.

Never specify upstairs stripper bolts. They are "verboten" because they can break off and smash the die. Spools or keepers in the upper pads are better.

Always use pilot holes to make sure that the strip is guided through the die set without any problems. It will save money if the holes are put through the scrap instead of the work.

If you want to form, form down. Prog dies do not like to form up.

Do not attempt to coin more than ⅓ of the thickness of 1010 steel or you will run the danger of getting cracks in the material.

2. This strip was carried through the die set by a two-sided carrier.

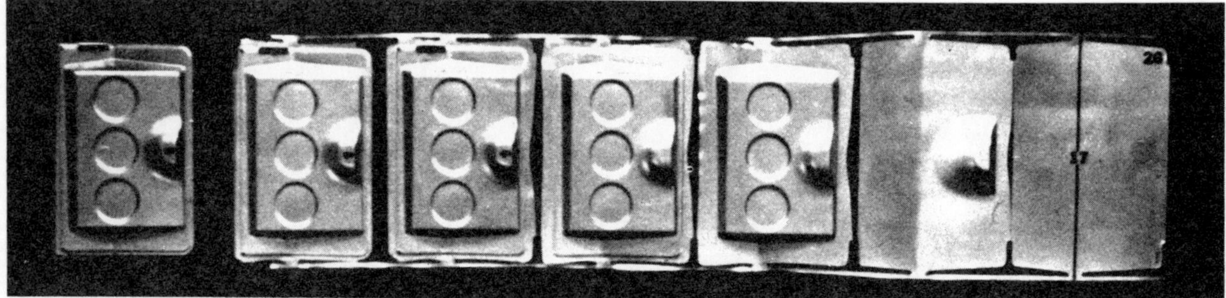

Carrying the strip

The two-sided carrier is the most common because it is the easiest to manage, **Figure 2**. Sometimes you can use a one-sided carrier (and there are times when the nature of the part requires it), but maintaining proper alignment of the strip in the die set can be a little tricky. The best one for saving money is the center carrier because it reduces scrap. Though it is a weak carrier, you can even use it for large parts (out of a strip, say 26″ dia) if you remember to use lifters and guides.

Whenever possible, the carrier should have a "stretch web" in it. A web is a flexible connector which allows the strip to adjust itself to the different positions made necessary by each die station, **Figure 3**. It allows the strip to hump at one station and straighten out at the next. It also ensures that if the die designer has made the die a few thousandths off pitch the strip can adjust for the error. Don't leave any corners on the web or it will not stretch evenly. If a web cannot be used, Capitol recommends that you provide an idle station which will permit the strip to flex.

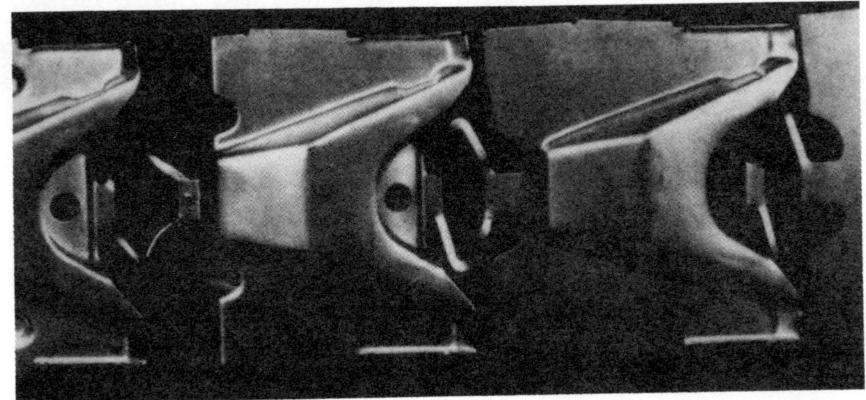

3. A flex web used in a center carrier. Note how much the web can be stretched without breaking.

The Advent of Progressive Coining

*If you still think of coining as a way to make nickels and dimes, take
a look at the applications and tooling discussed in this article.
Together, they spell big money – and a way to machine unmachinable parts*

DANIEL B. DALLAS
Editorial Director

Reprinted from Manufacturing Engineering, February 1974

MANY OF THE MORE DIFFICULT machining jobs can be handled by progressive coining. The big problem, however, is that too many manufacturers continue to think of coining as a form of embossing, as in the case of the dies used in the federal mints. As such, the technique is considered great for presenting past American Presidents in bas relief on pieces of highly malleable metal — but little else. But the imaginative firm interested in moving metal the way it's done in cold heading and cold forming operations can utilize coining as an effective way to cut costs and step up productivity in many hard-to-machine jobs.

One company that has done so with outstanding success is Eltee Inc., West Caldwell, N.J. Eltee is using coining as a process that's competitive with cold forming on numerous large jobs. It's also using coining in the production of miniaturized parts — since coining is the only technique that permits these jobs to be produced on a production basis. A case in point is the part shown in *Figure* 1.

COINING DOWN

Relay Base Design. This part is a relay base of the type used extensively in aerospace. For the most part, they're used in telemetry. They provide data on the physical condition of astronauts in orbit. Some are currently on the surface of the moon, providing information on moonquakes, solar storm, and the like. Thousands of them — produced to the dimensions shown in *Figure* 2 — are required.

From a manufacturing standpoint, several features are worth noting.

▶ RECESSES. Two recesses are required. The distance between them is 0.280 inch. Their lengths are 0.103 and their depths are 0.025 inch.

▶ WELD PROJECTION, NO. 1. Each recess has a weld projection at its center. Height of this projection is 0.003 inch x 45 deg.

▶ WELD PROJECTION, NO. 2. The second projection is located on the shoulder. Dimensions are 0.004 inch high x 45 deg.

▶ HOLES. Seven holes are required. Dimensions of these holes — 0.047 ± 0.001 — imply need for a shaving operation. It should be noted that the finish size of these holes is only 0.005 inch greater than the strip thickness. But the initial piercing operation is accomplished with 0.041-inch punches. Thus the piercing punches are smaller in diameter than strip thickness.

If this job were ten times as large as specified in *Figure* 2, it would still pose problems, particularly in hole punching. These problems are compounded by the height differential between the 0.010-inch flange and 0.017 inch height of the recess bottom. The solution to this problem — and to the others previously mentioned — is a progressive coining sequence.

The Strip. The strip shown in *Figure* 3 is an excellent example of progressive coining — in this case the coining required to produce the relay base of *Figures* 1 and 2. The operations performed on this strip can be summarized as follows: (1) notch to establish the rough blank outline, (2) pierce the holes, (3) shave the holes, (4) coin 0.025 inch deep to establish the recess and its 0.003-inch weld projection, (5) coin to full depth — to 0.010 inch thick — on the inside of the peripheral weld tab, (6) coin to full depth — to 0.010 inch thick — on the outside of the peripheral weld tab, and (7) blank finished part from strip. There are, of course, numerous idle and piloting stations in this sequence.

The Progressive Coining Die. *Figure* 4 shows the front elevation of the upper die and the plan view of the lower die steels. It will be noted that the sequence of die operations — after the initial outlining of the rough blank — begins at the point marked with a capital A.

The stripper in this die is a positive type, made of HCHC hardened to R$_c$ 60-62. Strip alignment begins at *a* in this strip when pilot type punches (Detail 20) pass through the stripper and enter the die. At *b*, a notching punch (Detail 21) defines the rough outline of the blank. At *c*, two more pilot-type punches establish strip alignment. The starting point of a strip would be bet-

ween *c* and *d*. The second hit would allow two pilots (Detail 22) to enter the notch that is blanked at *b*.

Two-in-one Die. This die is designed to produce two different parts, their only difference lying in their hold patterns. This difference is seen at *A*, where four punches produce four holes not shown in the strip or part. Removal of these punches permits production of the part under discussion. Accordingly, initial piercing begins at *B*.

The Complete Sequence. Using the upper and lower case letters indicated in *Figure* 4, the die sequence is as follows:

STATION *a*. Pilot for alignment
STATION *b*. Notch outline of blank.
STATION *c*. Pilot for alignment.
STATION *d*. Pilot in the notch pierced at Station *b*.
STATION *A*. Pierce four holes.
STATION *B*. Pierce eight holes.
STATION *C*. Pilot — otherwise idle.
STATION *D*. Shave holes pierced at Station *B*.
STATION *E*. Pilot — otherwise idle.
STATION *F*. Pilot and coin recesses to 0.025 inch deep.
STATION *G*. Pilot — otherwise idle.
STATION *H*. Pilot and coin inside of peripheral weld tab.
STATION *I*. Pilot — otherwise idle.
STATION *J*. Pilot and coin outside of peripheral weld tab.
STATION *K*. Pilot — otherwise idle.
STATION *L*. M and N. Idle.
STATION *O*. Blank part from strip,

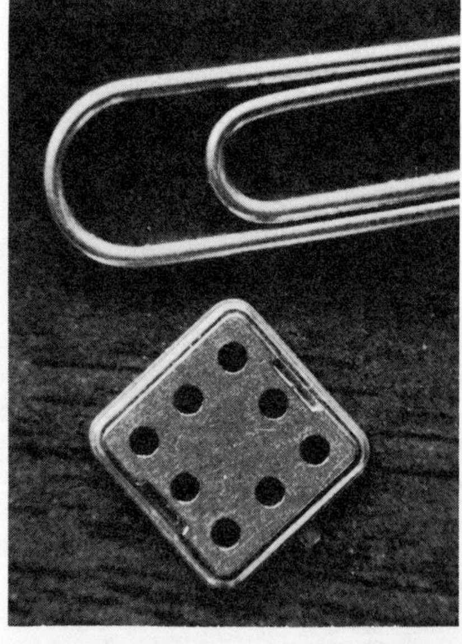

*1. COINED RELAY BASE
magnified by a factor of six.*

piloting from the bottom.

Pilots and Perforators. The number of pilots and piloting stations appears excessive. But more than simple alignment of the strip is involved in that the pilots also aid in stabilizing the part during the three coining and one cutoff operation.

Unfortunately, no data concerning punch and pilot material and hardness are available. The die buttons — held in place by 1/8-inch dowels — are high-speed steel hardened to R_c 60-62

Coining. Details 41, 42, and 43 are the first, second and third coining anvils respectively. These components are high-speed steel, hardened and ground to R_c 60-62. Each anvil contains a coining pad which also serves as a pilot retainer. Upon contact with the strip, these pads retract against disc spring pressure until the anvils contact the work. At this point, the pads bottom against the punch retainer, thus spanking the work.

Opposing the coining action of the anvils are three coining pads, denoted as Detail 19. (high-speed steel, R_c 60-62.

Compared to conventional sheet metal work, coining of the type performed here is a radical operation. But it works — in much the manner of a cold heading operation.

Lubricant is a must, of course. Eltee Inc. will not reveal the precise mixture used, but one of the basic ingredients is lard. All sulfur-bearing oils are ruled out for coining, since sulfur attacks both the work and the dies. Additionally, it becomes impregnated in the work, thus inhibiting sealing operations.

Temperature of the lubricant is also critical. Again, no details are available, but this much is known: The die, which normally operates at 80 spm, must be periodically halted to allow the lubricant to cool off.

The preceding example shows the potential of "coining down" in a progressive sequence. Coining down means moving the metal downward toward the lower die. This approach is preferred to coining up, which tends to deform the strip excessively. A significant drawback is that coining down generally requires "blanking up" to separate the finished part from the strip. This can pose problems in heavier gage metals.

COINING UP

Violations. An excellent example of "coining up" is shown in *Figure 5*. This is the base of a hermetically sealed connector for cable harnesses used in aerospace. Of interest in this design are the apparent violations of two generally accepted rules. The first violation is the gang punching of 34 holes — on a high production basis — through stock substantially thicker than hole diameter. As *Figures* 5 and 6 show, the maximum diameter in the group of 34 holes is 0.109 inch; stock thickness is 0.125 inch — 1/64 inch greater.

The second violation is the close proximity of the holes. An accepted rule of thumb is that two adjacent holes must be separated by 1.5 times stock thickness. These holes, pierced through 1/8 inch stock, are separated by a webbing that is less than 15 percent of stock thickness. Despite these apparent part design flaws, this coined hole header is being produced on a production basis without difficulty.

Strip Design. The operational sequence for this strip is shown in *Figure 6*. The lower die, which produces this strip, is shown in *Figure 7*. The sequence of events is as follows:

STATION 1. Two alignment pins in the upper shoe pass through holes in the stripper to correctly position the incoming stock. As in the previous example, a positive stripper is employed. This stripper provides guidance to punches, pilots, and alignment pins.

STATION 2. The strip — more specifically the part — is stamped from the bottom by the lower die. A solid punch mounted on the upper shoe provides the necessary pressure.

STATION 3. Two holes — 0.112 and 0.160-inch diameter — are pierced. Again the strip is straddled by two alignment pins. From this point on, the strip will be held in registration by pilots.

STATION a. Two 1/4-inch diameter holes are pierced in what will be the scrap between the blanks. Purpose: to aid in piloting the strip.

STATION 4. Ten 0.099-inch diameter holes are pierced.

STATION b. Pilots enter the two 1/4-inch diameter holes.

STATION 5. Eight 0.099-inch diameter holes are pierced. At the same time ten pilots enter the ten previously pierced 0.099-inch holes.

STATION c. Pilots enter the two 1/4-inch diameter holes.

STATION 6. Eight 0.099-inch diameter holes are pierced. At the same time 18 pilots enter the 18 previously pierced 0.099-inch holes. At this point the repetitive piloting in the 0.099-inch holes should be noted. The purpose is not to register the strip — the normal function of piloting. Rather, the multiple pilots are necessary to prevent hole collapse

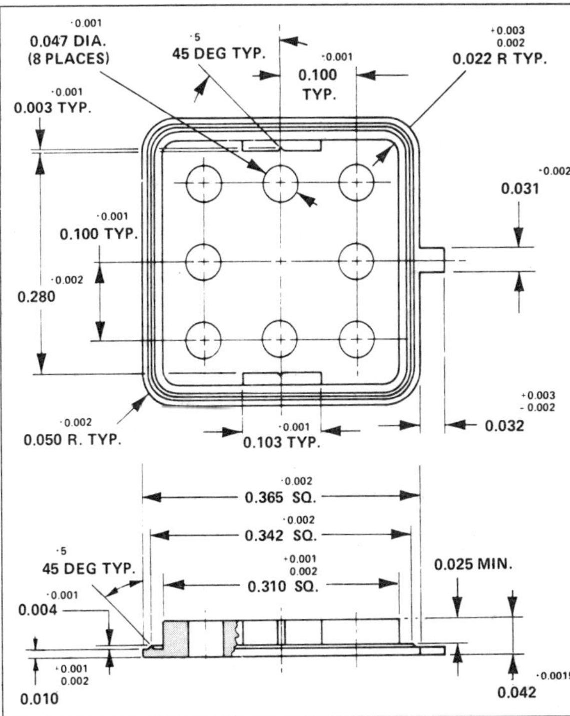

2. ENGINEERING DRAWING
of the relay base shown in
Figure 1 and in the strip below.

3. PROGRESSIVE SEQUENCE for the coining of relay bases.

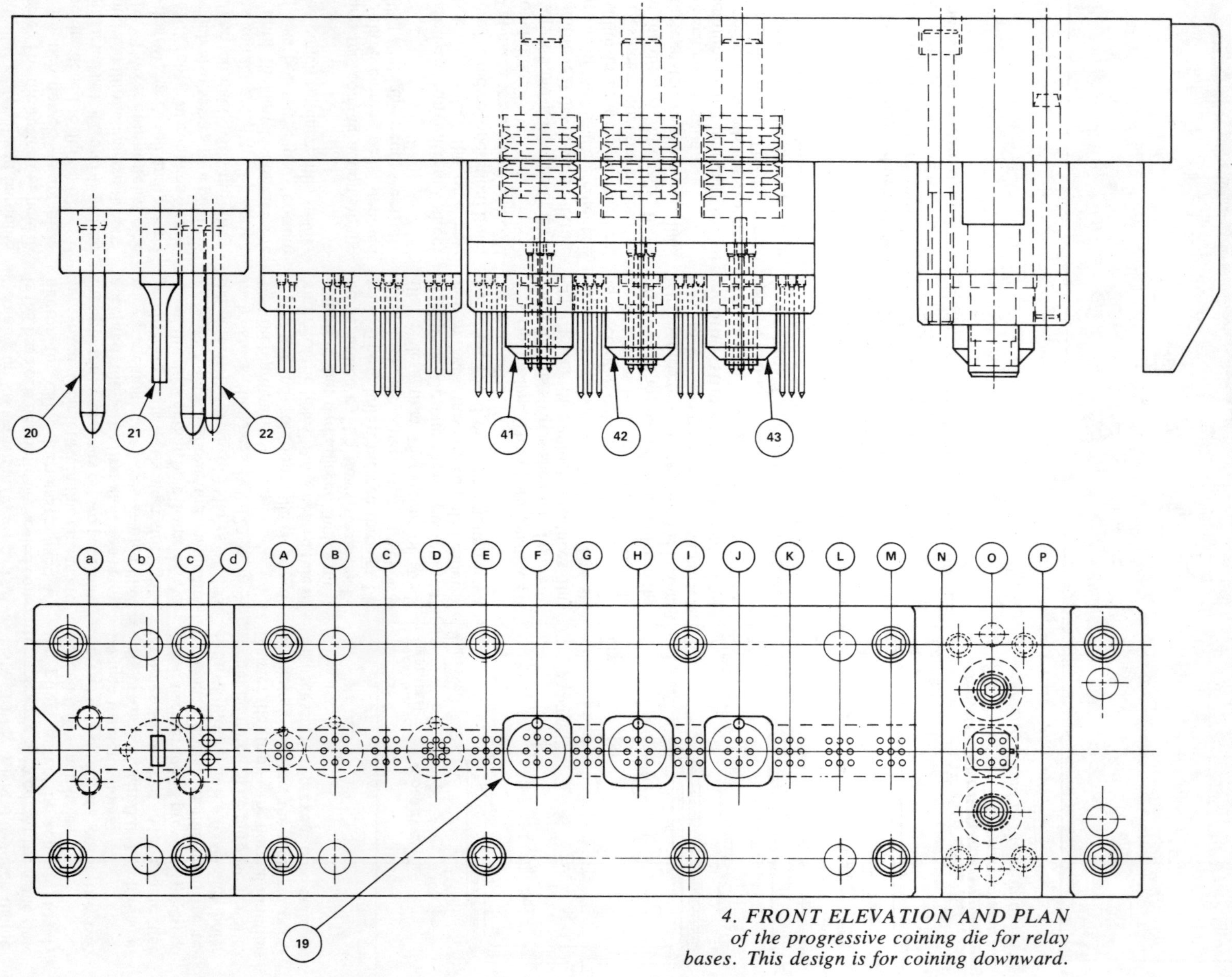

4. FRONT ELEVATION AND PLAN of the progressive coining die for relay bases. This design is for coining downward.

7. DIE LAYOUT for the connector job. The big problem (in addition to coining) is multiple piercing of holes.

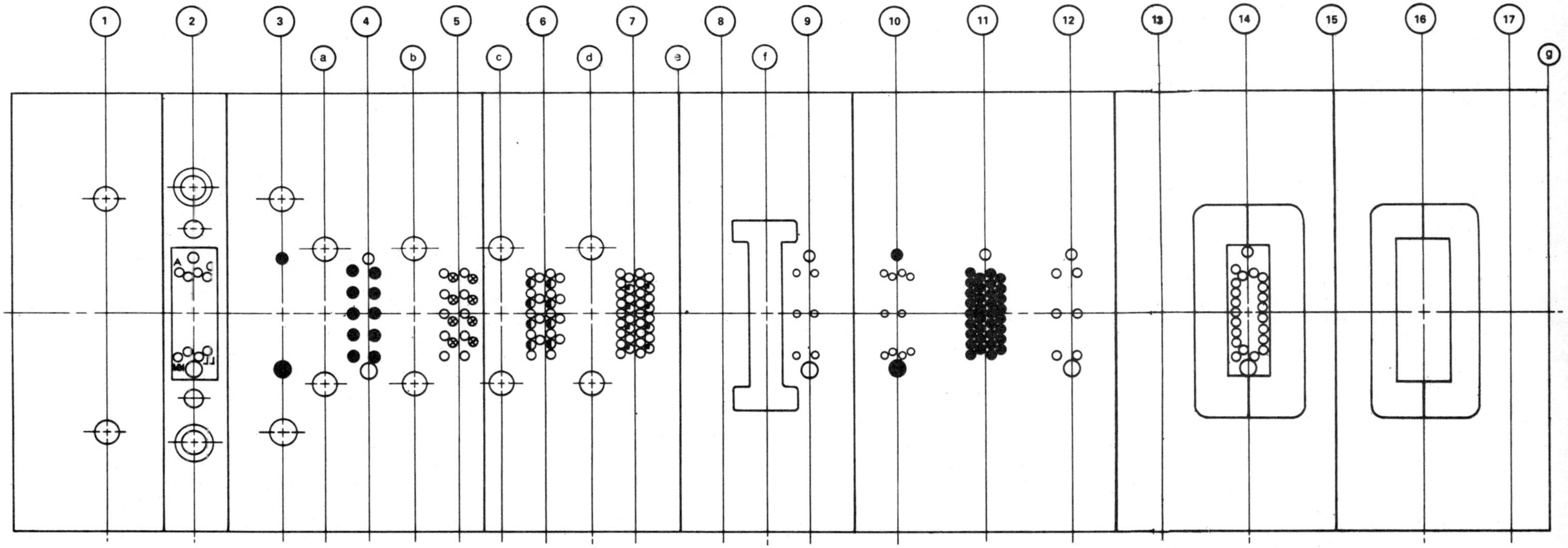

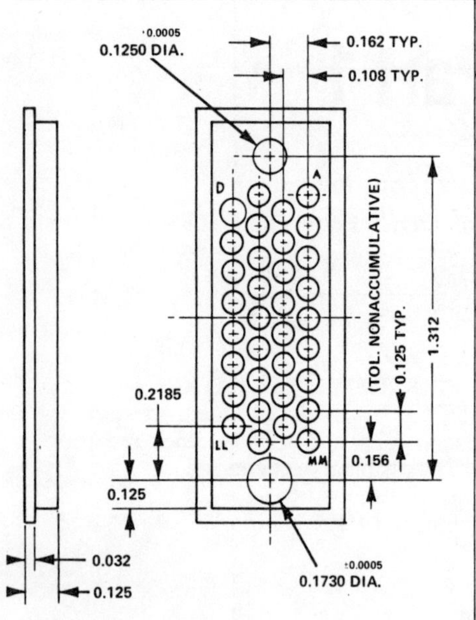

6. *COINING UP is done sequentially in this strip.*

5. *COINED CABLE harness connector used in aerospace applications.*

as new holes are pierced. It is this method of piloting that makes closely adjacent hole piercing possible.

STATION *d*. Pilots enter the two 1/4-inch diameter holes. As at stations *b* and *c*, it is this piloting that registers the strip.

STATION 7. Eight more 0.099-inch diameter holes are pierced. And pilots enter the 26 previously pierced 0.099-inch holes to prevent their collapse.

STATION 8. Idle.

STATION *f*. Blank slug to provide an outline of the rough blank.

STATION 9. Pilots enter six of the 0.099-inch holes as well as the 0.112 and 0.160-inch holes.

STATION 10. Pilots enter 10 of the 0.099-inch holes as shave punches size the center holes to 0.1250 and 0.173 inch.

STATION 11. Pilots enter the center holes and then 34 shave punches bring all of the 0.099-inch holes up to 0.109 inch.

STATION 12. Pilots enter the center holes and six of the 0.109-inch holes.

STATION 13. Idle.

STATION 14. Coining takes place. Before contact occurs, however, pilots enter all outside holes in the 34-hole cluster. Again, the objective is to prevent hole collapse during an extremely rigorous coining operation. The coining operation is shown in *Figure 8*. It will be noted that coining takes place against a spring-loaded pad. This pad seats on a backup plate in the lower shoe to provide the necessary part flatness. As the upper die ascends, the spring-loaded pin elevates the pad, thus ejecting the part.

STATION 15. Idle.

STATION 16. The part is blanked from the strip.

Two strippers are used in this die. The first stripper ends at a point halfway between stations 12 and 13. The second begins at a point halfway between stations 15 and 16. In other words, the coining operation, which takes place at station 14, is effected without a stripper.

PROGRESSIVE COINING TODAY

The preceding examples show the two basic methods used in progressive coining. As yet, this is not a large field of endeavor. Lars Johanson, president of Eltee Inc., claims that there are no more than six U. S. companies actively engaged in this type of work. One reason, perhaps, is a general fear of attempting to move metal in the solid, as is done in progressive coining. Another is that no company can be successful at coining until it has spent a certain amount of time in research and development.

But there's a payoff for those interested in trying. For instance, the connector base discussed in the second example is run in a straight-side 100-ton Minster. Its selling price is roughly 25¢. At 80 spm, the burden rate for the press is a matter of no more than academic interest. ∎

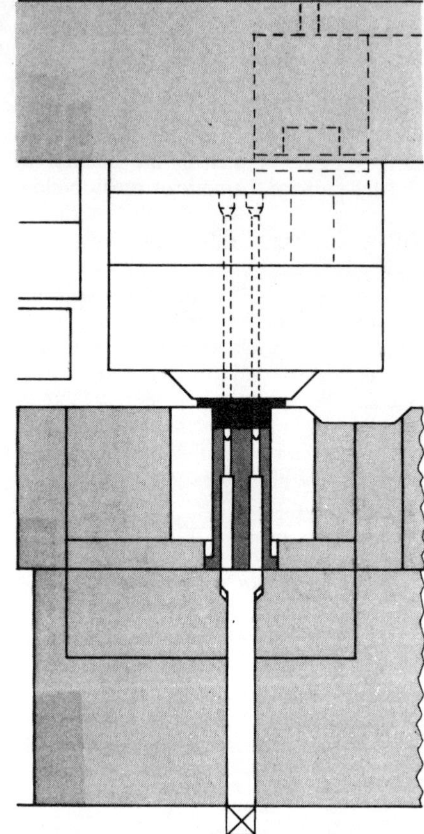

8. *COINING STATION is equipped with pilots to prevent hole distortion.*

METRICATING THE PRESSWORKING EQUATIONS: Part 1

*Metrication means conversion to the Systeme Internationale.
It's not as difficult a process as you may have been led to believe – as this
series of articles on the pressworking equations will show*

DANIEL B. DALLAS, *Editorial Director*

NOW THAT THE UNITED STATES is beginning the work of converting its industry to the metric system, the standard equations used in manufacturing engineering will also require conversion. It is important to note, however, that the United States — and the rest of the world, for that matter — is not simply converting to the metric system; it is converting to the SI (Systeme Internationale) system. Further, it can be said that there is no such thing as *the* metric system. There are, in fact, numerous metric systems, but the SI system is the one generally being accepted by the entire industrial world.

An excellent example of the differences in the various metric systems is seen in the manner in which presses are rated by tonnage. American press builders rate their presses by tonnage. Presses built for overseas use — and presses built by numerous European manufacturers — are rated in metric tons. But neither U.S. tons nor metric tons has any meaning with respect to force expressed in the SI system. Reason: in SI, force is measured in newtons.

Naturally, the customary use of pounds in measurement units of torque and pressure are dispensed with in the SI system. The average engineer — exposed to the "metric" system in his college chemistry and physics courses — isn't particularly concerned with this. He's ready, willing and (sometimes) able to express these terms in kilograms per square centimetre. The moment of truth comes when he's advised that, no — you can't use these units either; you have to use the SI units. It is often at this point that he begins to hope that the advent of SI will be delayed as long as possible — hopefully until he retires.

Still, there is nothing especially difficult about SI, as this series of articles on the metrication of the pressworking equations will show. By carefully studying the series — and by filing the articles for future study — the manufacturing engineer will soon come to an understanding of SI, at least as it pertains to pressworking.

THE STOCK REQUIREMENT EQUATION

A GOOD — RELATIVELY EASY — POINT at which to start is with the standard equation used to determine stock requirements in pressworking. The customary equation is,

$$W = C \times S \times P \times 7.3$$

in which,

W = weight of stock for 1000 parts in pounds
C = the dimension between blanks in inches
S = width of the stock in inches
P = weight of the stock in pounds per square foot (by gage).

Some equations used in Manufacturing Engineering are derived from the laws of physics. These laws, and the equations derived from them, apply in any system of units, but the units on the right side of the equation must always match the units on the left. Existing equations derived from physics can often be used unmodified if all values are inserted in consistent SI units.

Other equations are empirical. In many instances, these equations have unlike units on the opposite sides of the equal sign. Present empirical equations based on customary units require modification before they can be applied with metric values.

The dimensions in the preceding equations are:

$$\frac{\text{pounds}}{\text{thousand parts}} = \text{inches} \times \text{inches} \times \frac{\text{pounds}}{\text{square ft.}} \times \text{Constant.}$$

To obtain equal units, this equation should include a conversion factor from square inches to square feet of 144 in the denominator, or of 0.006944 in the numerator. Combining the square foot to square inch conversion factor with 1000, the thousand parts multiplier, a constant of 6.944 re-

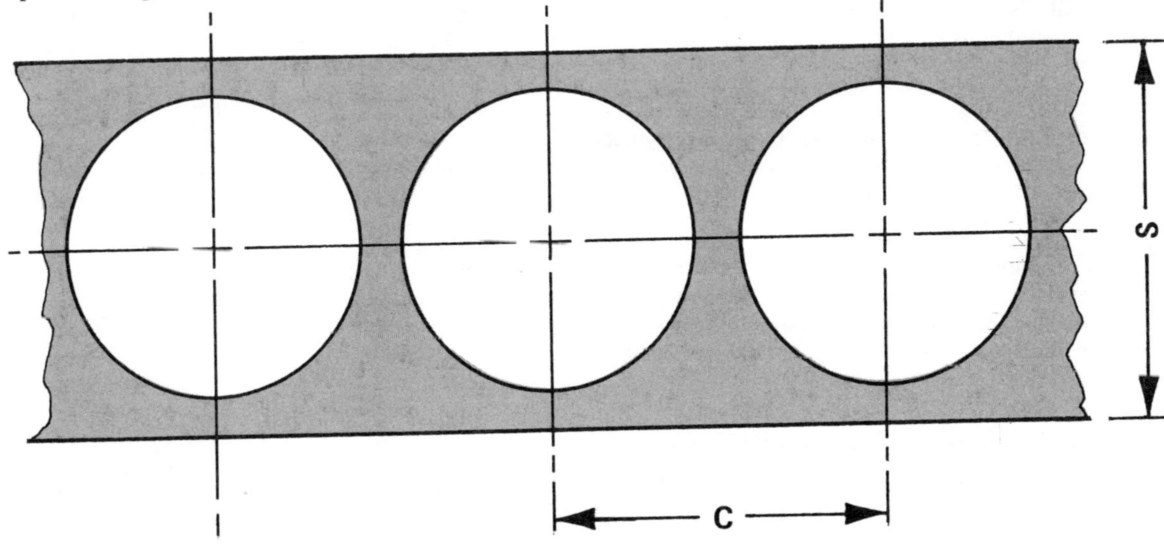

*1. STOCK REQUIREMENT
equation requires a new constant.*

sults. The actual factor of 7.3 in the customary equation includes a scrap allowance of approximately 5%.

Let us assume an example, illustrated in *Figure* 1, in which 2.6-inch discs are stamped from 3-inch, 10-gage stock. The interval between blanks is 2.8 inches. The equation can be solved quite easily once the value of P has been determined. From *Machinery's Handbook* (Twentieth Edition, Page 2292), the weight of one lineal foot of 10-gage, 3-inch flat rolled steel is 1.4352 lb. By multiplying by 4, we arrive at a value of 5.7408 lb/ft² for P.

From this, the weight of a thousand blanks can be determined by,

$$W = 2.8 \times 3 \times 5.7408 \times 7.3$$
$$= 352.023 \text{ lb.}$$

At this point, it is easy to make a soft conversion to determine the weight in SI values, remembering that,

1 pound equals 0.4535924 kilogram.

From this, the weight of the stock required for 1000 blanks is 159.67625 kilograms.

The question now arises: How is this equation solved if all dimensions are SI? The answer is that all inch dimensions will have to be in metres, weight will have to be in kilograms per square metre (by gage), and a new constant will have to be developed. Note:

1 pound per square foot equals 4.882429 kilograms per square metre.

Using the answer obtained in the last equation,

$$159.67625 = 2.8\,(0.0254) \times 3\,(0.0254) \times 5.7408\,(4.882429) \times \text{Constant.}$$

By simple algebra, the constant is equal to 1051.576. For our purposes, it should be rounded off to 1050, in which case the stock requirement equation becomes,

$$W = C \cdot S \cdot P \cdot 1050 \quad \ldots\ldots\ldots\ldots \textbf{(Eq. 1)}$$

with W in kilograms, and P in kilograms per square metre, and C and S in metres.

Note that the linear units are metres. Metric drawings normally show millimetres rather than metres. A relocation of the decimal point by three places to the left is required before values stated on drawings and in most specifications can be inserted into the equation.

One other difference should be noted. The "×" normally employed to indicate multiplication is replaced with a "period" centered between the terms. Use of a period instead of "×" to indicate multiplication is standard SI practice.

THE BLANKING FORCE EQUATION

BEFORE METRICATING THE BLANKING EQUATION, it would be well to review the precise meanings of force and pressure. Force is the *cause of acceleration*. Pressure is *force per unit of area*. In the U.S. Customary System, the two terms are often used interchangeably by some engineers who should know better. In many instances it really doesn't matter, however, but in SI the distinction is critical. In the SI system,

The unit of force is the newton.
The unit of pressure is the pascal.

In the blanking equation both terms are used — newtons to measure the force required to cut a blank, and pascals to indicate shear strength.

One pound of force equals 4.448222 newtons.
One pound per square inch equals 6.894757 kilopascals.

The equation for blanking force is,

$$F = S \times L \times t$$

in which,

F = force in pounds
S = shear strength in pounds per square inch
L = the length of the cut in inches
t = the material thickness in inches.

As an example, let us calculate the force required to blank

TABLE 1. SHEAR AND TENSILE STRENGTHS

Material	Shear Strength		Tensile Strength	
	psi	mega-pascals	psi	mega-pascals
Aluminum	8,000	55.2	—	—
SAE 3240	150,000	1034.2	105,000	723.9
SAE 4130	55,000	379.2	90,000	620.5
SAE 4130	65,000	448.2	100,000	689.5
SAE 4130	75,000	517.1	125,000	861.8
SAE 4130	90,000	620.5	150,000	1034.2
SAE 4130	105,000	723.9	180,000	1241.1
Stainless (18-8)	70,000	482.6	95,000	655.0
Steel (0.1 C)	45,000	310.3	60,000	413.7
Steel (0.25 C)	50,000	344.7	70,000	482.6
Steel (0.5 C)	70,000	482.6	95,000	655.0
Tin	6,000	41.4	5,000	34.5
Zinc	20,000	137.9	24,000	165.5

TABLE 2. INCH AND METRIC EQUIVALENTS OF U.S. GAGE NUMBERS

U.S. Gage No.	inch	mm	U.S. Gage No.	inch	mm
0000000	0.5	12.7	16	.0625	1.5875
000000	.46875	11.90625	17	.05625	1.42875
00000	.4375	11.1125	18	.05	1.27
0000	.40625	10.31875	19	.04375	1.11125
000	.375	9.525	20	.0375	0.9525
00	.34375	8.73125			
0	.3125	7.9375	21	.034375	.873125
			22	.03125	.79375
			23	.028125	.714375
1	.28125	7.14375	24	.025	.635
2	.265625	6.746875	25	.021875	.555625
3	.25	6.35			
4	.234375	5.953125	26	.01875	.47625
5	.21875	5.55625	27	.0171875	.4365625
			28	.015625	.396875
6	.203125	5.159375	29	.0140625	.3571875
7	.1875	4.7625	30	.0125	.3175
8	.171875	4.365625			
9	.15625	3.96875	31	.0109375	.2778125
10	.140625	3.571875	32	.01015625	.25796875
			33	.009375	.238125
			34	.00859375	.21828125
11	.125	3.175	35	.0078125	.1984375
12	.109375	2.778125			
13	.09375	2.38125	36	.00703125	.17859375
14	.078125	1.984375	37	.006640625	.168671875
15	.0703125	1.7859375	38	.00625	.15875

a 1-inch diameter hole in 20-gage aluminum. From TABLE 1, the shear strength of aluminum is 8000 psi. From TABLE 2, the decimal-inch equivalent of 20 gage is 0.0375. Using these values, the force required to blank a 1-inch diameter hole in 20-gage aluminum is,

$$F = 8000 \times \pi \times 0.0375$$
$$= 942.48 \text{ lb.}$$

Making a soft conversion (with one pound of force equaling 4.448222 newtons), the force required to blank the hole is 4192.35 newtons or — more properly — 4.192 kilonewtons.

Precisely the same answer is obtained if the equation is set up with shear strength in kilopascals and linear dimensions in metres. Thus,

$$F = S \cdot L \cdot t \quad \ldots\ldots\ldots\ldots\ldots\ldots\ldots\ldots\ldots\ldots\ldots\ldots\ldots \textbf{(Eq. 2)}$$
$$F = 8000\,(6.894757) \cdot \pi\,(0.0254) \cdot 0.0375\,(0.0254)$$
$$= 4.192\,35 \text{ kilonewtons.}$$

Obviously, it isn't necessary to carry the dimension as far to the right of the decimal as has been done in this example. But doing so enables us to explain still another facet of the SI system in that thousands are normally marked off by spaces, as has been done in this example. Had the answer been carried to only four places, however, the space would not have been used. The answer would have been given as 4.1923 kilonewtons.

The Gulf + Western E. W. Bliss Division has prepared an excellent nomogram of the blanking force equation. A metricated version of this nomogram, prepared by MANU-FACTURING ENGINEERING, is presented as *Figure* 2. To use this nomogram,

1. Find the total length of cut (or cuts) on Scale *B*
2. Find metal thickness on Scale *D*
3. Connect these two points with a straightedge to find *Area in Shear* on Scale *C*
4. From this point on *C* draw a line to the metal's shear strength on Scale *A*
5. Extending this line, read the force requirement in kilonewtons on Scale *E*.

THE STRIPPING FORCE EQUATION

THE FORCE REQUIRED TO STRIP A PUNCH is extremely difficult to determine, since it's influenced by the type of metal pierced, the area of metal in contact with the punch (which is to say the punch/die clearance), punch sharpness, spring positioning with respect to the punch, and so on. There is a stripping equation which does receive some use, however. Specifically,

$$F = L \times t \times 1.5$$

in which

F = stripping force in tons
L = the length of the cut in inches
t = material thickness in inches.

As was the case with the stock-requirement equation, it will be necessary to establish a new constant. This can be accomplished by solving an equation using typical values, then making a soft conversion of the answer to SI, and then reworking the equation with SI values, with the new constant as the unknown value.

To do this, let us use an example in which stock thickness is 3/8 inch and the length of cut is 9 inches. From these parameters, the tonnage requirement for stripping is,

$$F = 9.0 \times 0.375 \times 1.5$$
$$= 5.0625 \text{ tons}$$
$$= 45.038\ 427 \text{ kilonewtons.}$$

By reworking the equation in SI, using known values,

$$45.038\ 427 = 9\ (0.0254) \cdot 0.375\ (0.0254) \cdot Constant$$
$$20\ 684.415 = Constant$$

Since this value of the constant is unwieldly, and since the equation itself is empirical in the extreme, it is suggested that the constant be rounded off to a value of 20 600. The metricated stripping equation thus becomes,

$$F = L \cdot t \cdot 20\ 600 \quad \ldots\ldots\ldots\ldots\text{(Eq. 3)}$$

In this version, *F* is in kilonewtons, and *L* and *t* are in metres.

THE ENERGY EQUATIONS

IN BLANKING OPERATIONS it is often necessary to determine the amount of work done — or energy expended — in cutting through the workpiece metal. (For our purposes, work and energy are synonymous.) The rationale here is that energy requirements — and not the overall die size — often determine whether a certain die is operable in a given press.

In the U.S. Customary System, the equation used to make this determination is,

$$W = t \times \%\ penetration \times F \times 1.16$$

in which

W = work performed, measured in inch-tons
t = material thickness in inches
F = maximum force in tons

and 1.16 is an allowance for friction in the drive train. In other words, it's assumed that 16% of the work done is expended in overcoming frictional forces.

The value of *F* is actually the value obtained in the previously discussed blanking equation. Once this energy equation is metricated, *F* can be picked up from the nomogram, *Figure* 2. Naturally, a more precise answer can be obtained by solving for *F* in Equation 2.

For the moment, let us assume a situation in which the material thickness is 0.05 inch, the penetration is 30%, and the maximum force (as determined by Equation 2) is 15 tons. The question is: How much work is done in blanking the related part? The answer:

$$W = 0.05 \times 0.3 \times 15 \times 1.16$$
$$= 0.261 \text{ inch-tons.}$$

To metricate this equation, it is first necessary to note that in the SI system work is measured in joules.

Work is measured in joules.
One inch-ton equals 225.83 joules.

By soft conversion, the work performed in the preceding equation is 58.978 joules.

Precisely the same answer can be obtained with this equation by substituting values in which linear inch terms are expressed in metres and tonnage is expressed in newtons. Before making the substitution, it will be remembered that one pound of force equals 4.448 222 newtons. Therefore, the value of 15 tons is 133 446.6 newtons. Accordingly,

$$W = t \cdot \%\ penetration \cdot F \cdot 1.16 \quad \ldots\ldots\ldots\ldots\text{(Eq. 4)}$$
$$W = 0.05\ (0.0254) \cdot 0.3 \cdot 133\ 446.6 \cdot 1.16$$
$$= 58.978 \text{ joules.}$$

The Flywheel Equation. Having determined energy requirements, the next question is: Can the press deliver the necessary energy? (Again, energy and work are synonymous in the pressworking equations.)

At least two equations are available for determining the energy in a flywheel. The easiest one to work with is,

$$E = \frac{n^2 \times D^2 \times W}{5,250,000,000}$$

in which

E = available flywheel energy in inch-tons
n = revolutions per minute
D = diameter of flywheel in inches
W = weight of flywheel in pounds.

The value of *E* presupposes a 10% slowdown — the maximum allowable for continuous operation, as in the case of progressive die operations.

Let us assume a situation in which the flywheel weighs 1300 pounds. It measures 50 inches in diameter and it rotates at 300 rpm. The available energy in inch-tons is,

$$E = \frac{(300)^2 \times (50)^2 \times 1300}{5,250,000,000}$$
$$= \frac{90,000 \times 2500 \times 1300}{5,250,000,000}$$
$$= \frac{29,250}{525}$$
$$= 55.714 \text{ inch-tons.}$$

Again making a soft conversion by multiplying 55.714 inch-tons by 225.83 joules, we find that the available energy (at 10% slowdown) is 12.581 956 kilojoules.

To metricate this equation, it is necessary to express the flywheel diameter in metres, and the flywheel weight in kilograms, remembering that,

One pound equals 0.453 592 4 kilogram.

The value for revolutions per minute remains unchanged.

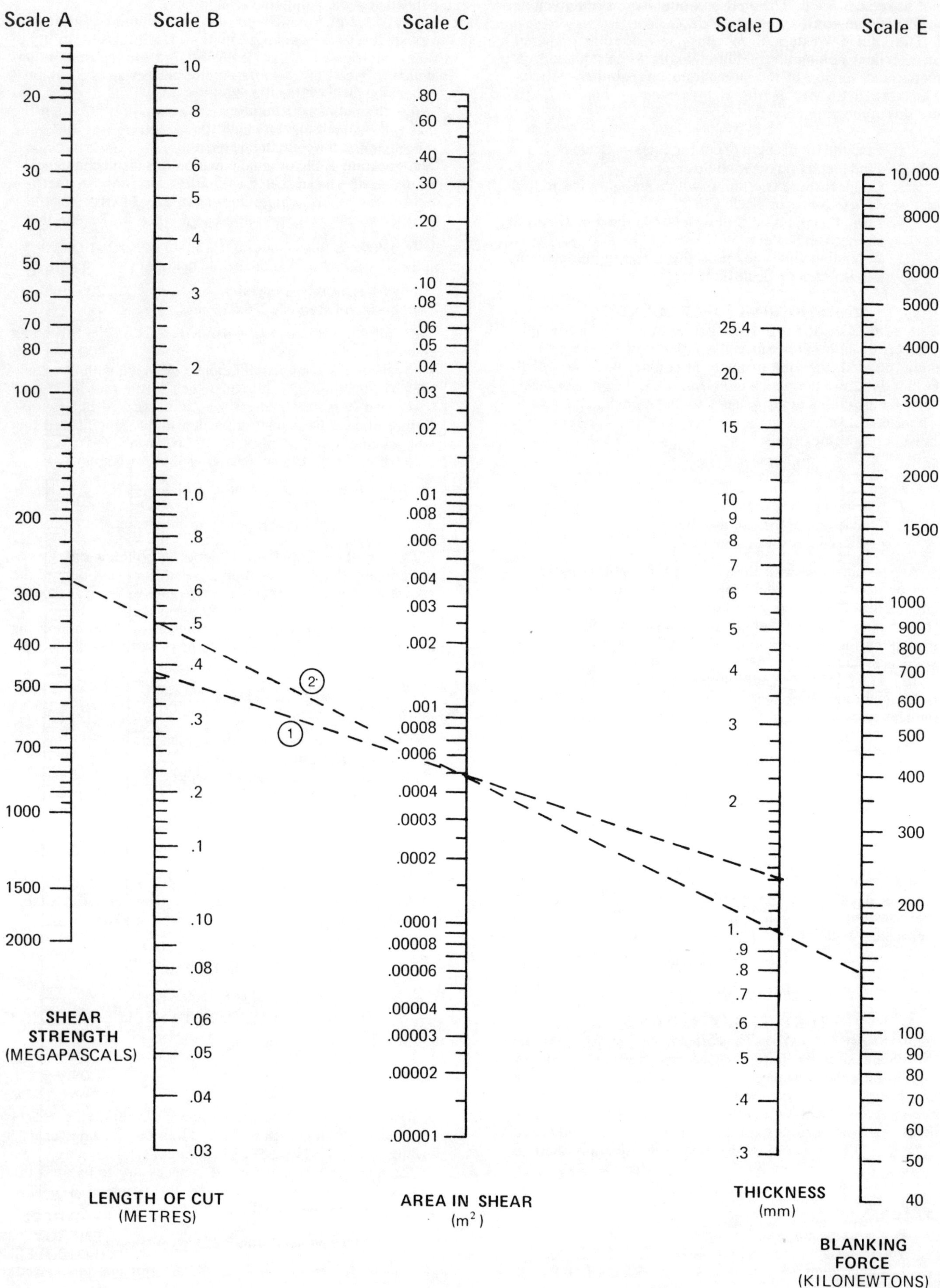

Scale A
Scale B
Scale C
Scale D
Scale E
SHEAR STRENGTH (MEGAPASCALS)
LENGTH OF CUT (METRES)
AREA IN SHEAR (m²)
THICKNESS (mm)
BLANKING FORCE (KILONEWTONS)

It's necessary to develop a new constant for the denominator, however. This can be done by substituting 12.581 956 kilojoules for E in the preceding equation, then converting inches to metres, and then solving for the New Constant. Thus,

$$12.581\,956 = \frac{(300)^2 \cdot (50 \cdot 0.0254)^2 \cdot (1300 \cdot 0.453\,592\,4)}{New\ Constant}$$

$$= \frac{(90\,000) \cdot (1.27)^2 \cdot (589.670\,12)}{New\ Constant}$$

$$New\ Constant = \frac{85\,597\,104}{12.581\,956}$$

$$= 6\,803\,163.5$$

No significant change in energy values is realized if the constant is rounded to 6 800 000. Accordingly, the metricated flywheel energy equation is,

$$E = \frac{n^2 \cdot D^2 \cdot W}{6\,800\,000} \quad \dots\dots\dots\dots\text{(Eq. 5)}$$

in which

E = flywheel energy in kilojoules
n = revolutions per minute
D = flywheel diameter in metres
W = flywheel weight in kilograms.

A nomogram of the original equation has also been developed by Gulf+Western's E. W. Bliss Division. A metricated version of this nomogram prepared by MANUFACTURING ENGINEERING, is presented as *Figure* 3. To use this

3. THE FLYWHEEL EQUATION

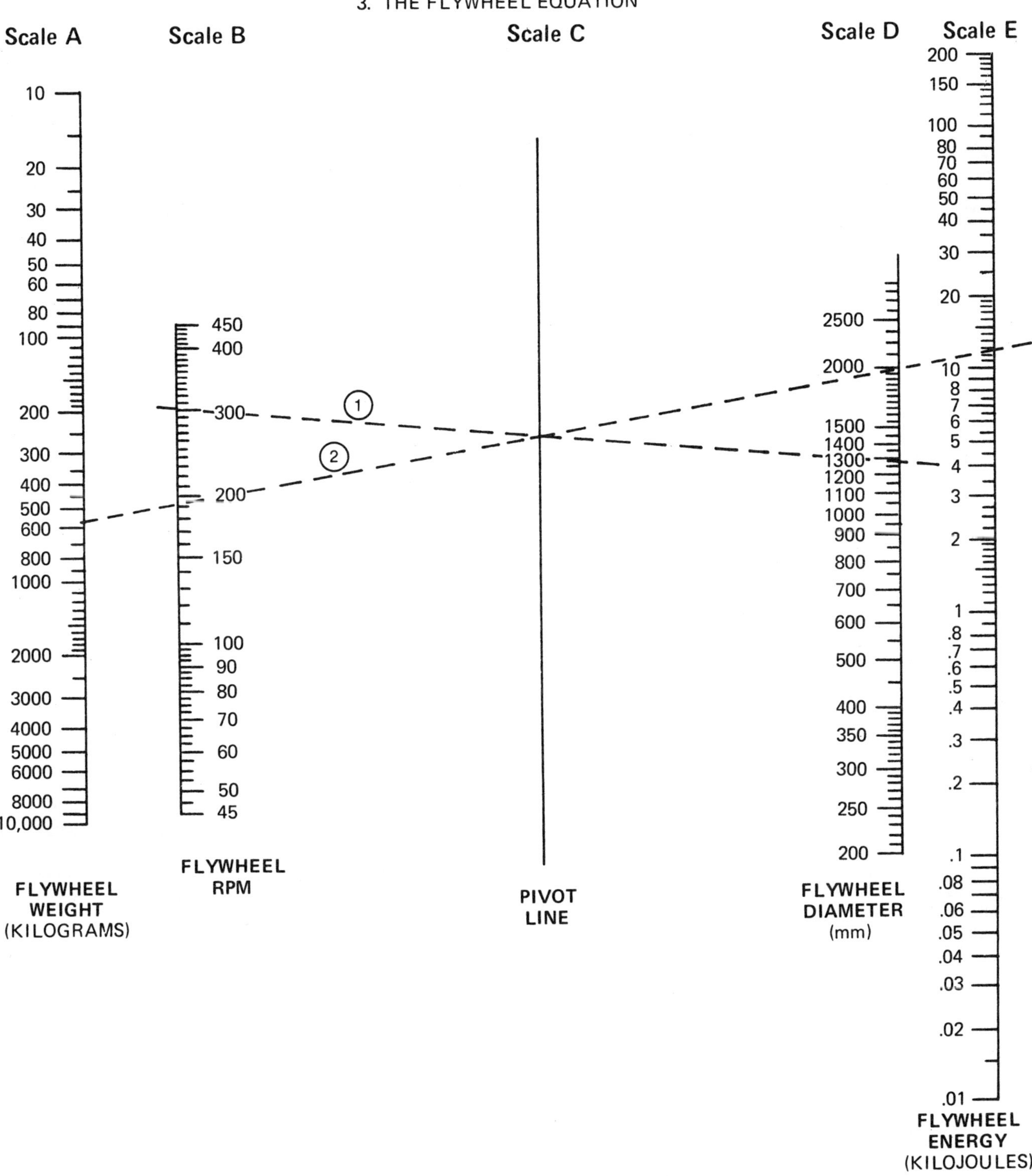

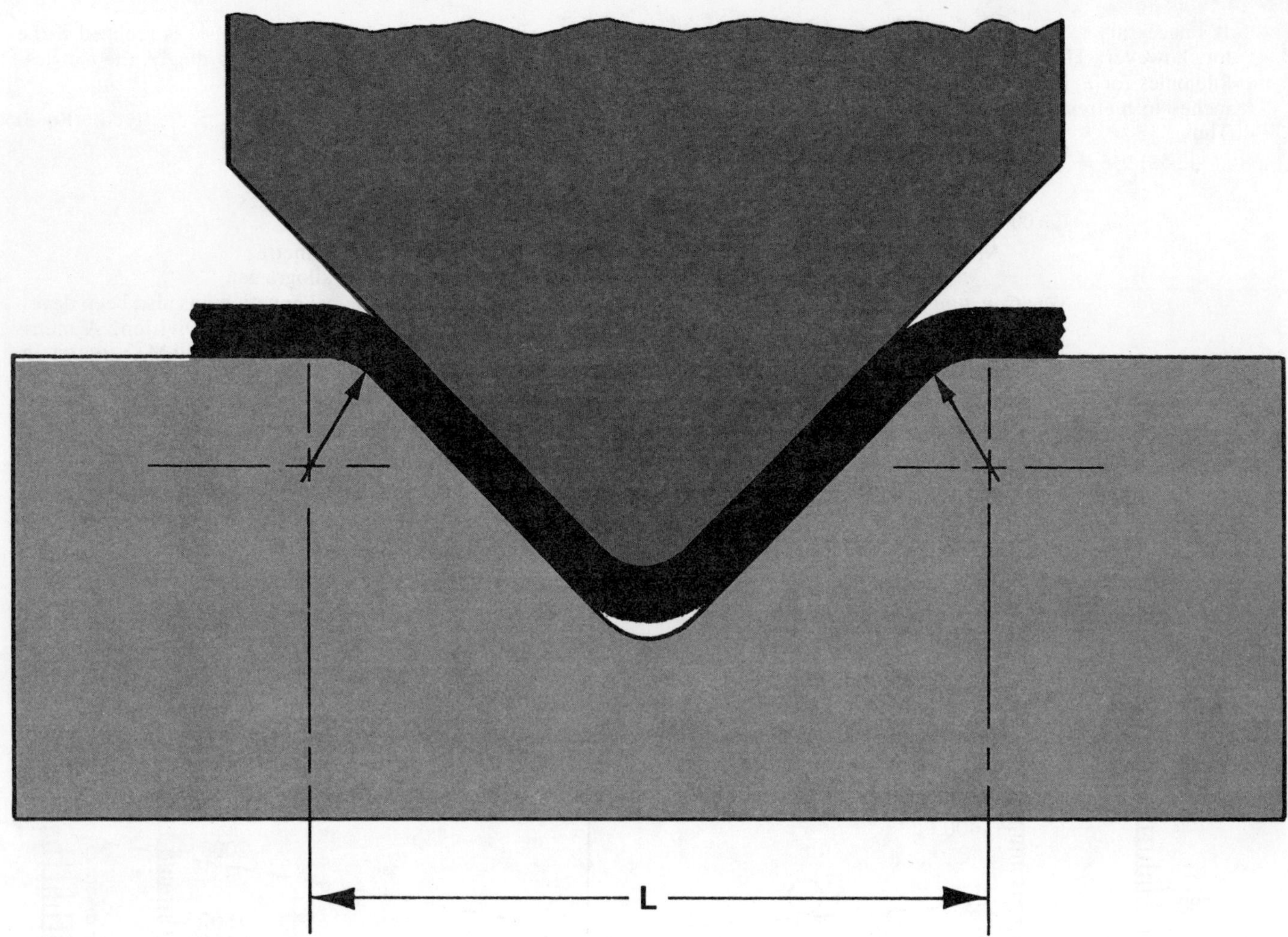

4. *TYPICAL V-BEND OPERATION.*

nomogram,

1. Find flywheel speed on Scale *B*
2. Find flywheel diameter on Scale *D*
3. Using a straightedge, draw a line from flywheel speed (*B*) to flywheel diameter (*D*)
4. Mark the intersection of this line with the pivot line (*C*)
5. Find the flywheel weight on Scale *A*
6. Using the straightedge, draw a line from this point on Scale *A* through the previously marked point on the pivot line (*C*)
7. The intersection of this line with Scale *E* indicates the available energy in joules.

THE BENDING EQUATIONS

TWO GENERAL TYPES OF BENDING are used in modern pressworking. One is the V-bend, which is used extensively in brake die operations as well as in stamping dies. The other is the wiping die. Both forms of bending have their equations — and the equations can be easily metricated.

The V-bend Equation. A typical V-bend operation is shown in *Figure* 4. The equation to determine bending force *F* is,

$$F = \frac{1.33\,SWt^2}{L}$$

in which

F = force in pounds
S = ultimate tensile strength in psi
W = the length of the bend in inches
t = stock thickness in inches
L = the dimension in inches between the radial centers of the bending die. (See *Figure* 4.)

As an example, let us calculate the force required to V-bend a piece of high-carbon steel, 3/16 inch thick. Its tensile strength is 120,000 psi. The bend, performed in a press brake, is 10 feet long. The dimension between the radial centers of the die is 3 1/2 inches. In other words, S = 120,000 psi; W = 120 inches; t = 0.1875; and L = 3.5 inches. The force required is,

$$F = \frac{1.33 \times 120,000 \times 120 \times 0.0351562}{3.5}$$

$$= \ 192,374.72 \text{ pounds.}$$

By soft conversion (multiplying the answer by the 4.448222 conversion factor), the force required to effect this bend is 855 725.46 newtons. This value is properly stated as 855.725 46 kilonewtons.

To effect a hard conversion, it is necessary to express *S* in kilopascals, and *W* and *t* and *L* in metres. First, however, it is necessary to solve for the Constant, as follows:

$$855.725\,46 =$$

$$\frac{120,000\,(6.894\,757) \cdot 120\,(0.0254) \cdot [0.1875\,(0.0254)]^2}{3.5\,(0.0254)}$$

$$\times \text{ Constant}$$

$$\text{Constant} = \frac{(0.0882) \cdot (855.725\,46)}{(827\,370) \cdot (3.048) \cdot (0.004\,762)^2}$$

$$\text{Constant} = \frac{76.073\,952}{56.993\,215}$$

Constant = 1.334 789, or, for short, 1.33, as in the original equation

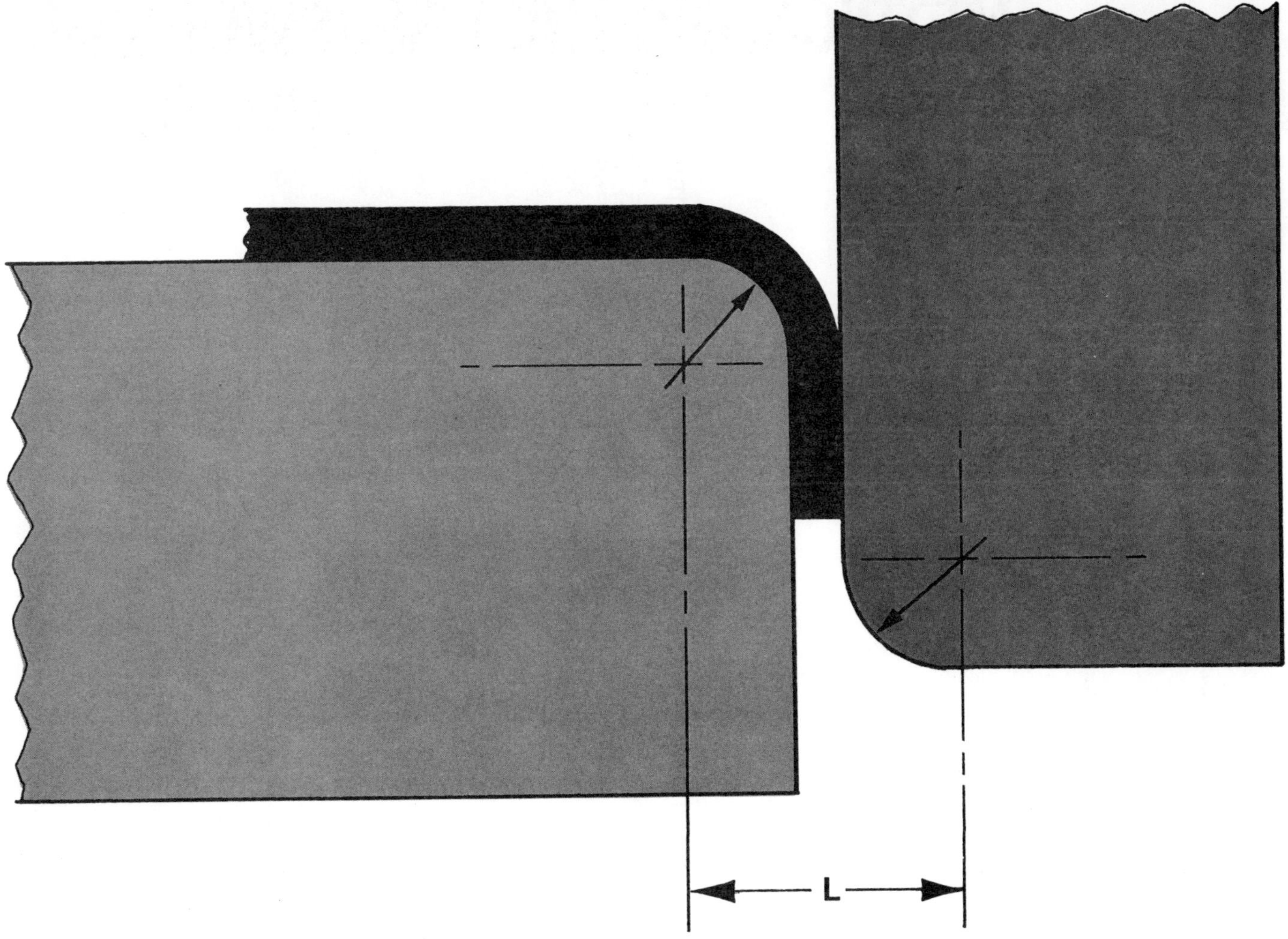

5. TYPICAL WIPE OPERATION.

$$F = \frac{1.33\,(120{,}000 \cdot 6.894\,757) \cdot (120)\,(0.0254) \cdot (0.1875 \cdot 0.0254)^2}{3.5 \cdot 0.0254}$$

$$= \frac{1.33\,(827\,370.84) \cdot (3.048) \cdot (0.000\,022\,6)}{0.0889}$$

$$= 852.655 \text{ kilonewtons.}$$

A discrepancy of 3.07046 kilonewtons between the soft and hard conversions is noted. This error, attributable to rounding of the constant, amounts to 69.027 pounds or 0.036%.

From the preceding work, the metricated V-bend equation is,

$$F = \frac{1.33\,SWt^2}{L} \quad \dots\dots\dots\dots\text{(Eq. 6)}$$

where F is in kilonewtons, S is in kilopascals, and W and L and t are in metres.

The Wiping Equation. Virtually the same situation is seen when solving the wiping force in SI. Referring to *Figure* 5, the wiping equation in U.S. Customary units is,

$$F = \frac{0.333\,SWt^2}{L}$$

where

F = force in pounds
S = tensile strength (psi)
W = width of the bend (inch)

t = metal thickness (inch)
L = the dimension between the radial centers (inch).

If we assume a tensile strength of 120,000 psi, a bend width of 4 inches, a stock thickness of 3/16 inch, and a die dimension L of 2.18 inches, the force equation becomes,

$$F = \frac{0.333\,(120{,}000) \times 4 \times (0.1875)^2}{2.18}$$

$$= 2577.69 \text{ pounds}$$
$$= 11.466\,142 \text{ kilonewtons (by soft conversion).}$$

Almost precisely the same force value can be obtained with this equation if tensile strength is expressed in kilopascals, and the linear dimensions in metres. Thus,

$$F = \frac{0.333\,SWt^2}{L} \quad \dots\dots\dots\dots\dots\dots\dots\dots\dots\text{(Eq. 7)}$$

$$F = \frac{0.333\,(120{,}000 \cdot 6.894\,757) \cdot (4 \cdot 0.0254) \cdot (0.1875 \cdot 0.0254)^2}{2.18 \cdot 0.0254}$$

$$= 11.44111 \text{ kilonewtons.}$$

This concludes Part 1 of this series. Part 2, to be published in March, will treat the draw die equations. Attention will be given to metrication of the equations for belt and motor energy, drawing reductions and drawing pressures. ∎

METRICATING THE PRESSWORKING EQUATIONS:

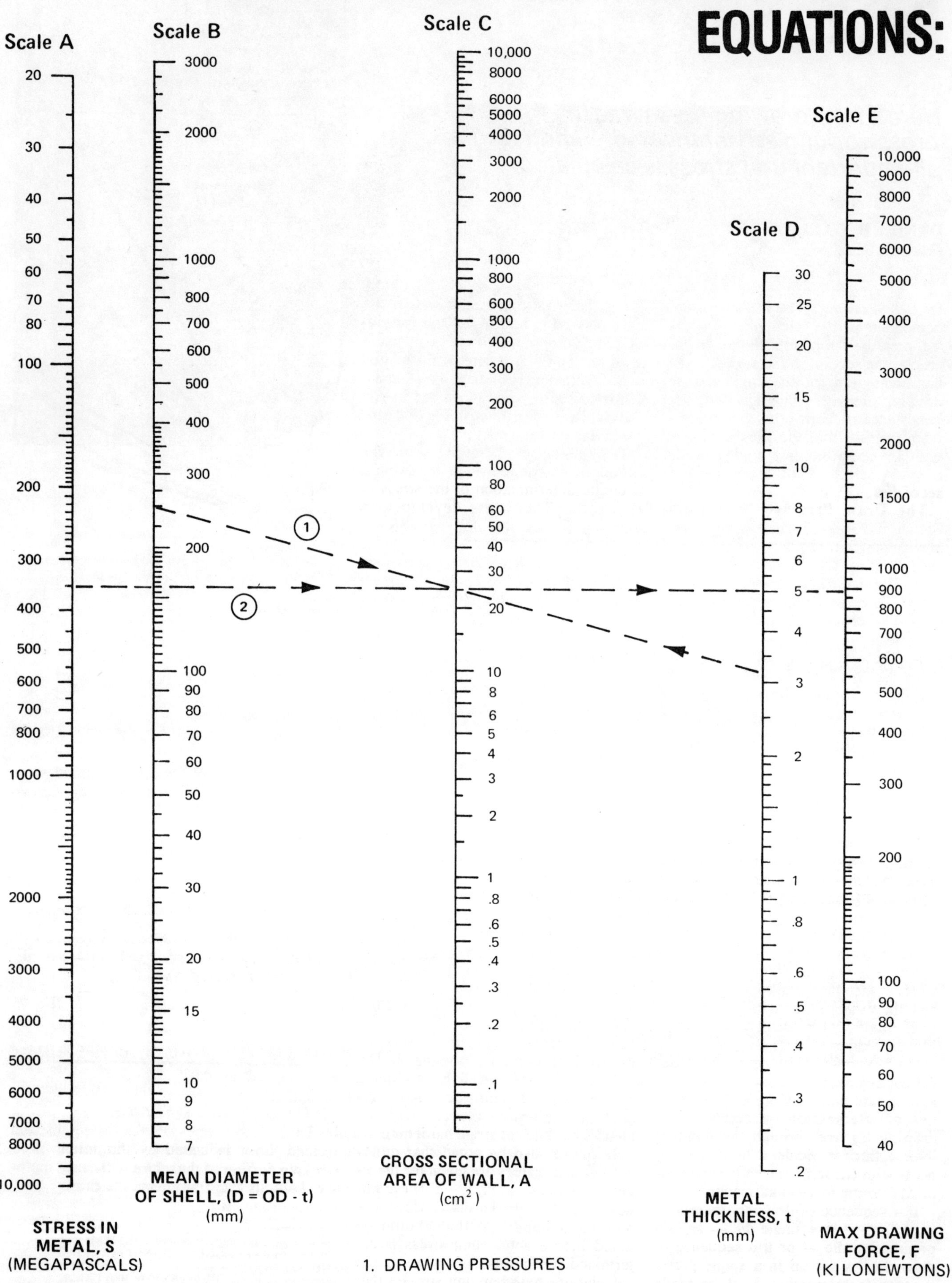

Part 2

Here's how drawing pressures in pressworking are metricated – and how the all-important unit stress is calculated

DANIEL B. DALLAS
Editorial Director

IN THE FIRST ARTICLE in this series a nomogram for the calculation of blanking pressures was presented. In this article, the second in the series, a similar nomogram for the calculation of drawing pressures is presented. The importance of both nomograms stems from the fact that the pressures dealt with are often the determining factors in press selection for a given die or set of dies.

The Draw Pressure Nomogram. Shown in *Figure* 1, the drawing pressure nomogram requires knowledge of the following:

▶ The stress in the metal
▶ The mean diameter of the shell
▶ The cross sectional area of the wall
▶ The metal thickness.

To use this information in calculating the maximum drawing force, the die engineer starts with Scale *B* — the mean diameter of the shell. This equals the OD of the shell minus metal thickness. Finding the appropriate value on Scale *B* — in millimetres — the die engineer then finds the metal thickness, also in millimetres, on Scale *D*. By connecting the appropriate points on Scales *B* and *D*, he is able to determine the cross sectional area by the point of intersection on Scale *C*. The operation of finding this point is denoted by the encircled numeral "1" on the nomogram.

Next, it is necessary to find the stress in the metal due to drawing. Just how the value of Stress in Metal (Scale *A*) is found will be dealt with shortly. For the moment, let us assume that the value is known. This being the case, the die engineer need only connect the appropriate point on Scale *A* with the line of intersection on Scale *C*. By drawing his line through the point of intersection on Scale *C* until it connects with Scale *E*, he finds the maximum drawing force in kilonewtons.

If a sequence of drawing operations is performed in a single die — as in a progressive die — or if a sequence of draw dies is set up in a single press, a separate calculation must be made for each operation or each die, as the case may be. The total of the individual calculations, which are easily made with this nomogram, are then matched against press capabilities. Press capabilities, in turn, are determined by the flywheel equation given in the first article in this series.

Determining the Stress. Use of the nomogram depends on a reasonably accurate determination of the stress in the metal. The stress developed in most metals used in drawing applications can be found in *Figures* 2 and 3. (Note: It is recognized that unit stress in *Figure* 3 should be noted in gigapascals. However, megapascals are used for consistency with *Figures* 1 and 2.)

To use these graphs, it is first necessary to determine the percent reduction between the blank and the drawn shell. This means the first draw — not the last draw if a sequence of draw operations is involved. If a sequence of draws is performed, the value of percent reduction is determined by the reduction between two successive draws. The implication here is that the drawing pressure must be calculated for the individual draws, since the actual unit stress is a variable and is not accumulative.

It will be noted that actual unit stress increases linearly with percent reduction. It will also be noted that each line representing a metal has both a light and dark portion. The dark portion starts with the yield point of the metal and terminates with the fracture point. In other words, the dark (or heavy) section of the line represents the working portion of the metal. The extension of each line all the way to the right — to the theoretical 100% reduction — provides the user with the metal's modulus of strain hardening.

It should also be noted that different terminal values are given for the aluminums shown in *Figure* 2. These values depend on the hardness of the aluminum, a condition that should be noted before actual unit stress is determined. The reason, of course, is that calculations based on unit stresses that

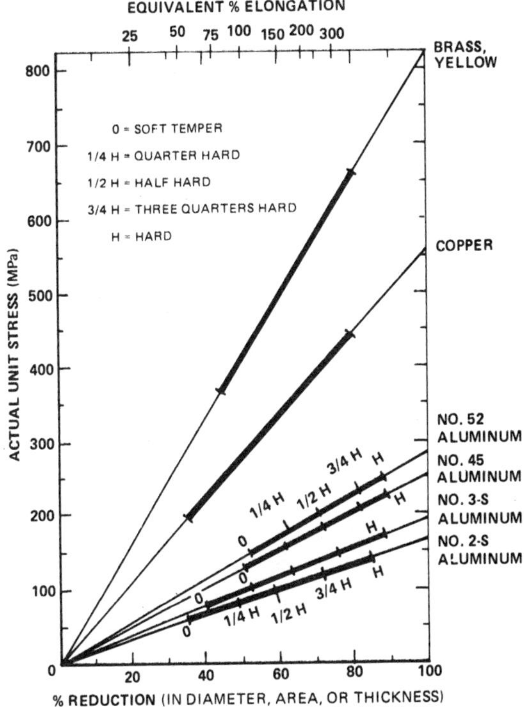

2. UNIT STRESS (NONFERROUS).

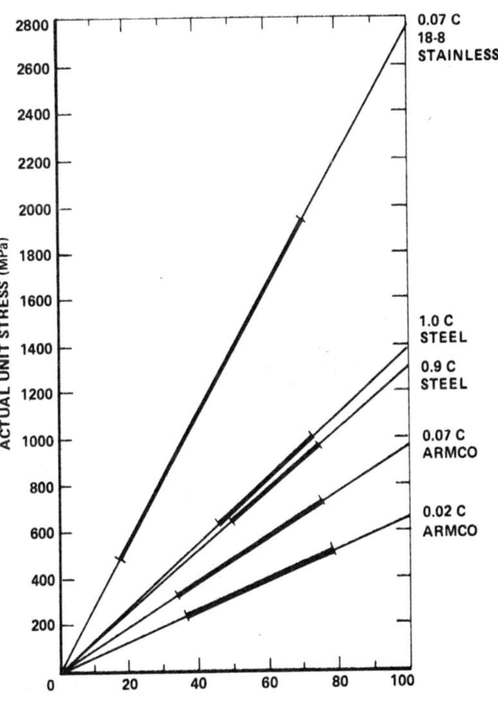

3. UNIT STRESS (FERROUS).

exceed those indicated by the actual working portion of the curve will cause the workpiece to fracture in the drawing operation. ■

The nomogram and graphs presented in this article were metricated by the editorial staff of MANUFACTURING ENGINEERING. The original nomogram and graphs were developed by Gult + Western's E. W. Bliss Division.

CHAPTER 4

DIE BUILDING

Revolution in Diemaking

By Siegfried Gruber
Vice President, Engineering
Argus Manufacturing Company

Using wire EDM over other known machine tools is much
more economical. The change of manufacturing procedures compared with other methods is revolutionary in itself because
wire EDM is finishing a punch and die opening, whereas, using
other methods, several machine tools have to be applied to finish a punch and die opening. Wire EDM is a fully automatic
process once it is set up. In other words, wire EDM finishes
its work by means of numerical control.

INTRODUCTION

The advantages of using wire EDM are: more economical,
very precise, and tremendous time saving.

    (a) Preparation of work piece requires only a start-
ing hole, large enough for a thin wire to pass
through.

    (b) By using a computer time sharing system, engineer-
ing procedures, which consist of a numerical control tape and plotter, are more rapidly achieved.

    (c) Contour cutting with wire EDM surpasses other
methods being used by about eight times.

MACHINE TOOL TECHNICAL DATA

The most unique machine tool, to help a tool and die maker
produce a tool in the fastest time possible, is the wire EDM
machine tool. About two and a half years ago we, at Argus, decided to look into wire EDM cutting process in conjunction with
our stamping operations, where we were in need of all sort of
cutting tools, like compound dies, progressive dies, and blanking dies. Since then we have acquired three wire EDM machines
which were made in Switzerland by Agietron Corp. One machine
tool is used for our own use and two are used to provide service to our customers.

The machine tool consists of (a) an X and Y movable table,
(b) a holding device for work pieces to be cut, (c) power
supply and (d) a control unit to activate the X-Y movement of
the machine tool by means of numerical control.

The EDM; or Electrical Discharge Machine system, consists
of (a) a tensioning device for wire electrode, (b) guiding
system, (c) flushing system (with deionized tap water being
used as the dielectric).   (Illustration No. 1).

FUNCTION OF MACHINE TOOL

The wire EDM tool can be compared in its accuracy to a jig
bore machine tool and in its function to a band saw.

I would like to illustrate in the following example how
easy it is to clamp a die block into the machine tool to be cut.
Picture a 6" X 6" X 1" thick die block with a starting hole
drilled through before heat treating. The block is clamped
with two clamps located on two sides of the square. The clamping area consists of approximately 1/2" per side to achieve the
maximum clearance area for the wire to move freely without ob-
struction and for clearance of the slug to drop through. One
side of the square die block is being indicated for parallelism
and the face being indicated to achieve perpendicularity. Two
outside edges are then brought into their actual location by
the use of an electronic sensing device which registers the
collision between the wire and the work piece in increments of
.0001". The same procedure can be used by picking up a hole.
The sensing device will register the dimensions at 90°, 180°,
270°, and 360° to pin point the center of the hole.

When the wire is inserted through the starting hole, the
flushing device is turned on, the wire is electrically charged,
the tape control is turned on, the wire EDM process can begin.
(Illustration No. 2)

WIRE

The cutting tool on the wire EDM is a commercial copper
wire in range of .008, .006 and .004. The material for wire
smaller than .004 in diameter is made of molybdenum. In other
words, the copper wire represents the cutting tool like an end
mill on a milling machine, or a grinding wheel on a grinder.

Commercial copper wire in its use as a cutting tool on a
wire EDM machine represents the most economical cutting device
in history compared to all tool machines. The cost of copper
wire in one hour of cutting time is 10¢. (Illustration No. 3)

TOLERANCES - WHAT CAN BE HELD. SIZES OF MATERIAL.
FINISH ACHIEVED.

It is essential for any machine tool used to produce die
openings and punches to have the capability of achieving and
maintaining very close tolerances. Our wire EDM equipment has
the capability to achieve very close tolerance requirements
like +.0002 in an area of 5.9" X 5.9". Clearances between
punch and die +.0001. Parallelism on a thickness of 2.875",
+.0002. Actual location from one opening to another opening,
+.0002. Surface finish, 15 to 20 microfinish. Maximum size
of a work piece on our equipment is 8" X 16" X 3" thick.

FEEDBACK OF PERFORMANCE OF TOOLS CUT BY WIRE TRAVEL SYSTEM

The excellent cutting finish achieved by wire EDM, with
low amperage, assures an excellent cutting tool. Dies previously manufactured by this process have been proven to give
tremendous performance. There are several factors which I can
mention why that is so. (1) Every punch is cut out of solid
prehardened tool steel block. It assures maximum heat treating
qualities even in the smallest cutting areas. Also, a small
radius is provided on every corner configuration. The same
radius will be cut into the die, therefore, an even cutting
clearance is achieved and an early breakdown of sharp edges
is prevented. (2) One hundred percent clearance requirements
through the whole contour are maintained even in the most
difficult shapes which assures an absolute even break in a
blanking operation.

ENGINEERING PROCESS - PROGRAMMING

Savings in time does not start with the cutting of the
wire EDM, it starts with the engineering process. In order to
manufacture a tape in a very precise and efficient manner, a
time-sharing computer is used for our wire EDM system. The
piece part drawing will be our guide to manufacture piece parts
in low quantities, openings for dies and punches, etc. Drawing
dimensions are transferred into the computer with the use of
computer language. Basically, all we are doing is describing
the piece part geometrically. All dimensions necessary to produce this part; like tensioning lines to circles, center points
of radii and angles of lines; are determined mathematically by
the computer. This computer will also provide us with a plotting generated by Hewlett-Packard plotter, to visually inspect
the contour. It also provides us with a tape used to activate
the wire EDM machine tool. As an example - a progressive tool
with "X" amount of notching and blanking stations will be generated into one tape. By doing so, a strip layout has been
created. Which also, by putting all this information into one
tape, creates a progressive tool in its entirety. One tape
can be used to manufacture punch and die. All dimensions are
programmed to the mean dimensions and an offset mechanism;
standard with this machine tool; lets us cut inside our outside
the mean dimension in increments of .0001 (.09999 +, .0999 -).
(Illustration No. 4)

INSPECTION OF PRODUCT CUT IN WIRE EDM MACHINE

No cut is being done at Argus Manufacturing Company without 100% inspection of the product to be cut. I know that 100%
inspection is almost unheard of, but it certainly can be done
with this equipment.

First the computer program is inspected by one other pro-
grammer. All dimensions appearing on this computer program
match the dimensioning of the piece part drawing figure by
figure. In other words, no piece part drawing dimension has to
be altered from its original.

Second inspection procedure occurs by using the machine
tool itself. The wire EDM machine tool is provided with a
duplicating feature in the back of the machine tool. A glass
which has been stained with ink is being scribed showing the
contour which we will cut.

Third - now the contour can be checked using a shadow-
graph.

     Before cutting the actual material the machine tool power
supply has to be set in reference to the thickness, the diameter of the wire, and kind of material. To assure dimensional quality, we are making a test cut to determine the slot
size created by wire electrical discharge machining. Slot
width, using .008 wire, range approximately to .0135. After
the product has been manufactured, the product itself is
checked in reference to finish and parallelism.

ECONOMY AND TIME SAVING OF WIRE EDM PROCESS

     It certainly can be said that wire EDM is a new and excit-
ing revolutionary process in its regards to manufacture of die
components, proto-types, extrusion dies, electrodes, gages,
machine parts and mold openings. It manufactures these products from start to finish without using grinding equipment,
milling equipment, saws, etc. In other words, the machine tool
produces a finished product without moving the product from one
machine tool to another. Wire EDM is a machine tool which
helps a diemaker do a job about five times as fast.

     Finally, let wire EDM do the work, and let tool makers do
the thinking.

Figure 1

Figure 2

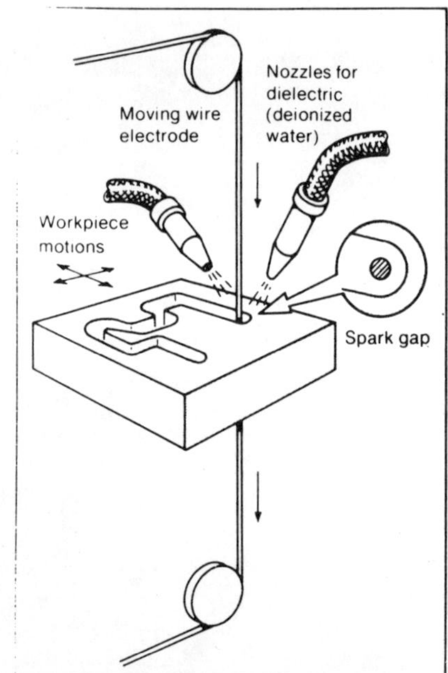

Contours are produced by moving the work relative to the wire electrode

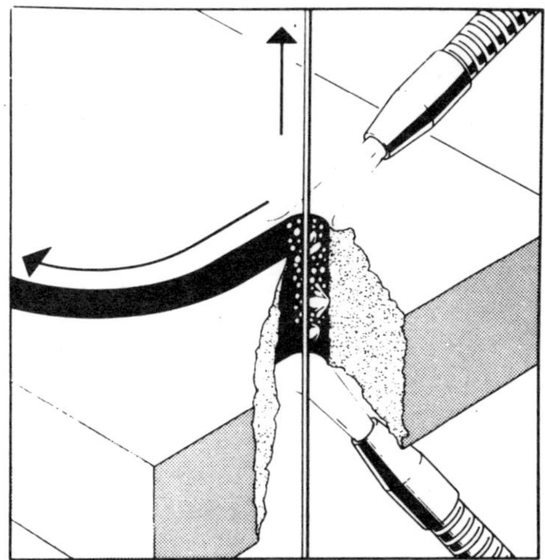

The principle of spark erosion cutting with a wire electrode.

Figure 4

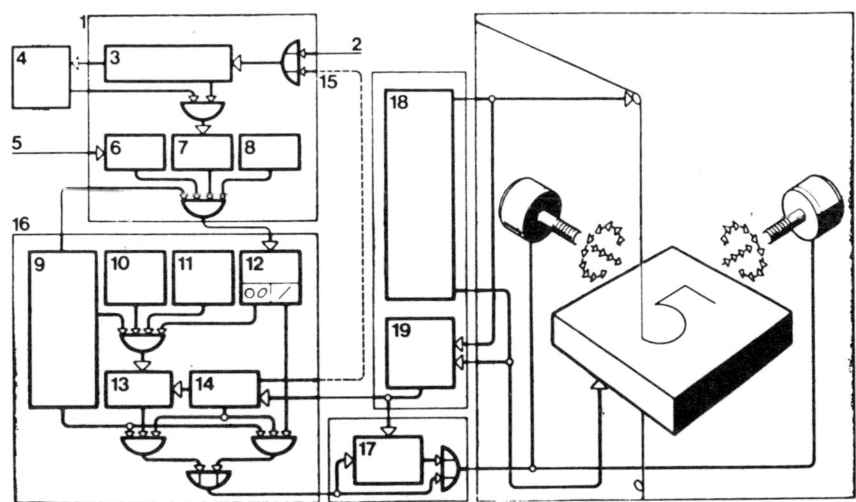

Block circuit diagram of AGIEMERIC numerical track control.

1 Data preparation
2 Start
3 Control programme for punched tape reader
4 Punched tape reader
5 Correction
6 Control programme for correction
7 Pre-memory
8 Control programme for decoding
9 Computer
10 Multiplication-division control programme
11 Fixed value memory
12 Intermediate result
13 Result
14 Control programme for linear interpolator
15 New block
16 Data processing
17 Reverse operation memory
18 Erosion power supply
19 Regulator device.

EDM: Making It Under the Wire

The high cost of building complex dies can be cut substantially through the use of traveling wire EDM. Here are a few pointers on the subject from Bowers Machine Company

Reprinted from Manufacturing Engineering, March 1979

THE STORY OF BOWERS MACHINE CO., Kalamazoo, MI, has an all-too-familiar scenario. The company started manufacturing tools and dies in 1928, thus anticipating the beginning of the Great Depression by one year.

Under the stewardship of its founder — a transplanted New Englander named Ernest Bowers — the company weathered the Depression, then moved into the heyday of the tool and die business: World War II. Bowers, who was trained in the craft at Greenlee Tap and Die, brought a certain class to the company he had founded. It flourished throughout the 50s and most of the 60s.

However, the company began a decline in the late 60s. Ernest Bowers, now dead, had left the company to his son Bob, who was deactivated in 1969 with a crippling stroke. By the time Bob Bowers had recovered sufficiently to take the helm again, the company was — for all practical purposes — finished as a contender in the ever-diminishing field of contract diemaking.

But Bowers had an idea. Why not take a look at EDM diemaking, specifically, diemaking via the traveling wire?

Enter Agietron. After a certain amount of research into traveling wire EDM, Bowers settled upon an Agietron unit. It can be submitted that Bowers' research was conducted in the best possible way. He read and reread articles on the various types of EDM machines in technical magazines. He then made his purchasing decision without benefit of further counsel.

Whether Bob Bowers would have been just as happy with an EDM machine of different make is — and will remain — an unknown. He has effected a significant turnaround in his company's fortunes through wire EDM, and he is contemplating purchase of a second machine of the same make. The prevailing 14% interest rates are holding this purchasing decision in abeyance, but Bowers is certain that a second unit will soon appear on the shop floor. In the meantime, his experience with EDM provides some interesting insights into the business.

Shop Capabilities. In enumerating the advantages of wire EDM, Bob Bowers notes that, "One advantage that we didn't expect is the amplification of our capabilities with respect to size. Ours is a precision shop, which means that there are definite limitations to the size of the blocks of steel or carbide that we can handle. However, the 4 x 6 x 8" envelope of our EDM unit exceeds the size capabilities of much of our equipment. In brief, we're now able to handle far larger configurations — precision configurations — than previously."

Increased Accuracy. "Increased accuracy is still another advantage," says Bowers. "Much of the work we're now doing with wire EDM is detail replacement for jobs previously built. There was a time when we had to have the entire die returned to us in order to replace a single detail. You know — the old die fitup business. But everything we now build is dimensioned and on tape. If a customer a hundred miles away has a fractured detail, all he need do is pick up the phone and tell us its number. We'll pull the drawing and NC tape for that specific detail and have it on the way to the customer in as little as 24 hours."

Substitute Jig Grinder. "We have a Pratt & Whitney jig borer but no jig grinder — something that might normally work against you in detail work," says Bowers. "After all, you do have to pinpoint those dowel hole positions. Our solution is to drill the holes to a close layout, then harden the part. The dowel holes are subsequently finished with the wire. Dowel hole size accuracy? We'll give you a press fit, a gage fit or a slip fit — it's that accurate."

High Up-time. "One of the biggest advantages of wire EDM is the fact that you can work 24 hours a day, seven days a week," Bowers adds. "Moreover, you don't have to have an opera-

BOB BOWERS OF BOWERS MACHINE CO.
"One advantage we didn't expect is the amplification of our capabilities with respect to size."

KEN GILMORE
OF BOWERS MACHINE CO.
"The price to the customer is still a function of the marketplace. But our costs to design and build are definitely lower."

tor in attendance. During the day we run the machine through cuts that will take no more than a few hours or less to complete. But before we close up for the night, we load a long-cut piece in the machine. On weekends we'll load it with a part that may require two to three days to finish. Generally, it's a job that will be finished sometime Monday morning.''

Wire Breakage. What happens if the wire breaks at night or during the weekend? ''Wire breakage isn't a frequent happening, '' Bowers answers. ''But it does happen — and when it does, this is the solution.'' At this point, Bowers pulls a small radio receiver somewhat larger than a matchbox from his pocket.

''The minute the machine is deactivated for any reason whatsoever, this bleeper is activated,'' he says. ''All key personnel carry one. The minute the signal sounds one of us gets down to the plant and corrects the problem and restarts the machine. As for range, anyone within 20 miles of the plant will pick up the signal.''

Programming. The question of programming arises. Bowers handles his programming through MDSI. If you're curious as to his programming costs, he'll show you his invoices. They average around $500 monthly.

''Programming costs aren't exorbitant.'' says Bowers. ''But they're high enough to make me think of the advisability of getting our own computer. But I'm still interested in getting another machine, so the question I've got to answer is how do I spend my money judiciously? So we'll probably hold off on a computer until we have our second machine, at least.''

How about personnel training? ''Currently, we have two programmers,'' says Bowers. Each spent a week at MDSI and another week at the Agietron plant in Elgin, Illinois. This training was sufficient to give them the capabilities needed to program the machine and set it in operation. Naturally, both have learned a great deal from on-the-job practice, but two weeks training is enough to get a smart diemaker started.''

Skilled Help. Like most die shop owners, Bob Bowers laments the growing shortage of skilled diemakers. At the same time, he's quick to note that wire EDM has largely obviated the need for traditional skills.

''When we were building dies by conventional methods, our diemakers had to have two sets of skills. One pertained to the precision finishing of punches and dies,'' he says. ''The other pertained to assembly. Wire EDM has taken over the precision punches and dies; and a good machinist can handle the job of assembly. Naturally, it's easier to develop machinist skills than diemaker skills. But we still call the man on the bench a diemaker, although he's

TWO WIRE-CUT PARTS– a proposed blank (left) and a finish formed part. Problem: the proposed blank opened when formed and lost dimensional accuracy. Redesign with a bridge (right) enabled part to hold size when formed The important point: wire cutting of prototype blanks enables engineering to discover part errors prior to the fabrication of permanent tooling.

nowhere near the type of diemaker we once employed.''

To emphasize this point, Bowers notes that the shop employed 20 to 25 diemakers at the height of its previous success. Today, it's doing even more work — with four or five diemakers.

Leadtime. If wire EDM means shorter leadtime, the question is: How much shorter? ''Approximately 50%,'' is Bob Bowers' answer. ''We had one interesting case of a plant manager making a perfunctory call about one of his dies. How was it progressing? Would we meet the scheduled delivery date? We took a certain amount of pleasure in telling him that the job was already set up and running in his plant.

''Reduced leadtime stems partially from the previously noted fact that the wire is running on close to an around-the-clock basis. Further reductions are realized from the elimination of time-consuming precision grinding and the old bugaboo of heat-treat distortion. But still another factor enters the equation — the fact that you're now on computer.

At this point, Plant Manager Ken Gilmore interjects a few thoughts.

''Let's say a partprint comes in with the part shown in the conventional X-Y position,'' he hypothesizes. ''Let's also say that two parts — right-hand and left-hand — are required. Finally, let's say that you're going to design a two-out progressive die, but that the parts are going to be rotated 45° from partprint position.

''Normally, this means solving a lot

WIRE-CUT TEMPORARY TOOLING used to produce part shown in previous photo. Die and punch were cut from the same program. Knockout was sawed, as was urethane stripper.

PAWL PRODUCED ON A TEMPORARY WIRE-CUT DIE required broached teeth. Problem: the broach was at least a year away. Solution: a temporary laminated broach made on the wire machine. Finishing broach segment is shown here.

of trig problems, which also means checking and double checking to eliminate the last possible error.

"Our solution to this problem is to label all critical coordinate positions *A, B, C* and so on. Next, we feed the coordinate positions from a theoretical centerline to the MDSI computer. Now here's an example of how we really cut down on leadtime requirements. We ask the computer to rotate the panel 45°. The computer instantly responds with a printout of the new coordinate positions of the designated points. These dimensions are then transposed to the die drawing.

"All of this has nothing to do with the programming of the EDM wire machine," Gilmore concludes. "It's simply a matter of taking advantage of the fact that we have access to the computer because we use it for NC programming. Why not use it for all calculations?"

Punch Footings. Conventional die engineering usually calls for heavy footing on the punches. It would seem that this might be an area where wire EDM might prove deficient, since it makes a straight-through cut.

"This isn't a problem," says Gilmore. "Remember, you can offset the wire on either side of a theoretical contour. You're machining on one side of the contour to make the punch and on the other side to make the die. Now you can also machine on the inside of the contour to make a press-fit or gage-fit punch retainer, which is just exactly what we do.

"If we're making a heavy duty punch, we like to machine the retainer to a press fit. We then like to put one or more tapped holes in it for socket head screws that come down through the upper shoe.

"If the punch in question is relatively fragile, we simply key it to its retainer. In any event, the problem of producing a punch with footing in a wire machine is not a problem," Gilmore concludes.

Strippers. The offset capabilities of EDM wire makes the process a natural for machining punch-die clearances. Only one program is needed to produce the punch, the die, and the press-fit punch retainer. But how about the stripper?

"Generally speaking, the same program can be used to machine the stripper openings," says Gilmore. "But the question you have to answer is: Do you really want to tie up an expensive EDM machine to burn out a comparatively loose component like a stripper? If you're building a Class A ultraprecision die in which punch/stripper clearance is virtually zero, the answer is yes.

"But in most die work, we feel that the stripper is still a component in which the various openings are best done with a drill press, a mill, and a DoALL saw. After all, there are a lot of parts more demanding of the wire's accuracy and time than a stripper plate."

TWELVE STATION PROGRESSIVE DIE STRIP.
The design was facilitated through computer rotation of the
blank and by the elimination of sectionalized details.

Applications. Bowers Machine Co. is involved in a great deal of progressive die work, but it's finding a surprisingly large customer demand for short-run and prototype tooling.

"Traveling wire has opened up an entire new field in prototype work," says Bob Bowers. "A customer has an idea of what a part should be, but he doesn't want to commit himself to permanent tooling until the part has been proved in prototype production. In the old days he would often make just such a commitment, since he had no other choice. If the part didn't work out, there'd be a part correction by way of an engineering change in the die.

"Now, we're getting quite a few requests for wire-cut parts, perhaps a dozen parts at a time. After trying them out, the customer may make certain modifications in design — or he may say go ahead and order temporary tooling on the basis of the initial design.

"We'll wire cut the temporary tooling, assemble the die and make perhaps a hundred pieces. If these prove out in prototype work, we're generally ready to quote on permanent tooling."

The versatility of wire EDM — and of Ken Gilmore — is seen in the pawl application illustrated.

"This isn't a particularly difficult part," Gilmore notes. "The problem with this part lies in the need to broach the teeth. Stamped teeth wouldn't do because of the breakage. So our job was to build a temporary die in which all dimensions would be finished except the teeth; we had to leave broaching stock on the teeth.

"Our customer's problem was that he needed 2000 of these parts for an exhaustive prototype study, but he had over a year's wait before the broach would be finished. Even by stacking these parts and finishing the teeth with wire, the machine would have been tied up indefinitely and the machining costs would have been prohibitively high," says Gilmore.

At this point, Gilmore — who by all accounts doesn't know how good he really is — came up with a solution that is simplicity itself. He programmed the wire for the finished pawl contour, then machined a sequence of high-speed steel segments that could be fitted together as a temporary laminated broach. With each segment, the wire was progressively offset farther to provide the necessary sequence of lands. A temporary broach fixture was built and the customer's needed 2000 stamped and broached parts were delivered in a matter of several weeks.

Gilmore tends to dismiss the entire episode as something that's done as a matter of course. "Traveling wire EDM is like most machine tools in at least one respect," he says. "You start out with certain preconceived ideas of what can and cannot be done. After you've worked on it for a while, you see all sorts of possibilities opening up. The laminated broach was one of them."

There is still another application of the traveling wire that is proving quite lucrative for Bowers Machine. It's the production of electrodes for the more conventional or standard EDM machines. Bowers Machine doesn't use the traditional form of EDM — but many of its customers do.

As Bowers tells it, "Whatever their electrodes may look like — simple or complicated — they wear with use. If they're used on long run jobs, the demand for electrodes that are all identical in size can become substantial. There's no better way to manufacture them than with wire EDM, which is where we come in. We stack them up and cut them six or eight or even more at a time. So you could say that wire EDM is lending quite a bit of support to the traditional form of EDM."

Designing and Building a Die. Although Bowers Machine Co. is deeply involved in die detail and prototype work, it's still a die shop engaged in the design and building of dies. An excellent example of what the shop can do — using wire EDM — is seen in the photograph and plan drawing of the progressive die shown. The questions arise: How did the design and manufacture of this die differ from the conventional? And what were the procedures involved, starting with the initial partprint and ending with the delivered die?

Ken Gilmore responds: "First, the customer submitted a partprint and asked for a quotation. As this was a design and build job, there was no die drawing. Before we could proceed, however, we had to know shut height, bolster dimensions and bolster hole locations, press speed, and height of the press feed system.

"I started by laying out the blank in the flat. This required calculation of bend allowance, of course. Next, I wire cut a number of sample blanks, then bent them into the desired configurations."

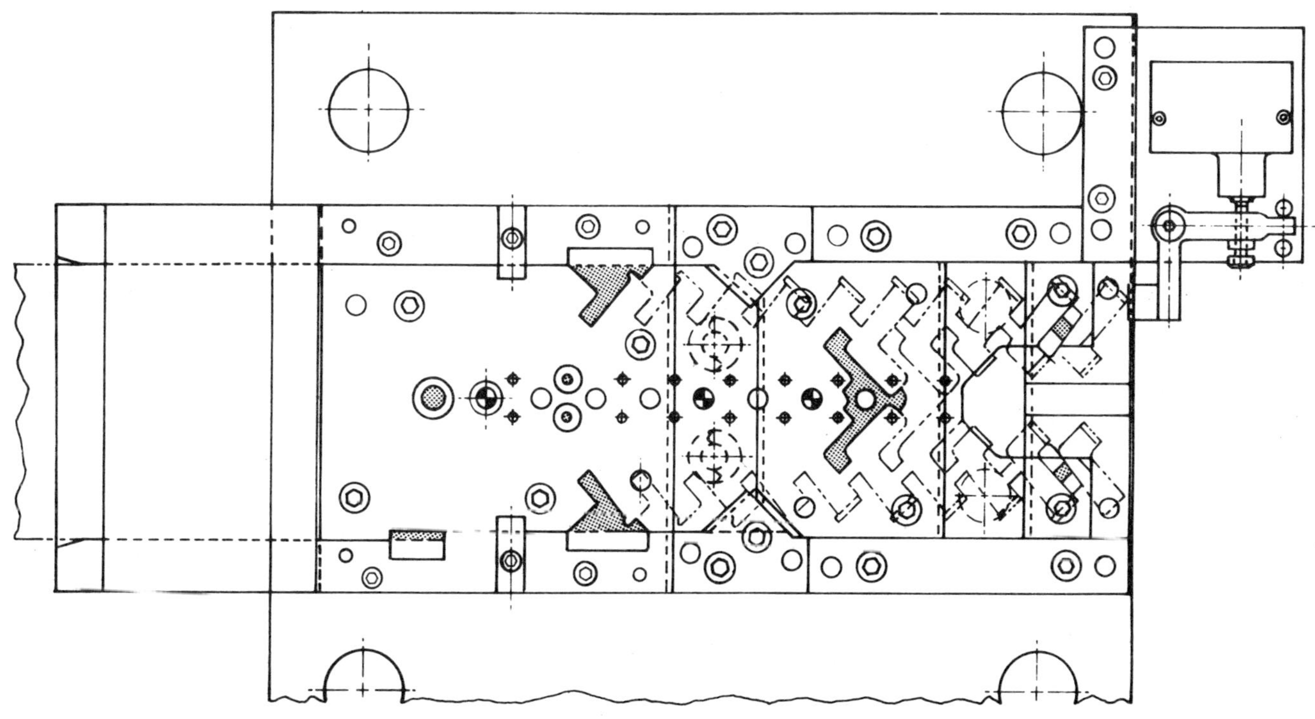

PLAN VIEW OF PROGRESSIVE DIE
for the strip shown. Of primary importance is the reduction of die sections
to the absolute minimum – something that wire EDM makes possible. **199**

Significantly, this is one of the first deviations from conventional practice when using wire EDM: the blank size is stored in computer memory — not in a template having varying degrees of accuracy.

Gilmore continues: "As is often the case, the amounts calculated for the bends were inaccurate, but it was easy to make the necessary corrections. Further, I could now visualize the die layout, which I then drew as an undimensioned plan view. At this point we were ready to submit a quotation, which resulted in our receiving a contract to build the die.

"We now started the process of finish design. I now reprogrammed the part and asked the computer to rotate it through 45°. With the new coordinate information thus obtained, it was easy — and fast — to go from the preliminary sketch to the finished plan and sectional views."

QUESTION: *What was the total time required for design?* The answer, according to Gilmore, is that, "I work on a number of jobs at the same time. About all I can tell you is that within a week and a half of making the initial rough layout for purposes of quotation, the total design was completed."

QUESTION: *What was the total design and build time for this job?* "This was one of our very first EDM wire dies, so we were still on the learning curve. I'd say we spent close to 300 man-hours and some 50 EDM machine hours. This wasn't bad for a die of this type, but we'd do it much faster today."

QUESTION: *What was the price of this die to the customer?* "About $8000. We took a bath on this die, but then you always have to pay for experience."

QUESTION: *Which is more expensive —this die as made by wire EDM or this die made conventionally?* "The price to the customer is still a function of the marketplace. But our costs to design and build are definitely lower. But there are a number of other considerations involved. First of all, this admittedly isn't a difficult progressive — but where would you find a diemaker today that could build it? A second consideration of importance is that of detail replacement. This isn't part of the initial price of the die, but it's always an important component in the cost of production. In this area, a wire-cut detail is far cheaper than one fitted by hand, and it can be produced much faster. So the answer is that to the die producer and to his customer, wire cut dies always end up being far less expensive, regardless of initial price."

FINAL COMMENTS? At this point Bob Bowers breaks in: "Just one. This shop, like many before it, was built on the qualities of craftsmanship that were built up over generations. Wouldn't it be wonderful if those old timers — and here I guess I'm thinking of my father in particular — could see how their vast body of diemaking knowledge is now a computer function? I wonder what their reactions would be if they could see the wire doing in a matter of hours what they did in a matter of weeks." ∎

How to make steel-rule dies

FOR SHORT RUNS, prototype production, or even 100,000 parts a year, you can build your own dies easily and cut costs significantly. Steel-rule dies can be made in ten simple steps, using a minimum of special equipment. The resulting blanking dies handle materials from 0.015″ to ⅜″ thick, holding overall dimensions within 0.015″ and placing holes within 0.005″. The photos show a sample workpiece revealing the many configurations that can be blanked with this tooling.

The Metlform die process, as developed by J. A. Richards Co., eliminates the use of a solid female die section and saves up to 90 percent in time and material costs. In this system, a ribbon of steel mounted in a wood holder becomes the cutting edge. Richards equipment will help you make the dies, and the company offers a training program with a "cycle time" of four days to teach your men how to do it. They even supply the special densified wood, a hard European beech known as lignistone. And, if you don't want to build your own, the company will make dies for you, still at a low cost.

Humble beginning

You start with the lignistone die block, scribing a layout on the reverse side of the high-density plywood board. Next, bend the steel rule to follow the layout lines and match the pattern. Then use special broach blades to jigsaw slots in the lignistone, following the layout pattern but leaving bridges for support. For this, Richards makes an Electromatic die saw, a universal machine for processing the Ligniform die blocks. It serves as a jigsaw, die filer, drill and circular saw.

Next, you notch the rule to match the bridges left in the wood, and then harden the rule in a small heat-treat oven. Finally, insert the rule in the wood to partially complete the upper die component. You generally have a 0.025″ interference fit to help hold the rule in place. Note that the steel rule has a sharp edge on it at this time.

To make the lower half of the die set, mount the punch material in the lower die plate. Attach the steel-rule die component to the upper die shoe in the press, and use it to strike a line on the punch material, using the sharp edge. Next, use a band saw to cut the punch outline, employing a blade angle of about 5 degrees to leave 1/64″ to 1/32″ from the line.

Machine a land on the rule and place urethane strippers in the rule section. Then "shear in" the punch so that it will have zero clearance. Finally file or grind the clearance on the punch to suit the material being blanked.

For thin stock, merely flame harden the edge of the punch. For blanking thick material, however, harden the punch all the way through. You can use air hardened or oil hardened tool steel for the punch.

The steel rule itself is a high-carbon oil-hardening steel, ranging from 6 to 12 points. A 10-point (0.140″) rule suits most jobs, but the sample shown uses a 12-point rule, 1¾″ high, to stamp ⅜″ mild steel.

Demonstration Metlform die. It blanks louvers, extrusions and knockouts, a sampling of the many operations possible. It also embosses names.

The Electromatic machine employs the circular saw to cut die blocks to size, and it incorporates a drill head for starting holes in the wood. Richards hand and hydraulic units bend the rule, and can serve elsewhere for prototype bending in model shops, etc. They handle flat and round materials. A special abrasive cutoff saw is available to cut steel rules to length, and there's even a small heat-treating oven for the rule if you don't have your own. It's truly a systems approach to die building. ∎

Drawings show the use of pierce punches and Multithane stripper pads. Spring ejector components may serve in the lower die plate. Clearance of rule at bottom of stroke over punch plate should be 20 percent of stock thickness.

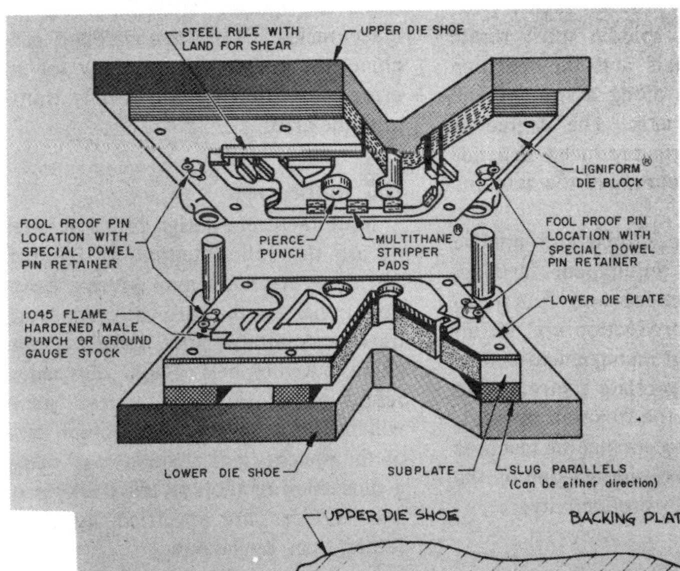

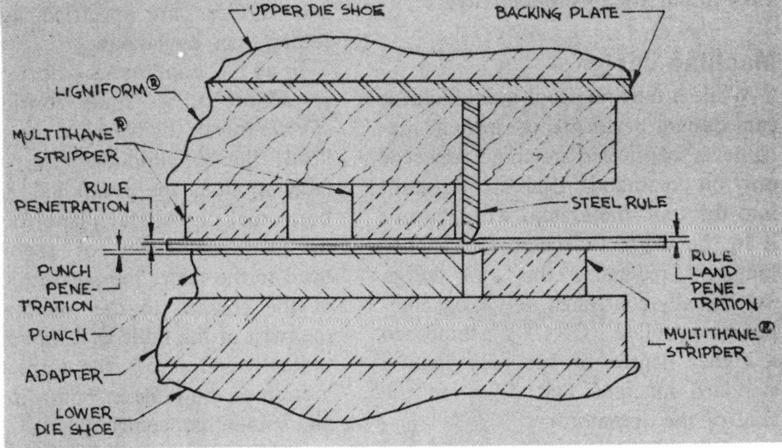

WHILE DOING editorial research on this story, how and where to get the documentation, I became acquainted with Joe Deuer of Main Machine. He pulled a letter from his files that states the case for the system explained here. The letter was to Harding Hugo of Danly Machine. Its excerpts are self-explanatory:

"This blank (sample enclosed) was produced on your Quik-Die machine; we here feel it could be one of the toughest parts that the process could be called on to produce. The outside of the part is not so difficult, but the small notch area, being only 0.06″ wide and 0.09″ long, is something else. And of course that end gets formed up later.

"We quoted the job and got it for permanent tooling and the production of the part. The problem had been to give the customer a pilot run of 150 pieces one week after receipt of the order.

"You know how concerned we were, and that we thought at first we would have to give the customer hand-filed parts. This of course was ridiculous and your process or others just had to be the answer.

"While trying the Quik-Die process we were concerned with what seemed to be literally impossible—to get any support for the small notched area with the Quik-Site material. With this doubt we went ahead with conventional development work while at the same time had a man make the blanking punch for your process.

"The punch blank development and

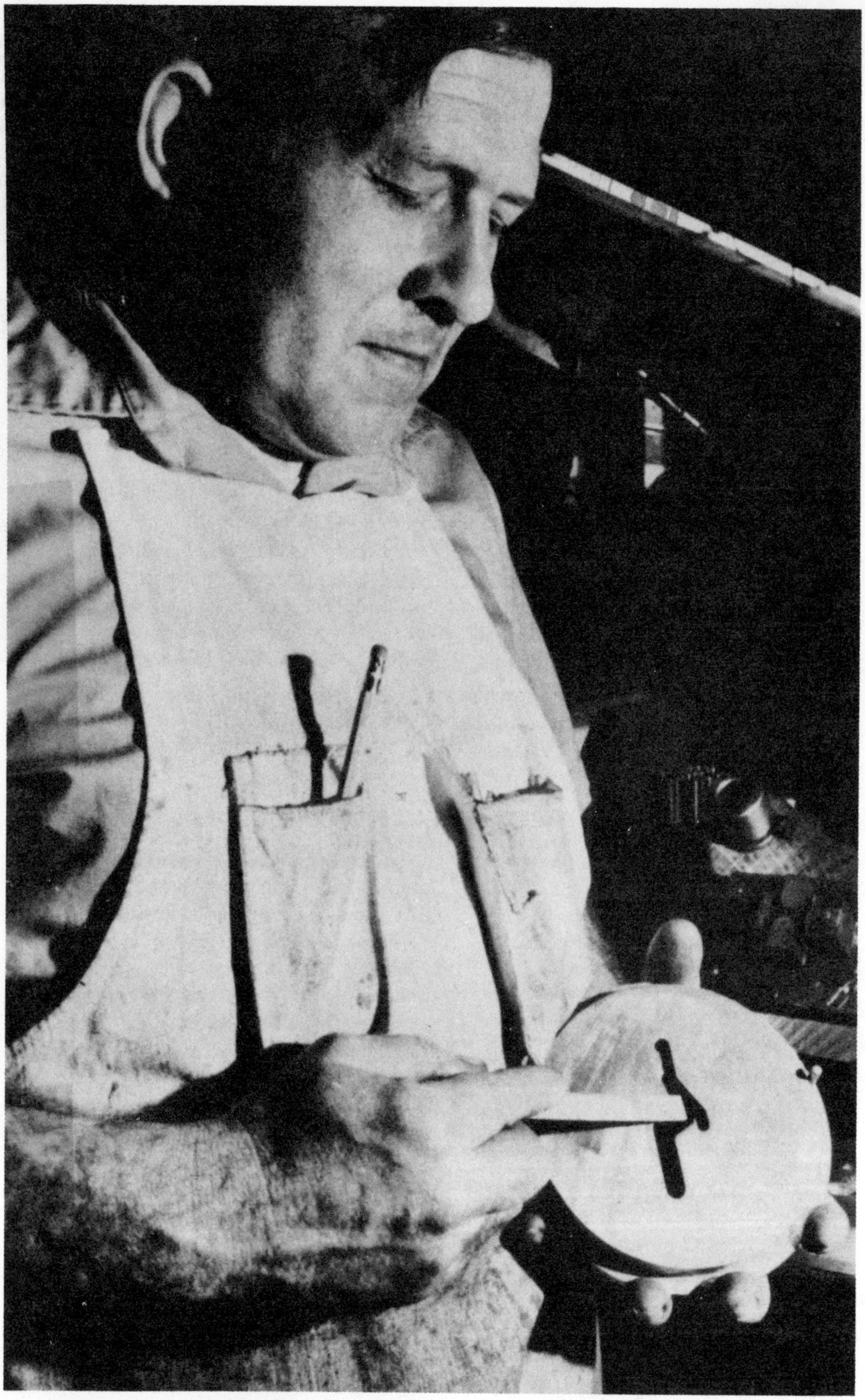

1. Pointing to a worrisome small projection in the die-cavity contour, Dick Irons worked it out. Story is a legend of a successful contract for making the part shown—inexpensively and in almost no turnaround time.

Another die system

Reprinted from Tooling & Production, September 1972

Quik-Die is Quick

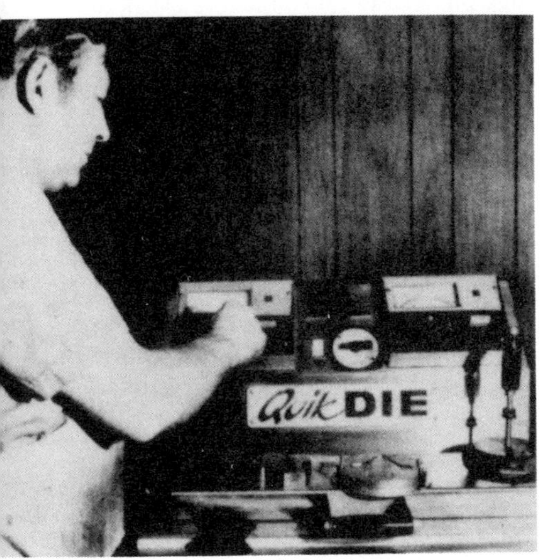

2. A punch that's necessary in any die, a couple of simple settings, and in a short time a usable die cavity. Melting pot and casting pot are both covered here.

3. A working die being used to punch the "H" shaped hole in the part in the foreground. As the story tells, production figures can be fairly high, but to increase numbers even further, steel inserts may be cast in the Quik-Site, at time of casting die.

machining took 15 hours, and using your process for the first time it took 2½ hours to pour the die and mount it in the die set. The 2½ hours included mounting the punch.

"You may or may not be interested in our use of the die, because you were telling us all along what could be done. But we were so impressed with the results, and maybe this configuration is something even you have not seen done with a Quik-Die.

"The first run was of 0.045″ cold roll for 150 pieces; the second run was of 0.036″ cold roll for the same number. Parts were all good, but there was evidence of breakdown beginning in the worrisome notch area. We sharpened the die with a banjo grinder and made a third run of parts with stock 0.056″ thick, half-hard. The die held.

"The way things went we felt this die could give 10,000 pieces with a few sharpenings. If more parts were needed we could easily, and quickly,

pour a new die around the same punch.

"We are pleased to tell you of our success, because it was great to find a way out, and a process we can continue to use. Everything doesn't always go this well."

How it's done

Dick Irons, toolroom foreman at Main Machine, cast a die to an existing punch while I watched. It seems easy to put a die in service within 3 hours after the punch is available.

Regardless of part configuration or complexity, the casting process takes only 15 minutes of operator attention with a total processing time of 2 hours. Removing the punch from the cast cavity and mounting it in a master die set requires only minutes; this includes application of pieces of rubber for strippers. Material cost is only $2.

The Quik-Site is a tough alloy similar to Kirksite. It melts at 820 F and is suitable for cutting die operations as well as drawing and forming.

Danly people say the "secret" is in the machine, which provides controlled cavity cooling from the inside so that the shrink is to the punch. This zero clearance obviously puts the cavity edge in pure shear—eliminating torque effect and/or edge rolling.

Already wide acceptance

Aerospace concerns have used this for small-lot production of airfoil shapes in exotic materials. The procedure is interesting: a plastic model of the part is coated with beeswax equivalent to stock thickness, and a metal-bearing plastic is poured over the model to produce a mating cavity. Dental plaster is poured in to make the male model; it in turn is used in the QD machine to make a QS tool cavity for drawing or forming the part. A mating punch is made on a template-controlled milling machine. Template is male model made from original.

When making large laminations with required electrical characteristics, the

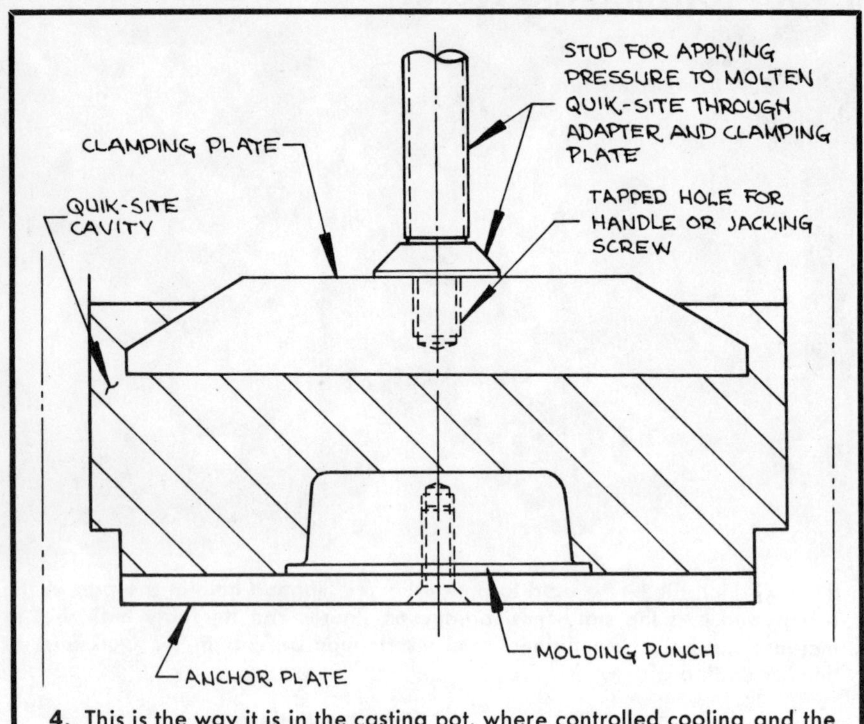

4. This is the way it is in the casting pot, where controlled cooling and the right material make the difference.

complicated openings can first be made this quick way and tested. Cost here is minimal compared with the permanent tooling made in carbide.

The zero cutting clearance, we are told, permits blanking or perforating thin metals or nonmetals. Cavities made this way have blanked asbestos, paper, cardboard, foil and many gasket materials.

The electronics industries appear to have embraced the use of the process too; instances of this have been in blanking glass-impregnated phenolic boards or Mylar at less cost than the typical saw-and-file method.

And of course the contract shops are reveling in the flexibility and quick turnaround time afforded them by this procedure. Large companies enjoy an advantage sometimes of unlimited resources to grab off a contract. Now this makes it easier for the little guy to get in there and "faceoff" with the competition. ■

William E. Hoffman

Reprinted from Modern Machine Shop, August 1970

Will your forming die work?

By Ken Gettelman

Any new stretch or deep draw die has one huge imponderable in addition to the money and human effort poured into it. Will it function properly without rupturing or tearing the metal during the stamping process?

If the die does work on a tryout press is there enough of a safety margin available that it will work in the actual production press where conditions may be slightly different?

Is there a method of actually measuring the effect of different lubricants, materials, die features or press characteristics?

When a good stamping operation suddenly turns sour, is there some method of quickly diagnosing the problem?

These questions haunt die builders and users alike and it had traditionally required a deep blend of intuition, common sense, experience and even hunch to come up with a working answer.

There is still no substitute for common sense but there is a powerful new tool to help answer the above questions. The circular grid will not only tell whether or not a die is likely to work, but will also tell what margin of safety there is in a stamping operation. In addition, it will enable the die builder or user to evaluate and study the effect of different variables found in the operation.

New Development

The circular grid system, originally conceived by Dr. Stuart Keeler of the National Steel Corporation, features a pattern of very small circles in a tight formation lightly etched onto a forming blank. After drawing or stretching, the deformed circles are measured to obtain a wealth of information. By quickly establishing deformation patterns and applying them against known strain curves it is not only possible to predict the probability of a successful stamping operation, but also to isolate and study individual elements such as die design, lubrication, materials, and press conditions.

In brief, the circular grid system will tell whether any portion of a formed workpiece is close to failure limits or well within its capability of accepting the deformation strain. The system is easy to apply, use and understand. This is no small consideration in a press where blank failures are occurring and something must be done in a hurry. If failure does occur it will be much easier to find the reason for it. The circular grid study can be applied to either stretch or draw operations or any combination of the two.

Idea Refinement

The process of marking some kind of a grid on a blank and then measuring its deformation after stamping is not new. For years, stamping specialists have scribed a pattern of one-inch squares on a test sheet and then formed it. The scribe marks were studied to observe the deformation. Hopefully, there was useful information that could be derived. Most often, the square with the greatest deformation was measured and its increase in area was related to some kind of a severity rating to obtain a degree of knowledge about the metal forming operation.

The square system has two major drawbacks. First, the scribed lines are not close enough for detailed study of a very localized area where blank failure might occur. This is not a major problem on large formed parts with no severe deformation. However, most failures take place in a very localized area. Secondly, the very act of scribing might so weaken a blank that failure would take place along the scribed lines.

Squares are used for the simple reason that they can be very easily made with a series of straight horizontal and vertical lines scribed either by hand or by machine. The etched, circular pattern does not weaken the test blank and it enables the study of a very localized area.

How It Works

The circular grid employs a pattern of small circles in which each circle is only one-fifth to one-tenth of an inch in diameter with a plus or minus one percent tolerance. The circles are closely spaced with approximately 5,000 one-tenth-inch-diameter circles on a nine-inch-square blank. With such a tight pattern of small circles any portion of a stamping, including very localized areas such as sharp radii or character lines where tearing and rupture often occur, may be measured and studied.

Obviously, hand scribing such a pattern would be impossible if each circle had to be made individually. Fortunately, there is a quick method of applying such a pattern over the entire surface of a test blank.

Electrochemical marking is used with simple, inexpensive, hand equipment and a preprinted electrical stencil, both of which are commercially available. The stencil is placed on a cleaned blank. A felt pad, soaked in electrolyte, is placed on top of the stencil. The pad is backed with an electrode which in turn is connected to one lead of a 14-volt power supply. The other lead is connected to the blank. The result is an electrically etched pattern with a depth of mark reaching only 0.000050 inch or even less, depending on etching time. The pattern cannot be removed by oil, chemicals, or ordinary rubbing as the workpiece material is pulled over the die surface. Application takes from 15 to 60 seconds.

The etched blank is put in the die and formed. At those areas where a strain occurs in the formed material, the etched circles will be deformed as shown in Figure 1. Both the degree and type of strain may be accurately de-

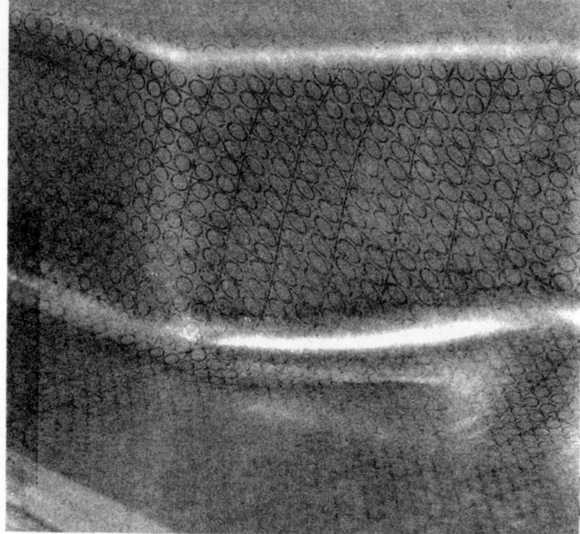

Fig. 1—When metal is formed, strains are created. The ellipses indicate the degree and type of strain. The most severe or peak strains are indicated by the most elongated ellipses. Note the severe strain in the lip area of the bumper jack (top photograph). Each circle was originally 1/10 inch in diameter.

termined by measuring the resulting axes of the deformed ellipses.

When metal is strained in a forming die, the actual strain that takes place may not be purely tension. Interestingly enough, there may be some severe compression strains. Metal is pushed as well as pulled and the particular type of strain is an important factor in determining the formability of a given material.

Since the circle is nondirectional, it is uniquely suited to indicating both the tension and compression strains as illustrated in Figure 2. *If there is a combination tension/compression strain taking place, the circle will be deformed into an ellipse with the major axis of the ellipse longer than the original circle diameter and the minor axis shorter in length. The long axis indicates the tension strain and the short axis the compression strain.*

If the ellipse that is formed has both its major and minor axes enlarged from the original circle diameter, then the material is in a tension/tension type strain condition. In a given forming operation it makes no difference as to type of strain condition so long as it is within safe limits. *Most low-carbon forming steels are able to take a greater compression/tension strain combination without failure than a tension/tension type.*

All materials have a safe strain limit before tearing or fracture takes place and the curve has been worked out for the basic low-carbon steels. Since the concept is quite new, strain curves are being developed for other materials. However, preliminary investigation indicates that materials most commonly formed have curves similar to the basic low-carbon steel. The strain curve illustrated in Figure 3 is an annealed and lightly skin-passed, low-carbon steel. First note the tension/compression side of the graph. If, after forming, the most severely deformed etched circle has an increase in length of 100 percent along the major axis of the resulting ellipse and the minor axis has a 40 percent reduction, the material is still within safe forming limits. It can be seen that there is a safe area, a critical area just bordering on **failure**, and a definite fail area. **The same** holds true for the tension/tension

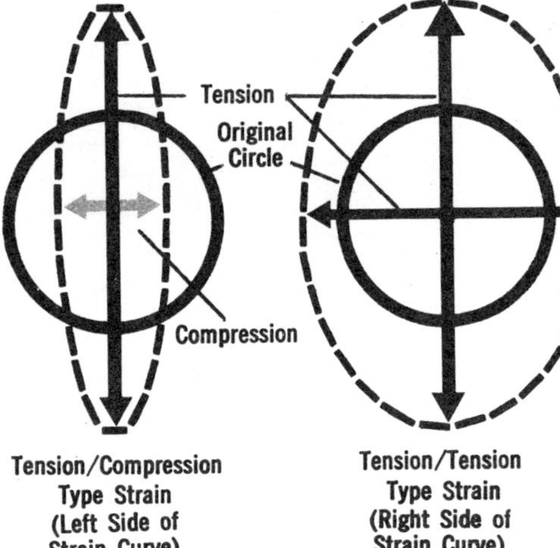

Fig. 2—Drawing operations will normally form an ellipse whose major axis is greater than the diameter of the etched circle. This is a tension type strain. When the minor axis is shorter than the original circle there is a compression strain. If both the major and minor axes are enlarged the strain is a tension/tension type. The latter usually results from stretch forming.

Fig. 3—Here is a failure curve for annealed and lightly skin-passed, low carbon steel. The band separates the safe from the fail conditions. The vertical axis is the largest principal strain found in the surface of the stamping (major axis of the ellipse). The horizontal axis is the principal surface strain perpendicular to the largest strain (minor axis of the ellipse).

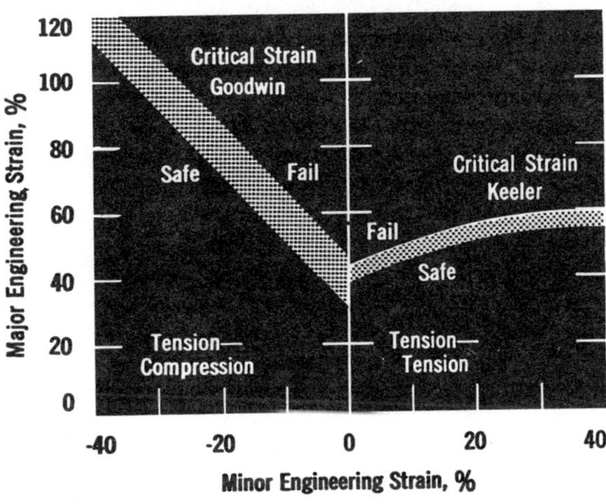

side of the curve. It should be noted that the tension/compression was worked out by Mr. G. M. Goodwin of Chrysler Corporation and the tension/tension by Dr. Keeler.

It is obvious that if the most severe strain deformation taking place is well within the safe area of the curve —whether tension/tension or tension/compression—there is no concern about possible material failure in the actual forming process. If, however, the worst condition is very near the critical or fail point there may be trouble.

Stretch and Draw

Basically, two types of dies are used to form metal. A stretch die holds the material in retaining rings and all deformation takes place by stretching the metal on the inside of the ring. A draw die allows all of the metal to flow between the male and female die components. Stretch dies generally set up purely tension/tension strains and draw dies normally create tension/compression strains although there may be various combinations in either type of die. It all depends on the particular die configuration being used.

Item By Item Analysis

It is very simple to grid a piece and then form it. The most severe strain conditions are measured and plotted as shown in Figure 4. Notice how the circular grid system has been used to isolate and study the effects of material, lubricant and die pad pressure.

By plotting the various strain points, a very good indication can be made as to forming success or failure. A single use of the grid circles and the

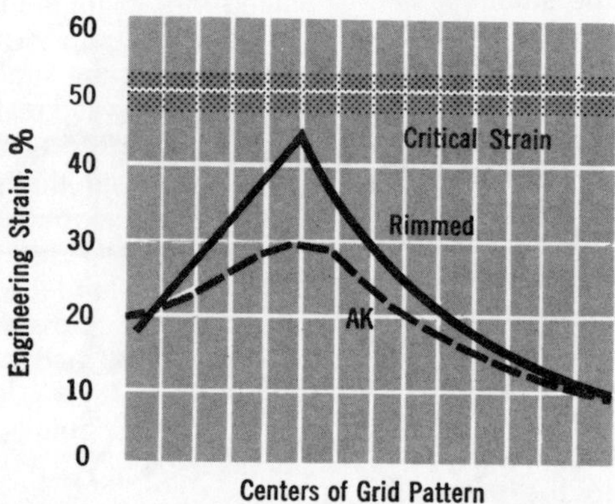

Strain distribution in an automotive fender. The critical strain level begins at 47 percent. Peak strain for stampings made from the rimmed steel is 45 percent, allowing a safety factor of only 2 percent. Slight variations in material properties or press conditions can lead to breakage during forming. The aluminum-killed steel, with its 31 percent peak strain and 16 percent safety factor, would be more satisfactory in this case.

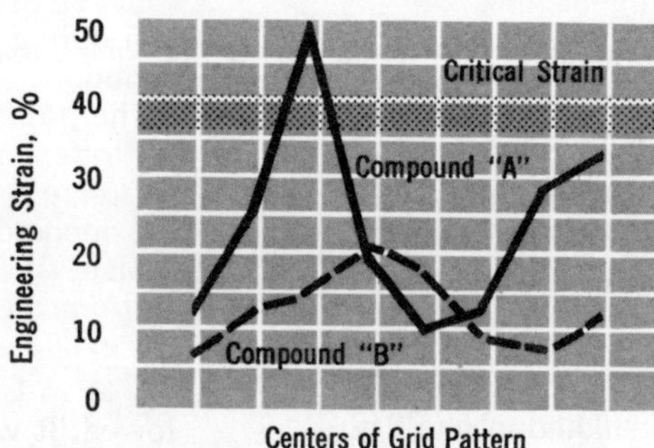

Two different lubricants were used in forming an automotive hood panel. Strains were determined from an electrochemically marked grid of 0.2-inch-diameter circles. The critical strain level was predicted from an empirical curve. Although both lubricants were considered "good," peak strains were different.

Increased pad pressure caused strain to increase during forming of an automotive panel. During Trial 1, satisfactory stampings were produced; however, during Trial 2, the strain level was somewhat higher and all the stampings broke.

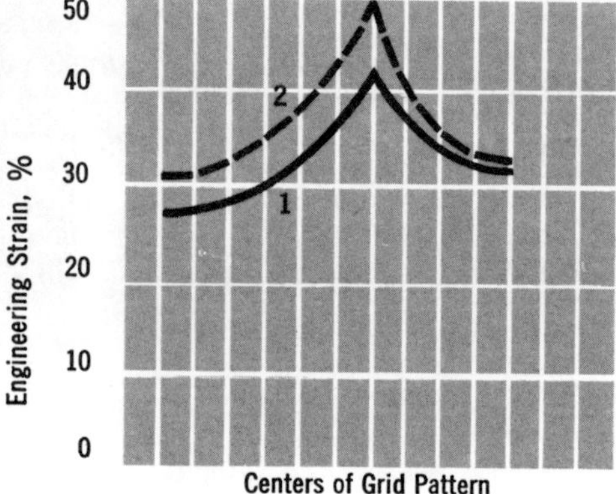

Fig. 4—These are examples of how material, lubrication, and die condition factors can be isolated and studied. The circular grid system makes these studies possible.

Keeler-Goodwin forming limit curve will not initially tell why a particular forming operation may be either successful or unsuccessful. By making a series of tests and varying only a single element, the significance of material properties, lubricant, die design, or press adjustment, may be determined. A test blank is formed and the strain curves plotted. Then the process can remain identical except for a change in lubricant. Thus, the lubricant factor can be determined. The same holds true for materials, press adjustment, and die design. If a situation is in the critical range, it may be determined that nothing more than a change in lubricant or a slight die adjustment will put the operation into the safe area.

Full application of the concept holds tremendous possibilities for improving a forming operation. For example, the use of the circular grid system is helpful in the selection of materials, particularly when stampings are critical. If the peak strain is near the critical level a change in material may be necessary. On the other hand if the peak strain is well below the critical level it may even be possible to substitute a less expensive material.

Lubricants are always an important factor and the circular grid system is an ideal tool for studying the effect of various lubricants. As Dr. Keeler points out, "There are a few broad guidelines for predicting the influence of lubrication on formability. Good lubrication aids in stretching material over a large radius punch.

Without lubrication, the straining of material near the pole of the punch is retarded. This means that increased straining is required away from the pole. As a result, strain distribution is more non-uniform and the resulting depth to failure is reduced. Good lubrication allows material near the pole to participate in the straining, thereby adding to the total depth of the stampings before breakage." Dr. Keeler's studies have shown that changes in lubricant can often make a substantial difference in the stamping operation even to the point of allowing the use of less expensive materials.

Die design is another important area that can be thoroughly studied with the circular grid pattern. Die modifications such as reducing the size of draw beads, increasing die radii, changing the size and shape of the blank, inserting lances, and so on, can improve the formability of a stamping to a much greater extent than a change in materials. Because of the multitude of variables in die design, there can be no hard and fast rules about improving the forming characteristics of an individual die with a specific modification. However, modifications can be tried and evaluated on an individual basis within a given die by using the circular grid system.

Dr. Keeler cited an example where a steel was formed with success one week and a week later all stampings broke. Since the material was stabi-lized no change in properties had occurred. Strain distribution pattern showed that the strain over the entire punch area was greater in the second trial than in the first. Further investigation showed that buckles or wrinkles had developed in the production steel during the interval between the two trials. In order to eliminate these buckles, the press operator had in-increased the pad pressure. This resulted in an overall increase in strain in all the stampings and caused the breakage.

Other Benefits

In addition to isolating and studying lubrication, die design, materials, and press conditions, the circular grid system provides an excellent training aid for apprenticeship programs. The gridded blank will immediately show the student the effects of changes in blank size, blank configuration, radii, pad pressure, and other variables. The effects of a large number of variables can be studied in one day.

Often, the final strain pattern does not yield sufficient information to detect a failure cause. By examining the grids after each stage of a multiple stage operation, or by stopping the formation of a stamping at various incremental depths, the cause of the strain can be better analyzed. The particular die set which generates the non-uniform strain distribution can be identified. Sometimes compressive strains are followed by tension strains. Different parts of the die in-teract with the stamping at different stages of its formation. All these characteristics can be studied utilizing the circular grid pattern.

When dies are replaced in a press, they are often reset differently from a prior production run. This can change the strain distribution and cause breakage. The grids will tell the location, type and magnitude of any such differences.

If a periodic test is made with a gridded blank it is even possible to detect changes that occur as an operation wears on. Impending trouble could be detected to permit modification during the next scheduled down period.

The possibilities are almost endless and offer many opportunities to the person in charge of stamping or forming operations.

Material for this article was secured from the presentation by Dr. Stuart Keeler of his paper "From Stretch To Draw" and discussions that followed. It was presented at a Society of Manufacturing Engineers seminar on metal stampings. The complete paper, MF69-513, is recommended reading and is available from the SME, 20501 Ford Road, Dearborn, Michigan 48128.　　**mms**

no scratches with urethane draw die

By utilizing a urethane punch in a draw die, the cost is reduced, there are no scratches on the polished material being formed and there may be no need for a steel draw ring.

By KNIGHT FARWELL

Urethane can be made to flow, but it will not compress. These qualities make it ideal for many types of forming dies — especially when polished stock is involved. For unlike die steel, urethane can form polished steel or aluminum without scratching the surface.

There are two basic draw die designs in which urethane may be used. The first makes use of a urethane male or female punch with a conventional steel draw ring to hold the workpiece blank. The ring is pressurized by urethane springs as shown in Figure 1.

The second design makes use of a steel or epoxy female die section and a male pad (or wafers in some instances) of urethane used as a combination punch and blank holder. This type is advantageous from several standpoints:

- Low cost
- No mating of parts is necessary
- Shaping of the urethane is usually eliminated
- The die can be efficiently operated in either a punch press or press brake, and
- Die set mounting is not mandatory.

A die of this nature for the drawing of the six-inch-diameter reflector shown in Figure 2 from pre-polished 0.030-inch-thick aluminum was recently made by Prescolite, a Division of U. S. Industries, Inc., El Dorado, Arkansas. The complete unit was mounted in a die set (for convenience of operation) on a 60-ton mechanical punch press.

The female (lower) half of the die was made of commercial steel with a 6 3/16-inch-diameter by 0.030-inch-deep nest at the upper edge of the section in order to properly locate the blank. A floating die button, striker plate, and spring cage were made as outlined in Figure 3.

The upper (male) die section was made from a section of eight-inch ID pipe welded to a one-inch-thick commercial quality steel plate. The urethane was placed in the pipe with no screws, clamps, tape or other holding method. The force fit between the urethane and pipe was sufficient to hold it in place.

A Tool-A-Thane UT-35 grade of urethane from Urethane Tooling and Engineering Corporation, 2800 Bernice Road, Lansing, Illinois 60438, was selected for several reasons:

1. The amount of urethane travel is approximately one inch, which (in the three-inch thickness) represents a one-third deflection. UT-35 will give optimum life up to 35% deflection.

2. In order to run the job on a 60-ton press, UT-35 had to be used as it has the lowest force requirement for the standard grades of urethane presently available. A force of thirty tons is developed at a one-inch deflection. The total force requirement for the part is the sum of the force developed as the urethane flows plus the forces to punch, emboss, and draw the part.

3. The urethane had to be soft enough not to mar the pre-polished aluminum surface in order to avoid a secondary polishing operation.

During initial setup, it was determined that the top of the urethane-punch had to be relieved in order to further reduce the tonnage as well as allow the proper forming of the dimple while still punching the hole.

Shortly after the beginning of the stroke, contact is made between the outer edge of the workpiece blank, located in the nest of the female die section, and the urethane punch pad. Blank holding pressure is developed before any piece part forming takes place. This action helps insure that there is absolutely no wrinkling of the blank. As the height of the urethane is further reduced, its flow is directed into the steel female cavity—the only place available. As a result, the workpiece is completely formed. The non-compressability of the urethane is a vital factor in the complete forming action. The cutout at the top of the urethane pad enables the flowing and forming action to occur with a lower strain level than would have been the situation with a solid section.

Prescolite reports that production, at a rate of 500 pieces per hour, has now passed 200,000 reflectors with another 25 to 50 thousand expected before replacement of the urethane pad is necessary.

Some breakdown of the urethane pad, as shown in Figure 4, is now being experienced around the outer edge of the punch but there has been no adverse effect on the part quality. As long as urethane will do the job, a slight breakdown or cutting is not harmful. The ultimate life of a urethane punch or die is affected by such considerations as material thickness being formed, temper of material, type of material, depth of draw, definition required in finished part and speed at which parts are cycled.

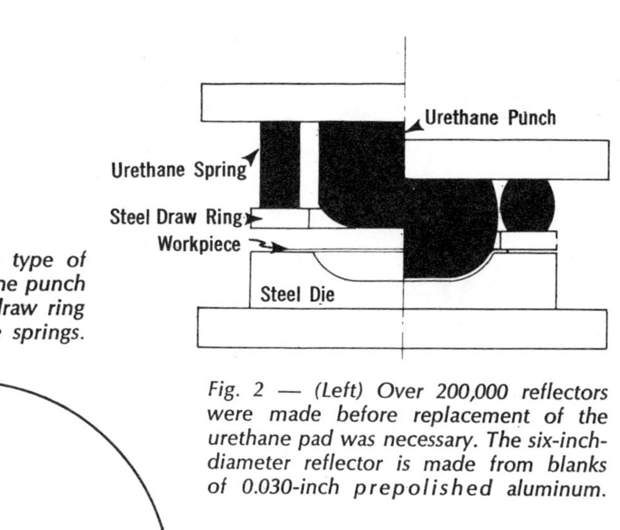

Fig. 1 — (Right) One type of draw die with a urethane punch makes use of a steel draw ring together with urethane springs.

Fig. 2 — (Left) Over 200,000 reflectors were made before replacement of the urethane pad was necessary. The six-inch-diameter reflector is made from blanks of 0.030-inch prepolished aluminum.

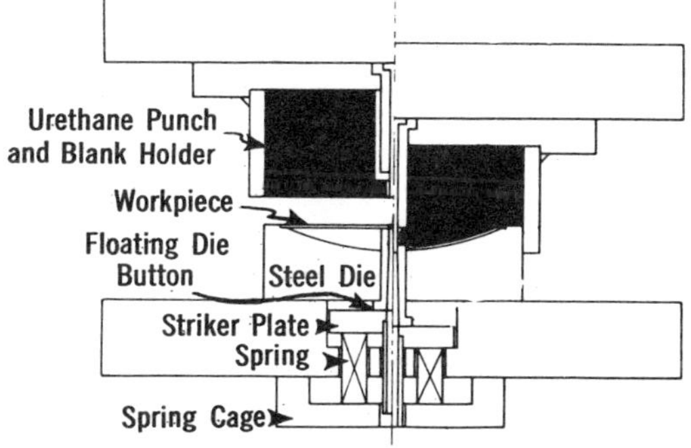

Fig. 4 — (Above) Although the urethane punch will show a slight tearing and breakdown at the edges after long use, it will continue to function efficiently.

Fig. 3 — (Right) Several advantages are gained by using a solid urethane punch member that both holds and forms. A 6 3/16-inch blank is placed in this die to form the reflector shown in Fig. 2. No trimming necessary.

Proper design will increase the life of a urethane punch or pad. Where possible, it is advisable to use the material in the cast condition, which is more resistant to cutting and breakdown than a machined surface.

The use of a urethane die has given Prescolite many advantages, of which the absence of part scratching is only one. By using the urethane as a blank holder, it was possible to use only a 6 3/16-inch-diameter blank. It draws to size and no trimming operation is necessary. The cost of the urethane punch material was only $48.75 — a substantial savings over the cost of a steel punch, which would have to be mated to the female section. Because the urethane is flexible, any variations in thickness of the aluminum stock will be automatically compensated for with no possible damage to the die.   **mms**

CHAPTER 5

MAINTENANCE AND REPAIR

Carbide Die Maintenance and Service

By John C. Vecchi
President
J. V. Manufacturing Co., Inc.

A carbide die is an expensive precision tool which, with proper care, will produce at least ten times as many pieces per grind as a high-carbon, high-chrome die of comparable quality. It cannot be overstressed that a well-planned maintenance program will make a significant contribution to higher production and a lower unit cost. In order for a die service man to perform his job properly and efficiently, he must have the proper die handling equipment and die servicing tools.

Sharpening Equipment

First to be considered is the grinder. It should be large enough to carry the weight of the carbide die, which will be somewhat heavier than a high-carbon, high-chrome die of similar size (see Figure 1).

A grinder which mounts a peripheral type (D1A1) wheel is much preferred to a machine having a vertical spindle and mounting a cup (D2A2) type wheel. The area of contact with a D1A1 type wheel is less, while visibility of the work is greater.

The wheel spindle should be sturdy and run as near perfectly true as possible with an uninterrupted flow of clean coolant to the face of the wheel at point of contact with the work as shown in Figure 2. The coolant will not only protect the die from damage by heat checking, but will also prolong the life of the diamond wheel.

The coolant flow will flush the work, remove the ground-off material and help reduce chatter caused by excessive wheel loading and glazing. A better finish is always obtained when the grinding is done wet. Coolants should be free of sulfur and have low alkalinity, since sulfur and high alkalinity will attack both the cobalt binder in the carbide and resins in the wheel bond.

The proper wheel is very important. Green grit wheels should not be used. They present too many possibilities of damage to the die. A resinoid bonded diamond wheel of the proper grit size and hardness with a Bakelite core has been found, through research and experience, to be the most efficient wheel. The use of a special reinforced Bakelite core will greatly add to wheel stability.

Use of wheels with a metal core must be avoided. The metal core has a tendency to go out of round because of localized heat, thus causing a chatter finish on the carbide.

It is advisable to use one wheel of 100 grit for roughing and one of 180 grit for removing the last 0.001 to 0.002 inch of carbide stock. The use of one 150-grit all-purpose wheel has also proved successful

Fig. 1—The grinding machine must be big enough to hold the added weight of the carbide die and it must be rigid enough to produce excellent grinding results.

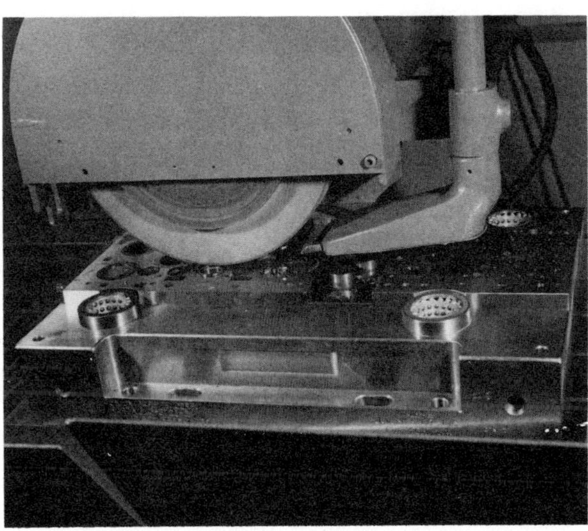

Fig. 2—There will be no sparks when properly grinding carbides. Notice how fluid is directed to the grinding area.

in most cases.

The wheels to be used include:

- For roughing: D100, R100, B7, 1/8, or equivalent.
- For final finishing: D180, R100, B7, 1/8, or equivalent.
- For all-purpose: D150, R100, B7, 1/8, or equivalent.

Speeds and Feeds

The wheel should be run at 4200 sfpm. A grinder with a D1A1 wheel, 16 inches x 1 inch, has met the requirements for die sharpening. However, such a large machine is not necessary for smaller dies. Likewise, it may be more economical to use an even larger machine with up to 3-inch-wide wheels for more massive dies.

Table speeds should be approximately 85 fpm for roughing, and faster speeds, if possible, for finishing. Cross feeds of 1/3 to 1/2 of the width of the wheel are best for roughing. Reducing the cross feed to 1/32 inch will produce a better finish on the final finishing passes.

Check For Run-Out

After the selection of the machine and the proper wheel, both the spindle and the wheel adapter should be checked with a dial indicator to insure true running. Flanges should be checked for burrs that may affect the proper wheel fit. When the wheel is mounted on the adapter, care should be taken that the size of the bore is sufficient to permit the wheel to pass freely up against the shoulder of the adapter. Diamond wheel manufacturers generally make the bore somewhat larger than the nominal size to permit the machine operator to adjust the path of the outside diameter of the wheel.

The outer flange should be brought into position and the flange screw lightly tightened. The wheel run-out should then be checked with a dial indicator, both on the side and on the grinding face. If the adapter flanges are free from burrs and dirt, the wheel should have no side play. To adjust the periphery, a small block of wood should be held against the high spot on the side of the wheel and the block should be tapped lightly with a hammer.

From the moment of trueing until it is completely worn out, the diamond wheel should never be removed from its wheel mount (or adapter).

Since diamond wheels are expensive, the price of a separate adapter for each wheel will soon be recovered by savings realized in less frequent dressings. The wheel should also be balanced, if necessary. An out-of-balance wheel can cause a rough surface finish and will have a shorter life because of more frequent dressings.

Dressing the Diamond Wheel

For the initial and all subsequent dressings of the diamond wheel, a Norton or similar brake-controlled dresser is recommended, similar to the one shown in Figure 3. The dresser should have a silicon-carbide wheel of the following specifications: 37C60Q5V Norton or equivalent (dimensions: 3 inches diameter, 1 inch face, 1/2 inch bore).

The dresser should be placed at some convenient place on the grinder table or the magnetic chuck, with its spindle parallel to the spindle of the grinder, and its wheel travel in line with the direction of travel of the diamond wheel. When the brake-controlled dresser is properly set, it will travel at approximately 1500 sfpm when driven by contact with the diamond wheel travelling at normal speed.

While many operators set the dressing wheel in motion by spinning it with a small wooden stick, it is preferable to use a knurled knob which has been added to the dresser spindle. This is a safety measure to prevent injury. The diamond wheel should then be brought into contact with the spinning dressing wheel. This is important—if it is not done, a flat will be ground on the dressing wheel before it starts to spin and damage to the diamond wheel could occur. The diamond wheel should be passed over the dressing wheel in line with the grinder spindle; that is, use cross feed only. All dressing is done without the use of a coolant.

Downfeeds from 0.0005 inch to 0.001 inch should be taken at each pass. After a total downfeed of approximately 0.005 inch, the diamond wheel should be stopped and examined for results. If necessary, the action should be repeated until the diamond wheel is dressed true and smooth.

When the diamond wheel is properly dressed, its color will be uniform over the entire face. The purpose of the dressing operation is to remove enough wheel bond to make the wheel run true and to expose sharp diamond particles which will give the desired cutting action. A sharp, true-running wheel will produce an excellent finish on the face of the punch and die.

Cleaning sticks, supplied by the wheel manufacturers, should frequently be used to keep the wheel "open." These cleaning sticks are generally about ½ inch x ½ inch x 6 inches. A Norton 37C500HV stick or its equivalent is the recommended type.

If a considerable number of diamond wheels are used, it will prove advantageous to develop a diamond wheel salvage program in cooperation with the diamond wheel or diamond grit supplier.

Inspection Prior To Grinding

When a carbide die is ready for sharpening, it should be cleaned of all oil, slugs, and other foreign material. It is necessary to remove stock guides and lifters from the die and push pins, pilots and other appurtenances from the punch as shown in Figure 4.

Fig. 3—Dressing of the wheel is done effectively with a brake type dressing unit.

The bottom of the die shoe and top of the punch holder should be inspected and all burrs removed to insure that the assembly will lay flat on the grinder's magnetic chuck.

To determine the amount to be ground from the punch and/or die when the die is being serviced, it is a good idea as shown in Figure 5 to examine the burrs on the last skeleton strip that was produced by the die in operation. Generally, when the strip and parts have uniform burr height throughout, a normal sharpening will suffice. If there is an exceptionally high burr in a localized area, it is advisable to check the cause. When checking the strip, it is best to examine the areas where holes, slots, and so on, are pierced rather than at succeeding stages because a sizeable burr will often be flattened by a spring stripper as the strip progresses through the die.

Generally, a chip or worn condition will be revealed on the punch or die, or both. If examination fails to show excessive wear or chipping, burr problems could be caused by excessive break clearances. Even if no uneven burr condition exists, it is advisable to examine the corner radii on punches and sections as shown in Figure 6 to see if a wear condition is developing. Unless an extreme wear condition exists, or the workpieces are adversely affected, it is not recommended that the wear area be ground off. However, after four or five normal sharpenings it is not unusual for punches with small corner radii, such as a cluster of rotor or stator punches, to be ground down equivalent to the amount of depth of entry into the die and material.

Usually, a 0.005- to 0.008-inch grind should be sufficient on both punch and die to restore the cutting edges to a like-new condition. Any area that is chipped or worn in excess of 0.008 inch should be raised by shimming, grinding off collars, removing step-a-head rings, or whatever, rather than grinding the entire punch or die the amount required to restore a chipped or localized worn area. If, however, the entire die is generally bad to depths greater than 0.008 inch, it is advisable to grind off the complete punch and die the required amount, rather than shimming. If there are special areas to be shimmed or raised, the work should be done and the components returned to the punch or die so that everything can be ground at once.

The stripper should be placed over the punches when sharpening the punch assembly. If it is a spring-type stripper, it should be placed on the remaining guide posts (only two guide posts should be removed) of the punch assembly and lowered over the punches to expose that portion of the punch to be ground off plus approximately 1/32 inch for a safety factor. A parallel or block of wood should be placed under the stripper to keep it at the desired height.

If a fixed-type stripper is part of the die, it should be placed over the punches with extreme care to the same distance below the punches as recommended with the spring-type stripper. Extreme caution must be exercised in removing a stripper when used for this purpose. A skeleton strip may also be used to support the punches during the grinding operation

Fig. 4—For inspection of the punch unit, guides, pins, and so on, are removed.

Fig. 5—*Examination of the last skeleton strip will tell much about the job ahead.*

Fig. 6—A close microscopic examination of all corners will show extent of wear.

but placing and removing the strip over slender punches is a hazardous operation and should only be used as a last resort for supporting the punches.

The above service methods apply to carbide dies used in straight piercing and blanking of flat parts. If the die is used to make parts with numerous forms, generally, the easiest and the best way to service the tool is to shim up all the cutting members, both punch and die, and grind them off to the existing height or level. In so doing, the form areas of the tool are not affected. This procedure does not in any way affect the shut height of the die, so the press setting does not need to be readjusted.

The determination of the amount of stock removal necessary for sharpening is usually established by the die service man or die grinder. After grinding, he will re-examine the die assembly and decide whether or not more stock removal is necessary. By focusing on the cutting edge, die-wear microscopes will measure die wear up to 0.020 inch to an accuracy of 0.001 inch. The examination is localized and it becomes quite tedious and time consuming to accurately determine the amount of wear throughout a die. However, in many shops this procedure has been used successfully by inexperienced people. Using the microscope is particularly helpful if it has been determined to leave the die slightly dull. If the die shows 0.008 inch wear, grinding off 0.006 inch will leave the die just dull enough to help retain the slugs in the die during the next production run. This type of sharpening philosophy is not considered the best, but it is used successfully in numerous stamping operations.

To sharpen a carbide die, it should be placed on a magnetic chuck with the grinder stops set so that the wheel will pass at least one inch beyond either end of the die. Cross index and table speeds should then be set and the wheel brought in contact with the work. There will be no sparks while grinding carbide with a diamond wheel.

Diamond wheel grinding is done more by sound than by sight. A light hissing sound indicates good wheel contact with the work. After the initial contact, the wheel should be raised approximately 0.002 inch, then returned to the cutting position by downfeeding 0.0002 inch per pass. For finish cuts, the downfeed should be decreased to 0.0001 inch, with the wheel permitted to make several free passes between each feed.

If the machine is stopped, the wheel-head should be raised at least 0.002 inch before re-establishing contact with the work. This will compensate for any change in table elevation caused by oil accumulation on the table bearings.

In order to establish the necessary depth of grind, some operators plunge the wheel through the area requiring the greatest stock removal to a point which will remove all signs of wear. They then remove the balance of the material conventionally until the original wheel path disappears from the die component.

If the total stock removal will exceed 0.010 inch, it is wise to remove as much of the shoe and die steel material as possible with an aluminum oxide wheel before using the diamond wheel. The grinding of steel with diamond wheels has a unique chemical reaction phenomenon which wears the diamond wheel away much quicker than when grinding carbide.

Fig. 7—After reconditioning, the entire die assembly unit should be demagnetized.

Therefore, excessive steel grinding should be avoided.

When the punch and die are sharpened, both members should be thoroughly cleaned. All sludge, grit, and other foreign material should be washed off all stripper guide bushings, guide posts, guide post bushings and grease cavities, as well as from all punch and die openings. The die and punch assemblies should be dried and demagnetized as illustrated in Figure 7.

All pilots, push pins, and other devices in the punch should be inserted and the stripper then replaced, making certain that all back-up screws are in place. The stripper may be drawn back when necessary by placing washers under the heads of the stripper bolts, provided the same total thickness of washers is under each bolt.

When drawing back the stripper, it is also important not to over-compress the stripper springs. Over-compression promotes early spring failure. Shorter springs can be inserted or washers removed from under the springs to give the correct spring pressure and deflection. When the stripper is drawn back, the pilots and push pins must also be reground to the correct length. Stripper-mounted pilots need not be reground unless they are worn undersize. Lifters and stock guides in the die are then replaced and grease, such as Lubriplate No. 130AA or equivalent, is then applied to all grease fittings.

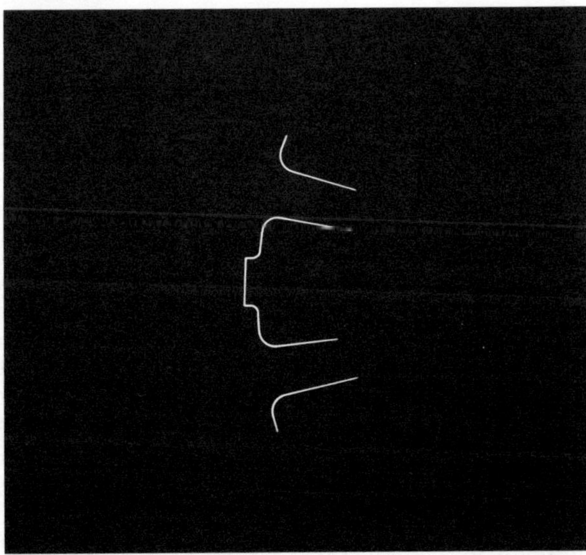

Fig. 8—The best test for final alignment of die components is the light check.

The alignment light check is the final procedure before returning the die to the press. Figure 8 shows what a serviceman sees when he checks alignment with a light. The punch should be entered in the die approximately 0.010 inch. Then, with a little experience, the die maker or serviceman can become familiar with the light view indicative of good alignment. If the light is perfectly uniform, the punch is centered in the die cavity. Other alignment check methods include cutting paper of various thicknesses or creating an impression on mylar or plastic material and then viewing it on a magnifying projector. These methods are acceptable in some cases, but the light check is usually more efficient, accurate and strongly recommended.

The pilots, push-off pins, stripper bolts, springs, and stock lifters from time to time will need attention when servicing a die. With a few exceptions, such as very thin material being punched or a blank push-button situation, it is good practice to draw the stripper back after the stripper over-hangs the punches 1/16 inch.

If the stripper is constructed with the normal-type socket shoulder screw, it is proper to put parallel uniform spacers (or collars) under the heads of the stripper bolts to draw the stripper a few thousandths. If the die is provided with a sleeve-type stripper bolt, the sleeve should be ground off the amount the stripper is to be drawn back. With this type of construction, there are no loose shims or collars that can be misplaced or fracture in die operation.

Generally, the pilots should extend above the stripper face one and a half times the material thickness. This rule of thumb is commonly used and is effective in most cases. There are other situations where the pilots cannot extend beyond the punch to this extent. In some dies the pilots may have to extend farther, particularly where there is forming or drawing to locate the strip prior to any other punch or die contact. The effective initial relationship should be maintained when the stripper is drawn back. In other words, if the stripper is drawn back a sixteenth, the pilots should be shortened by a sixteenth. However, if the die is equipped with stripper pilots that shoulder against the stripper, then little attention is required as the pilots will be withdrawn with the stripper.

Stripper springs should be of an adequate pressure to properly align the strip without pilot deflection unless the strip is totally misfed and the pilot has no hole in the stock for entry. Pilot springs will last for a long time if the press has a good feed and misfeeds are few and far between. If misfeeds happen quite often, then the pilot springs will lose their pressure and should be replaced periodically.

Push pins generally act as a counter-measure to slug pulling by breaking the seal between punch and slug. The push pins need not extend below the punches more than 1/32 to 1/16 inch. The extension initially established should be maintained when the die is serviced. It is not necessary to shorten push pins every time the die is ground, but after grinding 1/16 inch from the puches, the pins should be ground to the original overhang length. Push pins deflect their springs on every press stroke and the springs will have to be replaced periodically.

The stock lifters in the die will wear. If they are the type that are contained under the stock guide rails, they should be reground to establish their initial step height when they are worn to any amount. The lifters that shoulder against the die section, usually in the center of the die, should be shortened every time the die is serviced to maintain initial projection height above the die surface.

Before reinstallation in the press, the die and ram surfaces should be cleaned and stoned free of burrs and any projections. With rare exceptions, it is necessary to lubricate the stamping material before entering the die. A solenoid operated oiler that provides lubrication to the stock only when the press is running is a good type.

It is very important that the punch and die be adequately clamped into the press as shown in Figure 9. If at all possible, it is best to clamp the die and punch holder directly to the ram with screws extending through the die shoe into the press bed, or through the press ram into the punch holder. If this

Fig. 9—Adequate clamping is a must.

is not possible, then good clamping methods must be used so that the punch and die can be held securely and without danger of side pressures that would cause one or the other to move when the die assembly is in actual operation.

Slug Pulling

Slug pulling has been a major headache for all die users. So many ways have been employed to counteract slug pulling that it could be a special subject all its own. Push-off pins and air blasts are popular. One slug-retention method is a tight clearance between punch and die member. This is a short-sighted approach. It may work but the die will be pulled for servicing much more frequently than when proper cutting clearances are provided. Another effective slug-pulling prevention is a shear ground on the punch which creates a slug that becomes a piece of scrap. Circular punches can be crowned which creates a slug slightly larger than the die opening.

One of the most effective and simple ways to retain slugs is the diamond-paste treatment of cutting edges (refer to Figure 10.) This is not to be confused with dulling the cutting edge, which effectively prevents slug pulling but it also shortens the run, increases burr height, and decreases the overall die capabilities. The effective and accepted treatment is the application of diamond paste with a soft steel tool. The cutting edge can be changed so that slugs will be retained in the die and neither the burr height nor the eventual length of the die run will be adversely changed. The treatment is most effectively applied to round holes with a piece of diamond-paste-covered cold rolled steel or drill rod turned with an 18-degree included angle on the tip; something similar to a mechanical pencil point. The treatment very slightly enlarges the die cavity at the cutting edge which in

turn creates a larger slug. As the top edge of the die bushing does the cutting, the slightly larger slug will be retained when it is pushed down below the die face. There is danger of overdoing this process and it is recommended that the edges be given just a slight treatment and repeated if necessary, rather than initially going too far.

It is necessary from time to time to check the die components, such as guide posts and bushings, stripper guides, and so on, for wear (see Figure 11). With the conventional or solid type posts and bushings, bronze plated or otherwise, it is very easy to determine the clearance between guide posts and bushings simply by checking the ID of the bushings and the OD of the guide posts. When the clearance exceeds that between punch and die it is necessary to replace the worn posts and bushings. With the ball-type guide posts and bushings, it is a little more difficult to determine when the clearance becomes too large, but here is a method that works: Set the inverted die and punch on the die bench for an alignment check, and if the die can be pushed by hand to one side in excess of the clearance between punch and die per side, then it is necessary to replace the guide posts, bushings, and races. When replacing these parts, they should be of maximum length to provide maximum rigidity in operation. The ball-type guide posts and bushings should be lubricated with a light oil. Grease should never be used as it will retain dirt that will accelerate bushings, guide post, and ball wear.

A die cleaning cabinet, shown in Figure 12, is a handy, quick, effective way of cleaning the die after it has been in operation or on the grinder for die sharpening. An abundant flow of cleaning agent should be used with a brush to clean all areas of foreign material.

Fig. 10—A diamond paste treatment of the die section will prevent slug pulling.

Fig. 11—Guide posts and bushings are cleaned and checked.

Fig. 12—All components should be thoroughly cleaned, as part of the reconditioning process.

Die Storage

After a die has been in production or has been serviced, it is advisable to store it in a clean, uncluttered area that provides adequate die storage space and a method for ease of handling similar to that shown in Figure 13. It is advisable that the die be stored on a non-metallic surface. Never leave the die covered with oil or dirt for any length of time. There are numerous outside elements that cause carbide deterioration.

Certain lubricants will also affect carbide. Although it is a slow process, one that requires weeks, months, even years to cause damage, the deterioration will happen just as readily, if not more so, while the die is in storage as when it is in production. Therefore, always clean the die after completion of a production run, even when it does not require servicing prior to storage for future use.

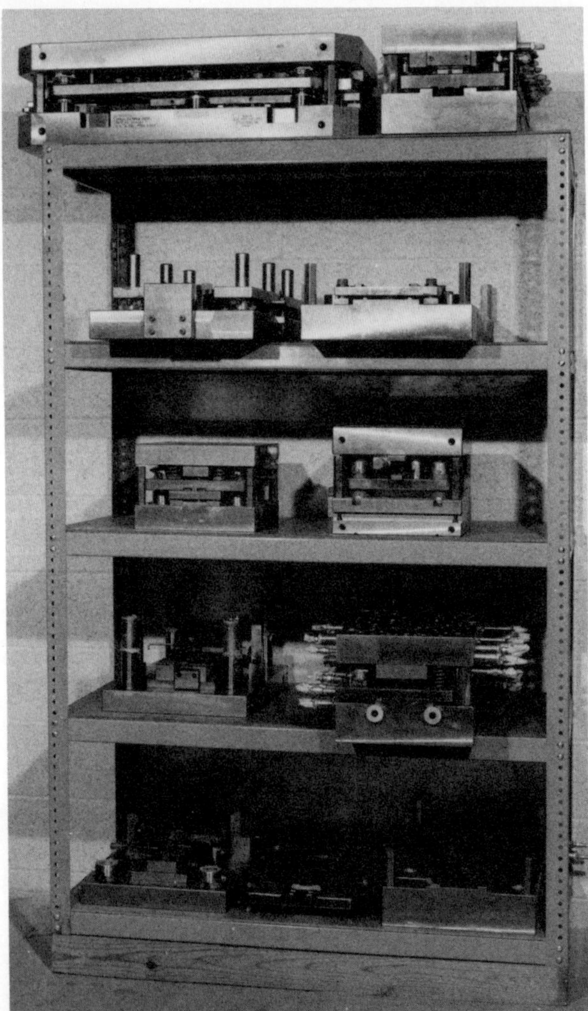

Fig. 13—When not in use, store carbide dies on wooden pallets in a clean area.

About the author

John C. Vecchi
President of J. V. Manufacturing Co., Inc.

John Vecchi has been working with carbide dies for 30 years. From 1949, Vecchi held various supervisory and management positions in the carbide die design, manufacturing and press working areas. In 1975 he formed the J. V. Manufacturing Co., Inc. A company dedicated to design and build the world's best carbide stamping dies with a customer rapport to insure maximum die performance.

224

Ironing of Cans—Tooling and Die Maintenance

By Cor Langewis, P.E.
President
Langewis Consulting & Engineering, Inc.

Ironing of a can wall can be described as an operation whereby the wall thickness of the shell is reduced and its surface smoothed.

In ironing, the key to success lies in the can material, the coolant and the shape and conditions of the ironing tooling. Discussed are tool construction, surface finishes required and forces generated during ironing. Recommendations for die construction and rework are presented.

INTRODUCTION

Most of the beverage cans produced in the U.S. are manufactured by the draw-and iron method. This method of seamless can making became commercial in the late fifties and has largely replaced the so-called three-piece can. It also opened the way for wide-spread self-manufacturing of cans by breweries and producers of soft drinks, etc.

The first steps in draw and iron (D&I) can making are blanking and cupping. In the subsequent ironing operation, the drawn cup is pushed through one or more ironing dies which have a smaller inside diameter than the outside diameter of the drawn cup. The wall thickness of the cup is thereby reduced (ironed out), and the height of the cup is increased. For beverage cans, it takes usually two or three ironing passes to elongate the wall to the desired length.

Forces generated during ironing are substantial and the tool finishes and die angle are of the upmost importance.

For a good process control, non-expanding dies are recommended. These dies can be manufactured with a controlled amount of interference fit between carbide insert and steel casing.

IRONING

A typical ironing operation is shown in Fig. 1, for 3004-H19 aluminum and a normal reduction, an average punch force of 4500 Lbs. has been measured. It should be obvious that the thin can wall cannot transmit this force from the bottom of the can to the ironing area.

For instance, a 211 size can being ironed in the final die might have a wall thickness of .005" and an inside diameter of 2.590". The cross-section of the can wall has an area of .0407 square inches. ($\pi/4$ x2.6^2-2.59^2)= .0407 inch^2.

At a maximum tensile strength of 42,000 PSI, the maximum force

the can wall can transmit is 42,000 x .0407 = 1709.4 Lbs--
which is much less than the measured ironing force of 4500 Lbs.
In other words, the can is not pulled through the ironing die
by the punch.

Fig. 2, shows an enlarged portion of a can wall being ironed.
Entrance angle of die = α_1 . The pressure P to deform the can
metal has a resultant, perpendicular to the punch surface, of
PC. This force times the coefficient of friction between the
can wall and the punch creates the friction force PF. The sum
of the tension in the can wall (WT) and this friction force PF
equals the punch force PP. The wall tension has a definite
limit - ·i.e. the ultimate strength of the can wall material -
and therefore the product of friction coefficient and the
force PC has to supply the largest part of the punch force PP,
especially when low strength metal, such as fully annealed
aluminum, is used. To decrease the wall tension WT, the product of force PC and the coefficient of friction between punch
and can has to be increased and this can be accomplished by:

   1.   Increase the coefficient of friction between can wall
        and punch;
   2.   Increase the force PC.

1.   FRICTION

For best ironing results, the punch should be kept dry and free
of oil or coolant. In practice this is, of course, impossible
or at least very difficult. Therefore, punches used for ironing aluminum are usually roughed up or "cross-hatched" to increase the friction between can and punch. Tin plate, with
its greater ultimate strength, can be ironed with rather
smooth punches.

2.   Increase Force PC

It can be seen that the resultant force PC depends on the semi-
cone angle of the die. If this angle (see Fig. 2) , is decreased to α_2 the resultant force PC becomes PC_1 and the
friction force becomes PF_1. The wall tension WT will become
zero if the semi cone angle of the die is reduced to a certain
value and the product of the resultant force PC and the coefficient of friction equals the punch force PP.

At first glance, this seems to be a very desirable condition--
if the tension in the can wall becomes zero, then the can wall
should never rupture during ironing. However, for best results
and even metal flow, some wall tension is desirable.

Without wall tension, only the friction between can wall and
punch "drags" the metal through the ironing die. Uneven surface finish of the punch and a spotty distribution of coolant
between can wall and punch will create an uneven metal flow,

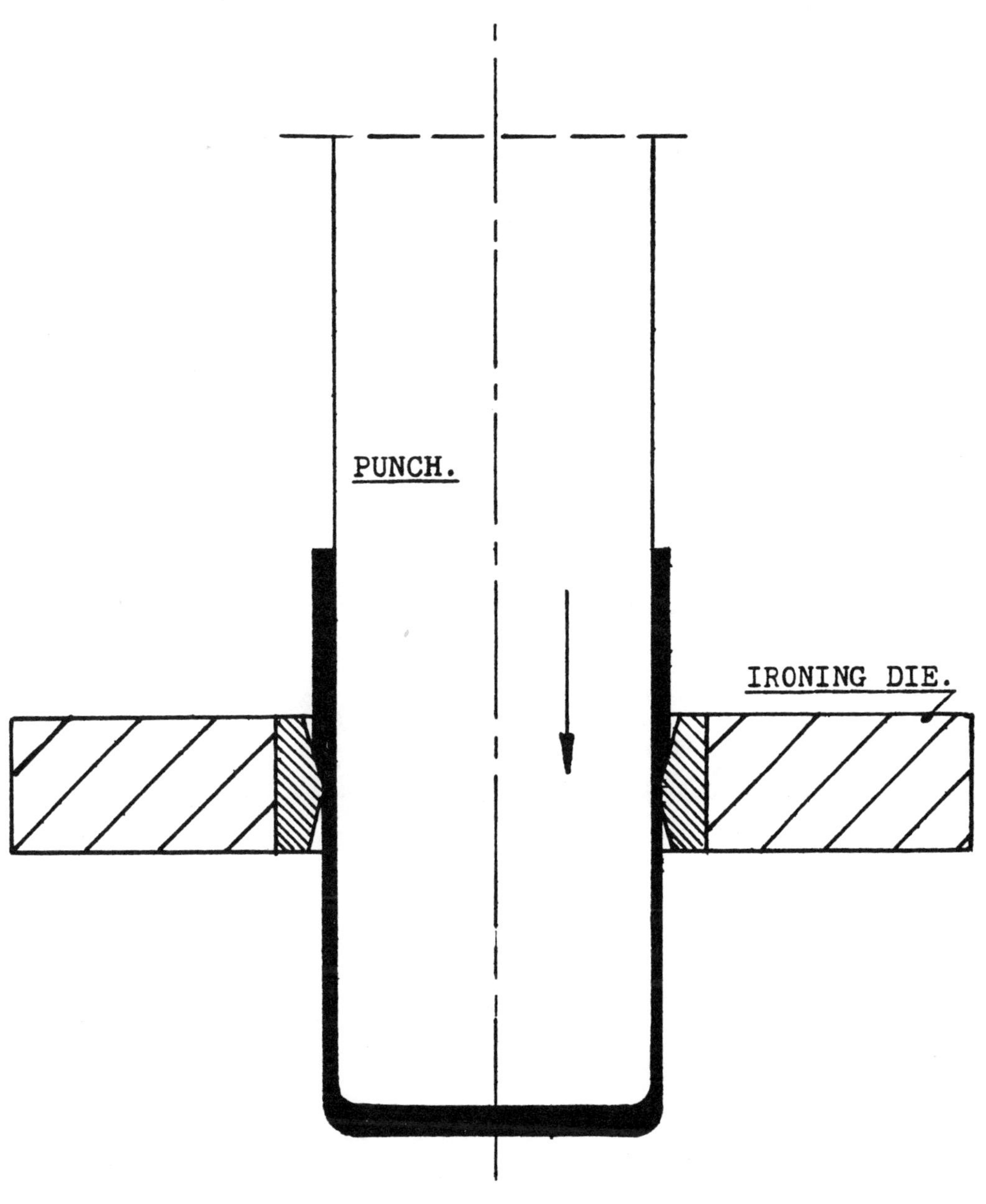

IRONING.

FIGURE 1.

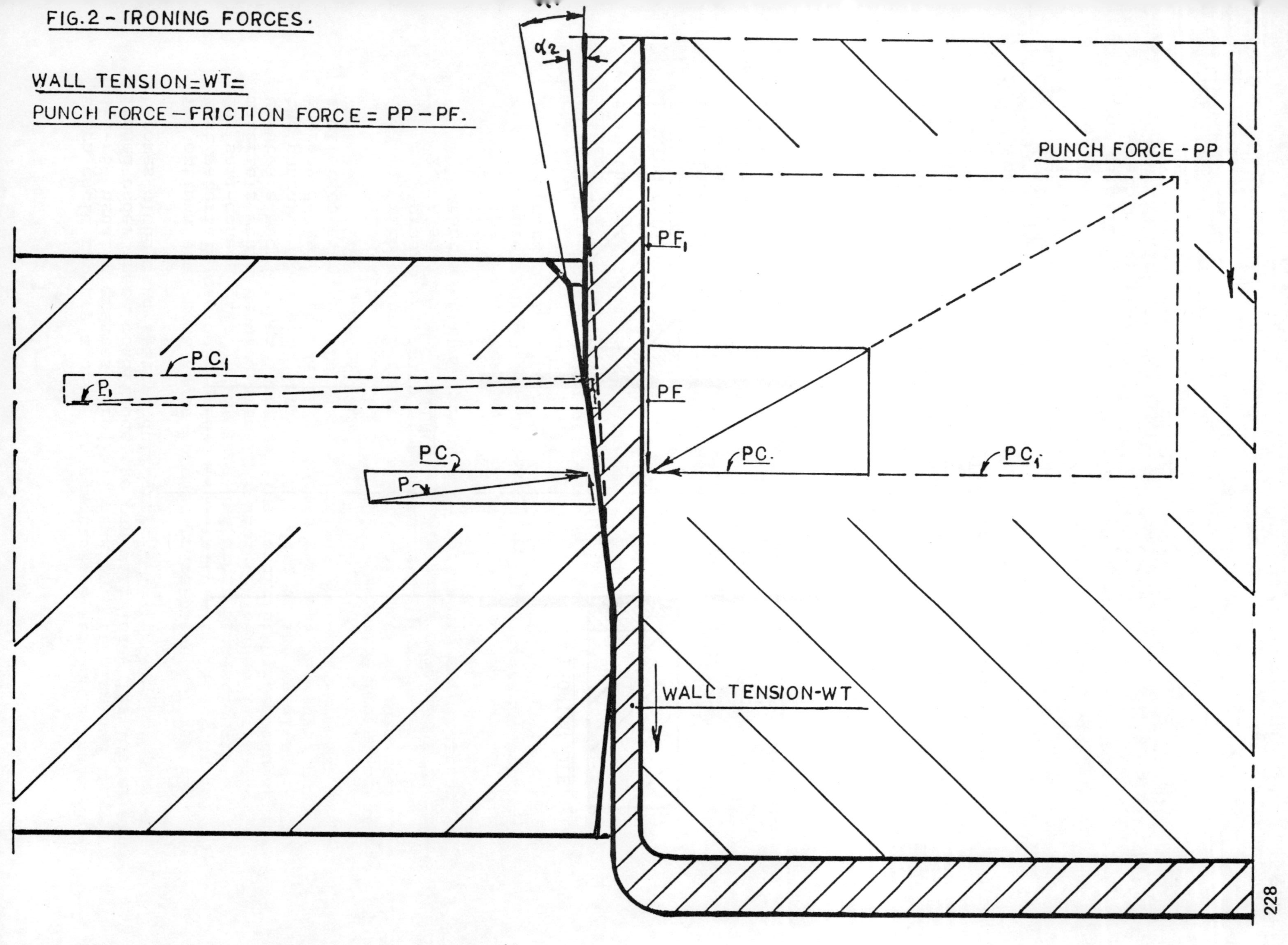

FIG.2 - IRONING FORCES.
WALL TENSION=WT=
PUNCH FORCE - FRICTION FORCE = PP - PF.
α_2
PUNCH FORCE - PP
PF_1
PC_1
P_1
PF
PC.
PC_1
PC_2
P_2
WALL TENSION-WT
228

which could rupture a can. In practice, a semi cone angle of
$4\frac{1}{2}$ - 5° usually creates zero wall tension. If aluminum is
ironed under these conditions the can wall loses its luster and
becomes cloudy and spotty, dull grey.

The optimum semi-cone angle appears to be between 7 and 8°.
Most materials can be ironed successfully with larger or smaller semi-cone angles. Tin plate and other high-strength
materials would require a smaller angle. The smaller angle
creates a much higher force PC (the resultant perpendicular to
the punch). This will increase the drag on the metal and
thereby decrease the wall tension in the can. However, the
smaller cone angle offers several disadvantages, such as;

1. The larger resultant PC_1 will press the ironed metal
    with greater force against the punch and this could
    create stripping problems.
2. A smaller semi cone die angle means more metal in con-
    tact with the ironing die--this means more heat and
    friction during ironing.
3. The larger resultant PC_1 will expand the ironing die
    more and close control of can wall thickness becomes
    more difficult.

In practice, more ironing dies are made with semi-cone angles
between 7 & 8° and the surface of the punch is finished to obtain the best results with the materials to be ironed.

<u>IRONING DIE CONSTRUCTION</u>

For close control of tooling and can wall thickness ironing
dies that do not expand during ironing are desirable. Carbide
inserts are shrunk in steel cases and if the interference fit
between case and carbide is large enough, the resulting shrink
pressure will exceed the maximum radial pressure generated by
the ironing process--consequently, the die will not expand
during ironing.

For instance, a die for a 211 size can might have a carbide insert of 3.2" diameter and a case of 6" diameter x 3/4" thick.
Cases are usually heat-treated to a hardness of $45R_c$ so they
can withstand rough handling without being nicked. Consequently, when the case has to be heated for the insertion of
the carbide insert, the case should not be made so hot that it
is annealed in the process. Assume therefore, a maximum case
temperature of 550^OF., then the 3.2" diameter hole will expand
.0093" - .010". For easy handling, the interference fit should
be about .007".

This interference fit will create a radial shrink pressure on
the carbide insert of <u>16,247 PSI.</u> The compressive stress in
the carbide will be <u>96,481 PSI</u>, and the maximum tensile stress
in the die case (at the steel/carbide interface) <u>29,164 PSI.</u>

These values are well within allowable limits.

If 3004-H19 aluminum is ironed and zero wall tension is assumed, then the radial expanding pressure on the die due to the resultant force PC (see Fig. 2) will be 14,242 PSI for a $7\frac{1}{2}^{\text{O}}$ semi cone angle.

The radial shrink pressure of 16,247 PSI exceeds this pressure and therefore the die will not expand during ironing. Also, if the hole in the die case and the outer surface of the carbide insert are ground with a 32 RMS finish, the ironing force PP will not push the carbide out and a straight insert without a shoulder can be used. (See Fig. 3A).

If a short carbide insert is to be used, it should preferably be located in the center of the die to avoid deformation of the casing. (See Fig. 3C).

For ironing tin plate, the above amount of shrink might be a borderline case and the die might expand somewhat during ironing. A thicker or a larger diameter case, or a larger semi cone angle would be required to make a non-expanding die for tin plate ironing.

DIE MAINTENANCE

During ironing, the metal tends to "bulge up", and it erodes the carbide surface of the cone angle. (See Fig. 4). If the land is not damaged due to can ruptures or hitting by the punch, rework of the semi cone angle might be the only work required. If the width of the land is reduced to the minimum set dimension, then the land should be ground as well. A back angle of 30 minutes on the land was found to be beneficial.

The width of the land should be kept between .015" and .030". Larger widths than .035" are not recommended.

Rework of the exit angle is very seldom, if ever, required. The corner between semi cone angle and ironing land should be kept sharp and hand polishing with diamond compound is definitely not a good practice! Use a 1000 grit diamond wheel for grinding only and never use emery cloth for clean-up.

Some water soluble coolants etch the cobalt binder out of carbide and dies should not be left wet and overnight in the grinder.

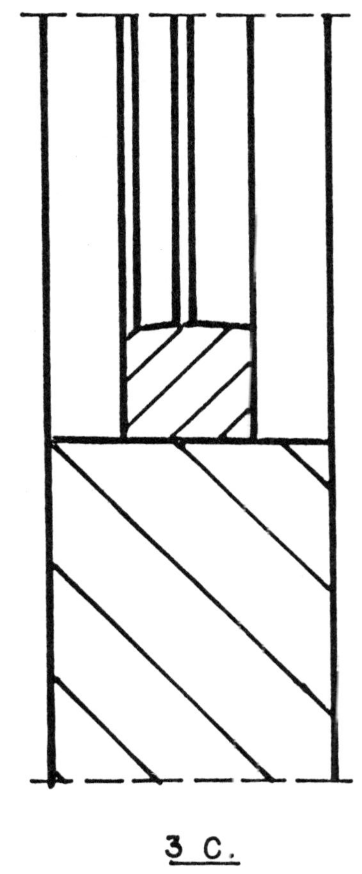

IRONING DIE DETAILS.

FIGURE #3.

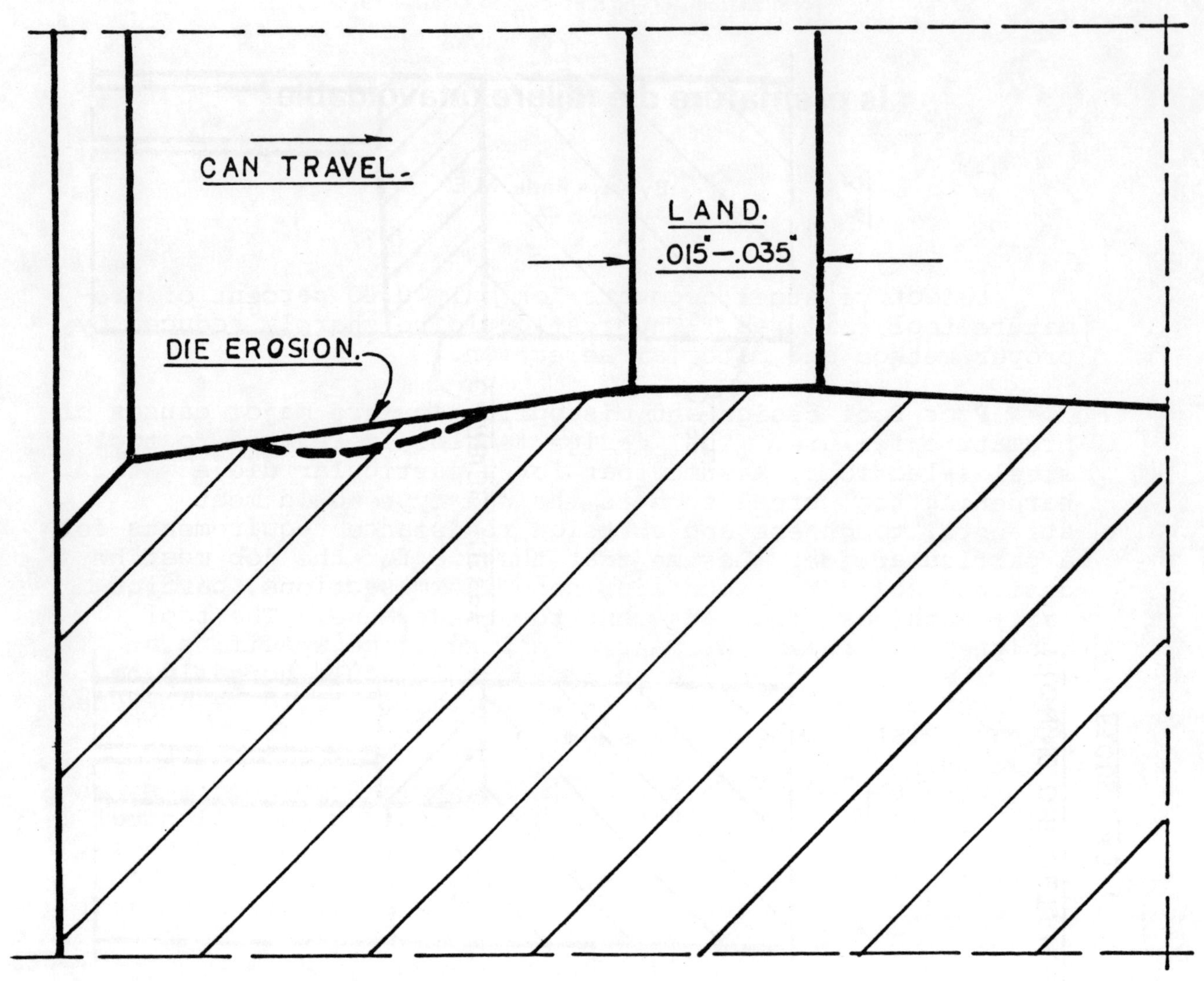

IRONING DIE DETAIL.

FIGURE 4.

Reprinted from Tooling & Production, October 1971

Is premature die failure unavoidable?

By Kaye Karle, M.E.
San Diego, CA

Defective steel accounts for only 0.20 percent of premature tool failures. The rest could be sharply reduced by proper method and material selection.

Poor tool design and misapplication are major causes of premature failure. Tool design is closely related to tool steel selection. Assume that for a particular die a water hardening tool steel such as the W-1 type would meet strength, toughness and abrasion resistance requirements for a particular job. Assume that the die for the job must be designed so that it contains nonumiform sections, particularly a thin section adjacent to a heavy one. The tool designer is risking premature failure if he specifies a water hardening steel, even though this would normally be his choice. It is improbable that the die could be hardened without distortion or cracking.

When the die must have thin or nonuniform sections, an oil hardening, or even better, an air hardening tool steel is best. Most tool designers consider strength, toughness, wear and abrasion resistance, machinability and size change in specifying die steel. Hardenability should receive just as much consideration.

Faulty Tool Designs

Premature failure is often due to faulty design. Cracked or broken dies can often be traced to sharp-cornered keyways. This type of breakage is usually blamed on fatigue failure caused by stress concentration at the corners, but it could be eliminated by specifying half-round keyways.

Tools, such as pneumatic tools that are subject to severe impacts, should always be designed with tapered shanks. Many such tools are designed with heavy fillets at the changes in section. These tools most often fail at that point. Tools with tapered shanks seldom break because of fatigue.

Heat treatment

Faulty heat treatment is probably responsible for a large percentage of premature tool steel failure. This is not always the fault of the heat treater. Often he is asked to treat a tool that would crack regardless of the care he took.

Almost all tool steel is supplied in the annealed condition. Faulty annealing by the producer does not usually create problems in normal hardening operations. An exception is improper annealing of a carbon or manganese oil hardening steel. This causes formation of a carbide network that cannot be removed by normal hardening processes. A tool made from this steel is sensitive to cracking in quenching and grinding. It will be brittle in service and may fail prematurely. These tools can be salvaged by normalizing from 1650 F to 1700 F, followed by a full anneal and rehardening.

Hot tooled or forged tool steels have a decarburized surface which should be removed prior to hardening; otherwise, it can cause cracking during hardening. If it does not cause cracking at this point, the tool will give poor performance because of inadequate surface hardness.

Tools that have been highly stressed by heavy machining or cold working should be stress relieved before hardening. Residual stresses from such operations, added to thermal stresses produced in hardening, may crack or distort the tool, or cause it to fail in service.

Premature failure can also be avoided by preheating tools made from materials quenched at temperatures above 1600 F. Preheating reduces the amount of time at which the material is kept at high temperature. This minimizes scaling and decarburization and makes it easier to heat uniformly.

The most critical operation in heat treatment is austenitizing and quenching. Many tool steel failures are caused by mistakes made during this part of the heat treatment cycle.

The most common causes of failure are overheating or underheating, removing from quench too soon, quenching too cold and not tempering immediately, using the wrong quenching medium, nonuniform quenching or insufficient volume of liquid quenching medium, rehardening without an intermediate anneal. Selection of the correct austenitizing and quenching method will eliminate many early failures.

A great deal of premature tool and die failure can be

traced to improper tempering. There is a correct time at
which to temper the tool after quenching. In almost all cases, this is as soon as the piece has cooled to the point
where it can be held in the bare hand. As-quenched tools may
contain extreme residual stresses. One of the main purposes
of tempering is to remove these stresses before they produce a
crack.

It is imperative that certain types of tool steels such
as the air hardening die steels, the hot-work steels and the
high-speed steels be double tempered. These steels are sluggish and retain austenite in the hardening operation. The
first temper will condition this austenite so that it will
transform to martensite when cooling. The second temper is to
temper the martensite.

CHAPTER 6

PRESSES

Don't Overlook Energy In Press Selection

Tonnage capacity is important, but energy requirements can also be critical in press selection and loading

CHARLES H. WICK
Machine Tool Editor

WHILE MECHANICAL PRESSES are rated for tonnage and energy, the tonnage capacity is too often the major or only consideration in their selection and use. Tonnage ratings are generally the most significant factor and can be the most destructive element, but energy capacity is also critical, particularly for continuous running and deep drawing applications. Unfortunately, the subject of energy is frequently misunderstood, neglected or ignored, and as a result, presses are often misused.

Tonnage Considerations. The rated tonnage capacity of a press is the maximum force (in tons) that should be exerted by the slide or ram-mounted die against a workpiece at a given distance above bottom of stroke. It is a relative measure of the torque capacity of the drive, and an actual measure of the structural capacity of the press frame. Industry standards for rated tonnage capacities are given in the accompanying table.

Drives of mechanical presses can provide a constant torque, but due to the mechanical advantage of the linkage, the forces transmitted through their clutches to rotating members (cranks or toggles) and reciprocating slides vary from a minimum at midstroke to infinity at the bottom of the stroke. Presses rated higher in the stroke require greater torque capacity in the drive members, and more flywheel energy. Tonnage curves available from press builders show the change in tonnage that the drive is capable of delivering at various distances above stroke bottom. These should be studied for each different application, especially when initial contact with the workpiece is high in the stroke, as in the case of a deep drawing operation.

Energy Needs. In addition to knowing that a press and its drive can provide adequate force (tons) at a certain distance above stroke bottom, it is equally important to determine whether the press has sufficient work (energy) capacity. This is the ability to deliver enough force through the distance required to make a particular part. The product of force times distance (working stroke) is the energy load on the press, usually expressed in inch-tons. As energy is expended in forming a part in the die, the flywheel slows down. If the amount of energy used is within the design limitations of the press drive, the motor will return the energy used to the flywheel, bringing it back to its starting speed before the press ram reaches the top of its stroke and starts the next stroke.

If not enough energy is available to form a specific part on the press, slowdown of the flywheel during the working portion of the press stroke becomes excessive. The results could be loss of press speed on each successive stroke, belt slippage and wear, and/or overloading of the main drive motor. If flywheel slowdown is severe enough due to the use of most or all the energy available during a stroke, motor overheating can cause thermal overloads in the motor starter to break the circuit and stop the press. Even more critical is the possibility of the press sticking at the bottom of the stroke. If this happens, considerable work is required to get the jam released.

Energy Available. A general rule-of-thumb is that a standard press will not have more work capacity than its tonnage rating times the distance from the bottom of stroke at which it is rated. For example, a 100-ton press rated at 1/4 inch from stroke bottom would usually be capable of providing only 25 inch-tons of energy to do work on a continuous basis, and maybe slightly less. For presses using multistation dies, the energy requirements at all stations must be added.

Kinetic energy stored in the rotating flywheel of the press is the major source of energy to perform work. The amount available is determined by the size and weight of the flywheel, and the speed at which it rotates. Total energy in a flywheel can be calculated from the following equation:

$$E_T = \frac{WR^2N^2}{140,766,740} \ \ldots\ldots (1)$$

where E_T = total energy (inch-tons), W = flywheel weight (pounds), N = flywheel speed (rpm), and R = radius of gyration (inches). Available energy for various slowdown percentage can be calculated from the formula:

$$E_A = E_T \times F_{SD} \cdots\cdots (2)$$

where E_A = available energy (inch-tons), and F_{SD} = slowdown factor, which is 0.36 for a 20% slowdown (slow, single-stroke presses), 0.28 for 15% slowdown, 0.19 for 10% slowdown, and 0.10 for 5% slowdown (high-speed presses).

If it is assumed that the tonnage is applied uniformly from the point of contact with the work, the tonnage rating T_R is limited by the available flywheel energy E_A, and can be determined from the equation:

$$T_R = \frac{E_A}{H} \cdots\cdots (3)$$

where E_A = available energy (inch-tons), determined from equation (2), and H = the distance above bottom-of-stroke where work is contacted (inches).

Since the flywheel energy increases as the square of its speed, fast rotation is desirable when large amounts of energy are required. However, flywheel rim speed is limited to about 6000 sfpm for cast iron, and 9000 sfpm for steel flywheels.

Energy curves vary with each different press, their shapes and values depending on flywheel weight and speed, type of drive, percentage slowdown allowable, press speed and stroke, and other factors. However, characteristic energy and tonnage curves are shown in the accompanying graph for a geared OBI (open-back inclinable press) operating at 30 spm and having a 20% allowable slowdown. Coordinates for this graph are given in percentages to divorce it from any specific press.

Assuming the press is rated at 100 tons and has a 10-inch stroke, tonnage is the controlling factor when the work is done between 0.35 inch above the bottom of the stroke (point A on the graph) and stroke bottom. However, when the work is engaged more than 0.35 inch above stroke bottom, energy becomes the limiting factor to prevent flywheel slowdown from exceeding the 20% allowable. Also, if the work is engaged 1 inch above the bottom of the stroke, the tonnage requirement should not exceed 36.5 tons (point B) to prevent excessive slowdown, even though the power train would, in fact, be capable of delivering 55 tons (point C).

Types of Drives. It is essential for users to inform press builders of the specific applications intended so that not only the proper size press can be supplied, but also one having the

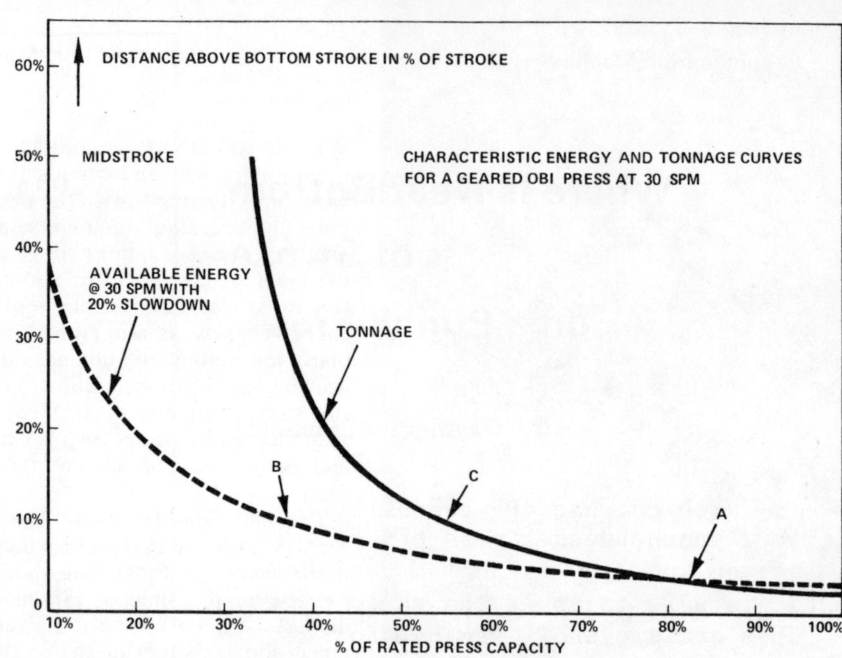

speed and energy characteristics needed for the job. Since blanking, piercing, and similar operations only require the application of force over a short distance, less energy is usually needed. Frequency of load application, however, is an important consideration, since higher press speeds reduce the time allowed for the flywheel to recover its speed. As a press speed approaches the area of 250 to 300 spm, the energy may be absorbed so rapidly that the flywheel is of limited value as an energy storage device. Also, if the press drive supplies power for automatic feeding as well as stamping, the energy drain may be high enough to keep the flywheel from recovering. In cases such as these, the press drive motor must be large enough to supply most of the energy requirements directly to the flywheel.

Deep drawing and similar operations require the force to be exerted over a considerable distance, and more energy is needed. Slower draw presses (usually geared) often have sufficient time between strokes for the fly-wheel to regain its speed, and relatively smaller motors can be used. In some applications, however, deep draws can cause flywheel slowdowns of 20% or more, and slip-type motors are required. The longer the working stroke required, the lower the pressure or tonnage must be at the start of the operation to prevent excessive flywheel slowdown or stalling.

Energy requirements are a major factor in selecting the type of drive for a mechanical press. The major types most commonly used are:

► **Flywheel-type (plain or nongeared) drives** have a flywheel mounted directly on the main shaft, and as a result they both rotate at the same speed. Such drives are used primarily for light blanking and piercing operations where energy requirements are comparatively small, and the press operates at relatively high speeds — from about 90 spm on large presses to 1000 spm or more on smaller ones.

► **Single-geared type drives** have the flywheel mounted on a separate driveshaft, with a single gear reduction used to transmit power to the main shaft. Presses with this type drive are generally used for shallow draw-work, forming and similar operations requiring more energy than available on flywheel-type presses, and operating at speeds of about 30 to 90 spm.

► **Double-geared type drives** have the flywheel mounted on a separate driveshaft, with a double gear reduction for transmitting power to the main shaft. Presses with this type drive are often used for deeper draw operations when more energy is needed, and the maximum allowable draw speed of the material to be formed limits the speed of the press slide. Presses with this type drive generally operate at 10 to 30 spm.

■

Industry Standards for Rated Tonnage Capacities at Specified Distances Above Bottom of Stroke

OBI Presses

Rated Capacity (tons)	Distance Above Bottom of Stroke (inch)	
	Nongeared	*Geared*
35 and Smaller	1/32	1/8
45 and Larger	1/16	1/4

Straight-sided Presses

Any Tonnage	*Nongeared Single-end Drive*	*Geared Single-end Drive*	*Geared Twin-end Drive*
	1/16	1/4	1/2

Reprinted from Machinery, May 1969

Where Is Near Bottom
of Stroke
on a Punch Press?

By Raymond A. Freeman

You can use the capabilities of your punch presses more effectively if you know how far up on the stroke it is safe to apply the rated tonnage. This article shows you how to find the answer.

Most press manufacturers rate the tonnage of their presses "near bottom of stroke." The question often arises: "Where is near bottom of stroke for a particular press?" Or, stated another way: "How far up on the stroke is it safe to apply the rated tonnage?"

To answer the question, we must analyze the geometry of the press mechanism and the forces involved. Such an analysis shows that "near bottom of stroke", which we shall call h for convenience, is a function of the stroke of the press, S; the diameter of the shaft at the journals, D; and the ratio of the length of the Pitman connection to the throw of the crankshaft, k.

The relationship is not a simple one and is best depicted graphically, as will be seen by the analysis that follows.

For any given press, however, the values of S, D and k are known, and h can readily be determined by reference to one of the graphs appearing with this article.

*GEOMETRIC RELATIONSHIPS between the essential connections in a conventional press are shown at right, top drawing. When the stroke of the press, S, the diameter of the shaft at the journals, D, and the ratio of the length of the Pitman connection to the throw of the crankshaft, k, are known, it is possible to calculate how far up on the stroke it is safe to apply the rated tonnage. The bottom illustration is a force diagram for this system.

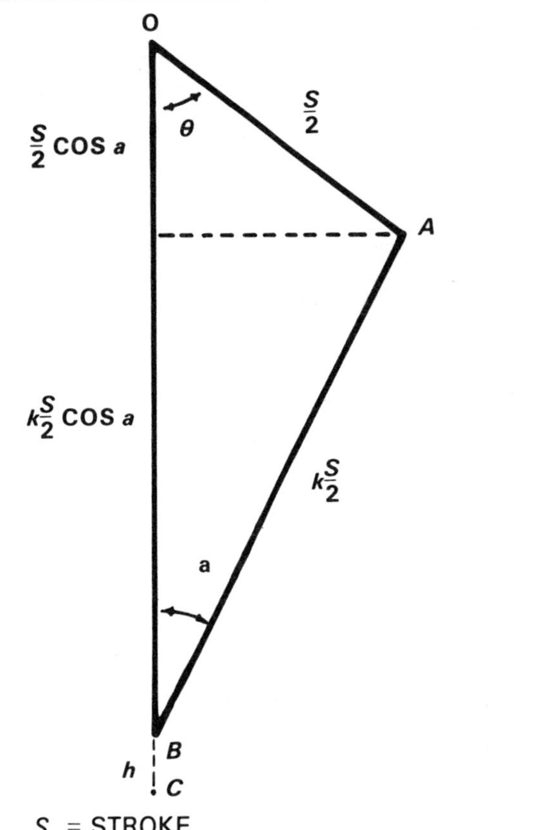

S = STROKE

OA = THROW OF CRANKSHAFT = $S/2$

AB = LENGTH OF PITMAN CONNECTION BETWEEN CRANKSHAFT AND **SLIDE** = $kS/2$

h = DISTANCE UP FROM BOTTOM OF STROKE = BC

k = RATIO OF LENGTH OF PITMAN TO THROW OF CRANKSHAFT = AB/OA

P = PRESS LOAD

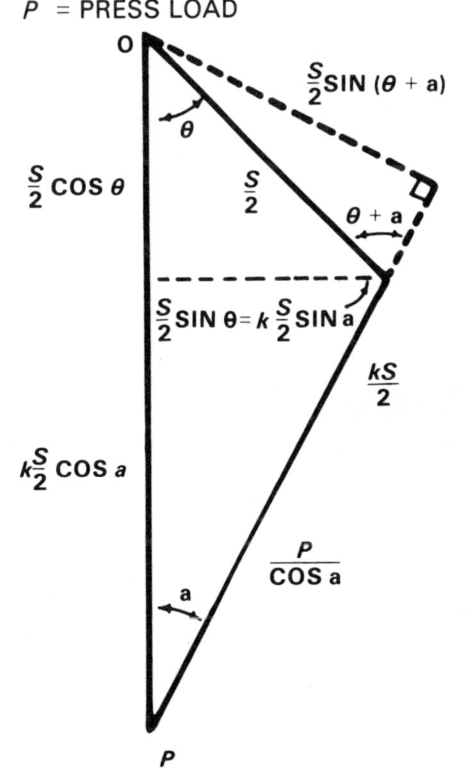

The geometric relationships between the essential connections on a conventional press are shown in the top drawing in Figure 1.

A force diagram for this system is shown at the bottom of Figure 1, where P is the press load:

Let the torque about point O = T; then

$$T = \frac{P}{\cos \alpha} \times \frac{S}{2} \sin (\theta + \alpha)$$

$$= \frac{PS}{2} \times \left[\frac{\sin \theta \cos \alpha + \cos \theta \sin \alpha}{\cos \alpha} \right]$$

$$= \frac{PS}{2} \left[\sin \theta + \frac{\cos \theta \sin \alpha}{\cos \alpha} \right]$$

Since $\dfrac{S}{2} \sin \theta = k \dfrac{S}{2} \sin \alpha$

$$\sin \theta = k \sin \alpha, \quad \sin \alpha = \frac{\sin \theta}{k}$$

$$\cos \alpha = \sqrt{1 - \frac{\sin^2 \theta}{k^2}} = \sqrt{\frac{k^2 - \sin^2 \theta}{k^2}}$$

Then:

$$\frac{\sin \alpha}{\cos \alpha} = \frac{\dfrac{\sin \theta}{k}}{\sqrt{\dfrac{k^2 - \sin^2 \theta}{k}}} = \frac{\sin \theta}{\sqrt{k^2 - \sin^2 \theta}}$$

$$T = \frac{PS}{2} \left[\sin \theta + \cos \theta \frac{\sin \theta}{\sqrt{k^2 - \sin^2 \theta}} \right] \ldots 1$$

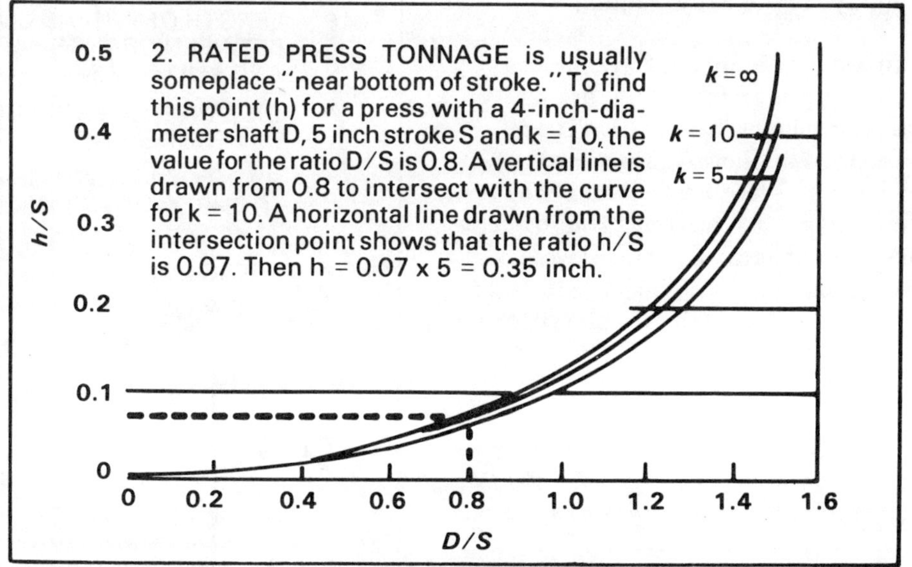

AUTHOR. Raymond A. Freeman is a vice-president of the V & O Press Co., Inc., Hudson, N.Y. He is a graduate of the Massachusetts Institute of Technology and has been a specialist in press design and application for V & O for 30 years. Hundreds of readers asked for additional copies of his article, "Rating Single-Crank OBI Presses," *Machinery*, January 1966.

We are concerned primarily with the condition where P equals the rated tonnage of the press. This rating is almost universally considered to be equal to 3 1/2 times the square of the diameter of the crankshaft at the journals = $3.5D^2$. Substituting this expression for P in Equation 1, the applied torque on the shaft at the rated press tonnage is:

$$T = \frac{3.5D^2 S}{2}\left[\sin\theta + \frac{\cos\theta\,\sin\theta}{k^2 - \sin^2\theta}\right] \quad ....2$$

If we consider the allowable shear stress on the shaft to be 12,000 psi, the allowable torque will then be:

$$T = 1.175D^3 \quad ....3$$

Then if we equate the allowable torque (Equation 3) to the applied torque at the rated capacity (Equation 2), we have:

$$1.175D^3 = 3.5D^2 \times \frac{S}{2}\left[\sin\theta + \frac{\cos\theta\,\sin\theta}{\sqrt{k^2 - \sin^2\theta}}\right]$$

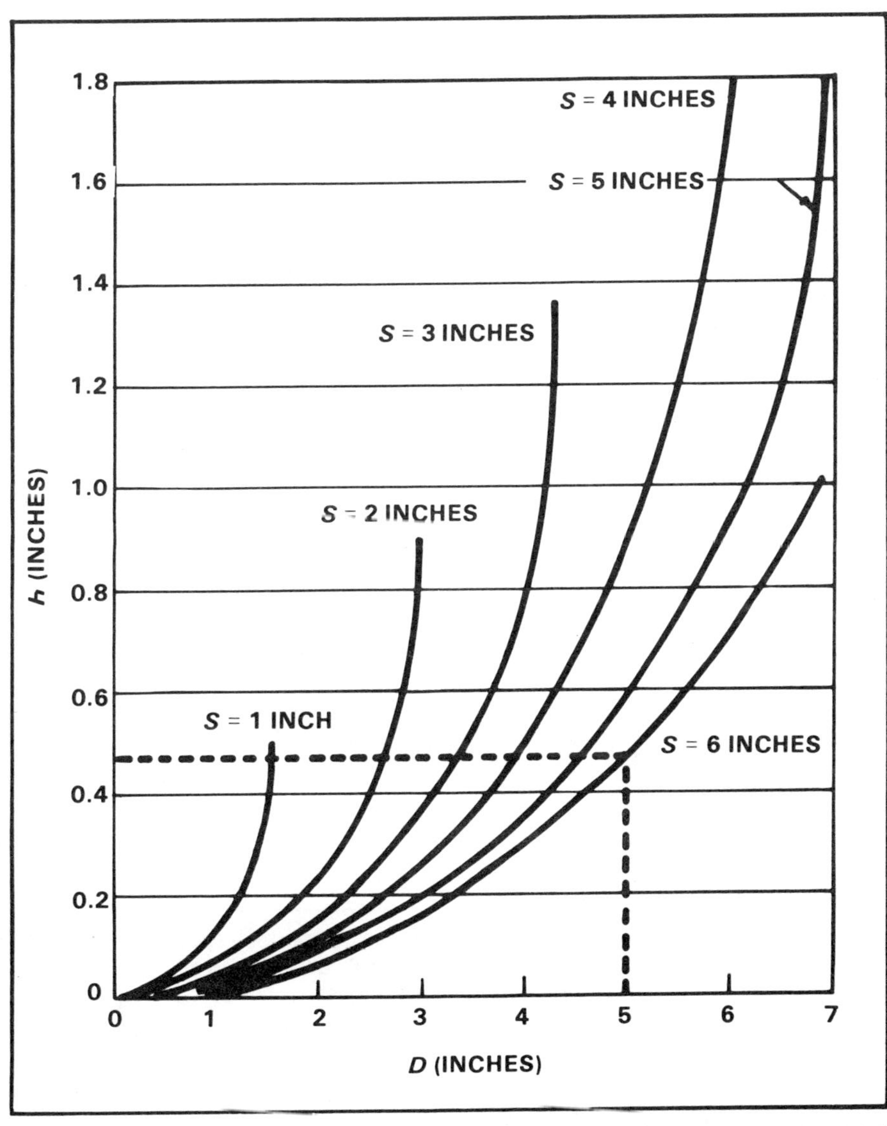

3. DISTANCE FROM BOTTOM OF STROKE where rated press tonnage is exerted can be easily found from these curves when press stroke S and diameter of shaft at journals D are known. In the example shown here, D was 5 inches and S was 6 inches. The value of h is around 0.45 inch. The curves in Figure 2 are more widely applicable.

From which we get:

$$\frac{D}{S} = 1.5 \sin\theta \left[1 + \frac{\cos\theta}{\sqrt{k^2 - \sin^2\theta}} \right] \quad \ldots\ldots 4$$

From Figure 1:

$$h = \frac{S}{2} + k\frac{S}{2} - \left[\frac{S}{2}\cos\theta + k\frac{S}{2}\cos\alpha \right]$$

And: $\quad \cos\alpha = \sqrt{1 - \dfrac{\sin^2\theta}{k^2}}$

Then:

$$h = \frac{S}{2}\left[1 + k - \cos\theta + k\sqrt{1 - \frac{\sin^2\theta}{k^2}} \right]$$

Which can be reduced to:

$$\frac{h}{S} = \frac{1 + k - \cos\theta - \sqrt{k^2 - \sin^2\theta}}{2}$$

Since k^2 is always many times larger than $\sin^2\theta$, the expressions

$$\sqrt{k^2 - \sin^2\theta} \quad \text{and} \quad k - \frac{\sin^2\theta}{2k}$$

may be considered equal. Then:

$$\frac{h}{S} = \frac{1 + k - \cos\theta - k + \dfrac{\sin^2\theta}{2k}}{2}$$

$$\frac{h}{S} = \frac{1 - \cos\theta}{2} + \frac{\sin^2\theta}{4k} \quad \ldots\ldots\ldots\ldots 5$$

By calculating both D/S from Equation 4, and h/S from Equation 5, for values of θ, the relationship between the ratios h/S and D/S for different values of k may be shown graphically, as in Figure 2. This graph may be used to determine the value of h as follows:

For a particular press with 4 inch diameter shaft, 5 inch stroke and k = 10, D/S = 0.80; from the curve for k = 10, the value of h/s is found to be 0.07. Then h = 0.07 (S) = 0.07 (5 inches) = 0.35 inch. For that particular press, then, the "near bottom of stroke" at rated press capacity occurs 0.35 inch above the bottom of the stroke.

The curves in Figure 2 present the relationship in terms of ratios, and may be used with any system of measurement, English or metric. From these curves, we may also construct a series of more useful curves for different strokes and values of k; from such curves the value of h may be obtained directly from known values of D and S. Such a set of curves for k = 10 is shown in Figure 2.

Either set of curves permits evaluation of the location of "near bottom of stroke." Those of Figure 3 simply permit a direct reading without the need for calculation but by the same token must be plotted for each different value of k and for specific values of S. The curves of Figure 2 are universal, and interpolation permits their use for any value of k.

It should be noted that the preceding calculations were based on the limits imposed by the static torsional strength of the crankshaft. Dynamic forces have not been considered in this analysis.

AUTHOR. Raymond A. Freeman is a vice-president of the V & O Press Co., Inc., Hudson, N.Y. He is a graduate of the Massachusetts Institute of Technology and has been a specialist in press design and application for V & O for 30 years. Hundreds of readers asked for additional copies of his article, "Rating Single-Crank OBI Presses", _Machinery_, January 1966.

Reprinted from Tooling & Production, February 1974

Proper Die Setup

by Nolan W. Rhea
Assistant Editor

William J. DeGain, president of Koppy Corp., offers some good advice on how to set up a prog die machine once you have your part and die set designed the way you want them. Follow this check list.

The first thing to do is make sure you have adequate press tonnage. With prog dies, 50 percent excess tonnage is not unreasonable. It reduces the number of parts you will have to reject.

Second, make sure your coil is in line with the press. If you stretch some 0.020"-dia piano wire between the two, you will see how they line up. To double check for parallelism, draw a wire across corners between the straightener and the feed.

Third, get the die in line. No one ever goes to school to learn a trade called "die setting," so watch your men closely. A badly set die will wear out much faster than one which has been set properly.

Fourth, occupy at least 70 percent of the table area, and make sure that the die area does not exceed the ram area.

Fifth, have as balanced a load as possible.

Sixth, use your stop blocks to check "shut" position. (Do not call them "crash" blocks. If they crash, you've done something wrong.) Put a tiny ball of clay or putty in a fold

of cellophane on top of each block. Closing the press will
cause no stress and the distance between ram and block can be
easily measured. Don't use paper matches; they will embed in
the blocks.

Seventh, lubricate the strip on both sides. Too little
lubrication will cause the strip to heat up. Too much will
cause sumps to form in the depressions of the lower die area.
These sumps can become hydraulic rams capable of cracking
the machine's crown.

Last, make a few test parts. Remember that the speed at
which you run the press will be determined by how fast you
can get rid of the scrap.

It is easy to blame the diemaker when something goes
wrong, but in four-fifths of the cases the real cuprit is
poor setup.

CHAPTER 7

SAFETY

SAFETY CONSIDERATIONS IN DIE DESIGN

JOSEPH W. HART, Director
Machine & Material Handling Services
Liberty Mutual Insurance Co.
Boston, MA

A major cause of failure of accident control programs in power press operations is poor planning and the resultant unsafe die design. This is a product of failure on the part of management to plan in advance for the safety of the operator and a desire to reduce the initial costs of tooling to a minimum. A frequent result of this short-sightedness is that it is difficult and expensive to adequately protect the operator from injury with a guard or safety device that is often an afterthought and thereby ineffective or inefficient.

The starting point in a power press safety program is in the planning stages when the die or tooling is being designed. Management must be aware of the necessity and the advantages of planning well in advance how the power press operator will be protected from the inherent hazards of press operations, rather than relying solely on the foreman and diesetter to keep the operator from injury yet still run the operation efficiently.

To illustrate the magnitude of the problem, an analysis of 300 serious power press accidents showed that about 80% of the accidents occurred on secondary operations and about 20% on primary-type operations. Approximately 95% of the serious secondary operation power press accidents occurred when the operator was loading or unloading the die or readjusting the part in the die. Nearly all of these accidents occurred during manual positioning or removal of the part or material at the point of operation.

It is on the secondary power press operations that the greatest problem exists. This is where most of the die design faults occur which allow serious exposures and the resulting high percentage of accidents which are frequently extremely serious. It is not difficult to recognize, then, that the attention given to the feeding and ejecting, or the loading and unloading, of dies is of major importance in the planning of safe power press operations.

Design Objectives

The first objective in safe power press operation is to design the operation so as to make it unnecessary for the operator to place his hands in the point of operation.

The second objective is to prevent the operator from placing his hands in the point of operation when the slide is descending. To achieve the first objective, the die designer must consider several important aspects of the operation. He should be able to answer these questions:

(1) How should the material be fed into the die without exposing the operator's hands in the danger area?

(2) How will the completed part be removed from the die without requiring the operator to reach into the danger area?

(3) How can scrap be removed from the die area without requiring the operator to reach in after it?

(4) What means will be provided for die lubrication other than having the operator reach in to do it manually?

(5) What features should be built in the dies to allow for safe handling of the dies during setting, transportation and storage?

The principal reason for placing the hand in the point of operation is to feed and eject stock. Fortunately, there are means by which a large percentage of press operations can be performed without placing a hand in the point of operation. These are:

(1) Automatic feeding, requiring no manual operations other than replacing of stock.

(2) Semi-automatic feeding by means of manually operated feeding devices, including sliding dies, dial feeds, chutes, push feeds.

(3) Hand tool feeding by means of tongs, suction cups, sticks, permitting hands to be kept away from the point of operation.

(4) Manual feeding of large pieces of such size or shape that when holding them the hands are positioned well away from the point of operation.

Automatic Feeding

There are available commercially several types of feed mechanisms designed to accommodate coiled stock or strip stock. These include roll feeds, hitch feeds and air-operated feeds. The stock is fed into the mechanism immediately adjacent to the die or as a part of the die. The stock is indexed forward into the point of operation of the die on each stroke of the press a predetermined and adjusted distance. The feeds are actuated by the ram or by action of a cam on the crankshaft. These feeds are usually designed so as to allow the press to operate on continuous cycling and thereby reach a high rate of production. A point of operation guard or device must be provided to prevent the operator from reaching into the danger area while the press is in operation. Other automatic loading and unloading devices such as transfer units and mechanical hands are also manufactured on a custom basis for long-run press operations.

Semi-Automatic Feeding

In the field of semi-automatic feeding, there is a wide variety of units which can be designed to facilitate die loading without exposure of the operator's hands. Dial feeds consist of a turntable mounted on a power press which allows the operator to load the part at one station or stations before the table indexes, bringing the part under the slide where the work is performed. The completed part is usually ejected by air, mechanically, or it is removed manually from the lower die as the table indexes again to the operator's position. Dial feeds are adaptable to many different secondary press operations using the same basic turntable. The turntable is indexed by action of the press ram or crankshaft.

Many secondary power press operations are adaptable to feeding methods which require manual handling but do not require the operator to reach into the danger area. One such method is the use of gravity feed in which the part is placed on a chute and allowed to slide down into the lower die. Pins, gages and stock guides should be provided to assure easy part nesting in the proper position. Open-back inclinable presses, because of their ability to be inclined, allow good application of this type of feed because the part can be placed into the die by gravity and ejected out the back of the press by gravity or air blowoff.

There are various adaptations of gravity chute feeds which can be used to assure only one part being placed in the die at a time. One of these uses a single-piece feeder. When oil or other lubricant on the parts causes the parts to stick in the chute or slide, wire or metal rods can be installed in the chute to allow the parts to slide into the die without sticking.

Probably the most common type of semi-automatic feeding is the follow or push feed which allows the operator to push the parts on a tray into the die by pushing one part behind the other or using a pusher to place the part into the die, thereby keeping the hands well out of the danger area. Dies with this type of feed are generally easily guarded with individual die guards. Even irregularly shaped parts can frequently be fed with a push feed to keep the hands out of the danger area.

The addition of a magazine on the push feed can increase the production rate considerably as well as minimize the manual handling of the blanks at the press operation. Blanks with various configurations are adaptable to magazine feeds. Magazine slide feeds can be adapted to a nearly automatic operation by actuating the slide through mechanical attachment to the press slide or crankshaft. The operator then need only keep the magazine filled with blanks to feed the press. Other methods of providing mechanical actuation of the slide feed include air and hydraulic cylinder control timed with slide movement.

Hand Tool Feeding

Finally, where parts must be placed in/or removed from the die manually, hand tools such as soft metal pliers, tongs, tweezers and suction cups are effective means of allowing the operator to work with the hands out of the danger area. If the part is flat it is sometimes desirable to provide clearance holes or slots to assist the operator when he is inserting or removing the work pieces with hand tools. Relieving of the dies to provide clearance will make the job of grasping the parts much safer and more efficient. The lower die can be automatically or manually moved from under the upper die through the use of a sliding bolster so that difficult-to-nest parts can be manually loaded outside the danger area.

Part Ejection

Dies should also be designed so that the part can be ejected without the operator reaching into the danger area. This is equally as important as providing safe means of loading the die. Some means is frequently necessary to first strip the part from the upper die or punch or to lift the part from the lower die before actually removing the part. The part can then be ejected by gravity with an inclined press or by other means such as air blowoff to force the part from the die.

Lift-out fingers in the upper die can remove cup-shaped parts from the lower die that tend to adhere or stick in the cavity. Spring strippers in the upper die or lower die also will loosen the part from the die. Knock-outs in the upper die are frequently used as a mechanical means of freeing the part to eliminate the need to reach in to pull the part from the upper die. On larger parts, air cylinder ejectors or lifters, as well as mechanical knockouts, are used to lift or eject the part from the die.

Shuttle-type extractors can be used to reach between the dies to catch the part as it is stripped or knocked from the upper die and then transfer it from the die area into a stock container. Unloaders of this type must be timed with slide travel and can be used on automatic or semi-automatic operations.

Scrap Removal

Designing an efficient and safe method of disposing of scrap is as important to accident prevention as loading and unloading of the part. Dies should be designed to shed all scrap with no handling required within the danger area. When scrap is produced in the die, the operator should not be required to reach through metal trimmings or into the danger area to remove the scrap. Adequate clearance must be provided for slug removal. If slugs cannot be shed from any side of the die or through the bolster, the die must be designed for risers under the lower shoe. Dies must be designed to permit visual inspection by the operator for proper scrap and slug shedding. Chutes can be added to direct the scrap outside the die area and beyond the bolster into a scrap container. The die can be relieved or sloped outwardly to prevent buildup of scrap which may tend to encourage the operator to reach in to remove the scrap manually.

A scrap cutter can be an integral part of the die or can be an attachment to the press. The scrap cutter should be designed to eliminate the pointed or needle-type scrap which can cause serious lacerations and puncture wounds. When rings of scrap are produced by trim operations, the rings should be cut in several segments or in sizes small enough to be efficiently and safely disposed of without requiring the operator to reach into or come unnecessarily close to the danger area.

Other Die Design Features

Guide pins or leader pins should be kept to the back of the die away from the operator. Pinch points can be created when the guide pins separate from the upper die shoe. Guide pin covers should be used to enclose the opening between the top of guide pins and the bushings in the upper die set. Helically wound flat steel springs or telescoping guards or covers are available commercially for enclosing this pinch point.

Stop blocks which limit the closing of certain dies to prevent die damage should be placed on the inside of the dies, facing each other, or to the rear of guide pins and away from the operator to minimize the potential pinch point between the blocks and the upper die.

Sufficient surface should be provided on the die sets for adequate bolting or clamping of the dies to the bolster or to the slide. Bolting is preferable to clamping. The die designer should know in advance the layout of the threaded holes in the bolster and in the slide so that bolting provisions may be made wherever possible.

Air, hydraulic, electrical and lubrication lines should be located so they will not be cut by sharp-edged panels of stock or parts being processed. Hand clearances must be provided between air cylinders and other accessory equipment on the presses to avoid pinch points between the die and the slide and other equipment. Pinch points under stripper plates or pressure pads in the upper and lower dies should be provided with steel angle-type guards unless other adequate die or press enclosure guards are provided. Cam springs should be guarded to enclose the pinch points and to contain flying rods that may break in tension. Pins used in dies for locating or nesting purposes should be as low as possible to facilitate easy part removal.

When safety blocks are used, especially on large dies, pads should be provided or areas in the dies should be provided for placement of the safety blocks. Safety blocks should be used whenever it is necessary to reach into the dies to adjust, repair or set dies in the press. Spring clips, clamps and die extensions to eliminate the need for the operator to hold the part in the die should be provided where possible. Safety devices, two-hand controls or guards should then be employed to keep the hands out of danger areas during downstroke of the slide.

There are several aspects of large die press operations that warrant special consideration. In many instances when mechanical functions of feeding or extracting the part from the die break down or malfunction, the mechanism is removed for repairs and production on the press continues by hand feed and extraction. Therefore, the designer should engineer into the die alternate means of safely loading and unloading the part by hand. This can be done by providing indentations or die cavities where the part may be handled safely by hand even if the die inadvertently closes. Where hand feed clearance cannot be provided, tabs which project beyond the die closure line should remain as part of the parent panel for ease in manual handling. If tabs cannot be engineered as part of the panel, handling holes should be punched into the panel in the blanking stage for later feeding and extraction by metal hook held in the operator's hand.

Dies should be designed with the operator in mind. Excessive bending or reaching may create operator irritation, discomfort, and fatigue, factors which lead to a greater number of accidents. Any unusual or special die setting instructions should be stamped or otherwise indicated on the die for the information of the die setter and supervisor.

Die Handling

Several features can be provided in the dies to facilitate safe die handling. Chain slots and flanges should be provided with 1½″ radius around the slots. Cast or inserted pin die hooks can be used on dies, especially on dies weighing more than 15 tons. Die hook holes can be provided in die shoes. Sharp edges in die sections should be beveled or chamfered. Eye bolt holes in dies and in all heavy individual die parts can be provided where eye bolt lifting is used. It is advisable to mark the center of gravity of some dies on top of the upper die to facilitate safe die lifting.

SUMMARY

Since press room safety begins on the die designer's drawing board, the designer's foremost consideration must be the operator's safety. To design safety into dies effectively requires frequent communication between the operator, supervisor, safety function, die engineer and die designer. The philosophy of everyone concerned with operation of power presses should be that every cycle is a potential hazard. In line with this, the policy of "no hands in the dies" should be followed. Every reasonable effort should be expended to make it unnecessary for the operator to reach into the danger area. Then, to prevent any part of the operator's body from coming into injurious contact with the die during the closing press stroke, point of operation guards and devices must be used.

INDEX

Draw quality steel, 29
Drawing compounds, 36-38
Drawing lubricants, 37
Drawing operations, 101-102
Drawn parts, 11-15
Drawn shells, 9, 19
Drills, 92
Ductility, 29

E

EDM, See: Electrical Discharge Machining
Electrical Discharge Machining, 189-195, 196-200
Elongation of material, 99
Energy equation, 179

F

Fatigue behavior, 27-28
Faulty tool designs, 233
Fine blanking, 104
Finish barrel, 94
Finishing, 99, 225-232
Finishing departments, 38
Flanging, 164-169
Flat blanks, 101
Flat stampings, 3-8, 16-18
Flex rolling, 142
Flywheel equation, 181
Form complexity, 50, 71, 74, 76, 78
Formed boxes, 9
Formed angles, 71
Formed stamping, 9-10, 18
Forming, 92, 99, 100, 164-169
Forming dies, 161, 205-209
Forming modes, 24
Free-standing bosses, 45
French stop, 139

G

Government, 22-28
Grinding, 18, 105, 216-218

H

Hand tool feeding, 248-249
Hardness, 161
Heat treatments, 233-234
Honing, 29
Human error, 112

I

Idea refinement, 205-206
Improved stock threading, 135-136
Induction vacuum furnaces, 161
Inspection, 93
Internal diameter matrix button, 108
Ironing, 225-232

J

Jamming, 116
Jig borer, 132, 196
Jig grinder, 85, 105, 196
Jig-ground hole, 105

L

Leadtime, 159-163

M

Machine tool technical data, 190
Management, 150-151
Manifold system, 165
Matched set tooling, 112-113
Material engineer, 35
Mating of parts, 210
Metal edge stiffeners, 45
Metal stamping, 106-120
Metallurgist, 29, 30, 31
Metrication, 177-183
Molding punch, 204
Multithane stripper pads, 201

N

NC tapes, 121-123
Nitrogen cylinders, 164-169
Notching, 113

P

Painting, 99
Part drawings, 108
Part ejection, 138-139, 249
"Parting tool," 4
Perforators, 173
Phosphate coating plating systems, 38
Piercing, 100
Pilots, 170
Pilot breakage, 141-142
Planishing, 18
Plastic, 40-45
Polishing, 99
Postmolding operations, 45
Premature wear, 111
Presses
    brakes, 210
    double-action toggie, 32
    downtime, 109
    energy, 239-240
    maintenance, 37
    multi-station, 103
    repair, 35
    stoppages, 130-131
    triple-action toggle, 32, 103, 104
Product design, 34, 99
Production rates, 15
Productivity, 127-152
Progressive coining, 172-176
Progressive dies
    coputerized die design, 121-123
    cost estimating, 80-83, 87-95
    design tips, 170-171
    feeds, 102-103
    plan view, 199
    positive-return type cams, 101
    truck instrument panels, 169
Punch footings, 198
Punch point, 111
Punching, 100

Q

Quality control, 163
Quality control inspector, 34
Quik die, 202-204

L (continued)

Load control, 106-120
Low carbon sheet steel, 22
Lubricants, 36-38

READER REVIEW

Now that you have reviewed **Pressworking: Stampings And Dies,** your comments below would be greatly appreciated. Consider the book's completeness, relevance, preciseness and any improvements that should be made.

Comments

(Optional)
Name _______________

Company Name _______________

Address _______________

City _______________

State _______________ Zip _______________

***Discounts. . .** on the informative books of the Manufacturing Update Series are only part of the benefits of Membership in the Society of Manufacturing Engineers.

An SME Member can take advantage of numerous opportunities to fulfill continuing education goals. The opportunities include:

- Chapter programs providing a foundation for involvement and participation.
- Educational programs including seminars, clinics, programmed learning courses and videotapes.
- Publications including Manufacturing Engineering, the SME Newsletter, Technical Digest and a wide variety of books.
- Manufacturing Engineering Certification Institute recognizing engineers and technologists for their technical knowledge acquired through years of experience.
- Conferences and expositions which enable engineers to see, compare and consider the newest manufacturing equipment and technology.

For more information, fill in the space below and mail. Reply postage will be paid by SME.

Please send me Membership information.

Name _______________

Company _______________

Address _______________

City _______________

State _______________ Zip _______________

PRESSWORKING: STAMPINGS AND DIES

Please send me _______ copies of **Pressworking: Stamping and Dies** at the current price.*

Name _______________

Company Name _______________

Address _______________

City _______________ State _______________ Zip _______________

☐ SME Member Number   ☐ Non-Member

Check One

☐ Payment enclosed   ☐ Purchase order enclosed

Check One

*For current price information, please contact the SME Publications Sales Department at 313-271-1500.

BUSINESS REPLY CARD

FIRST CLASS   PERMIT NO. 1103   DEARBORN, MICH., U.S.A.

POSTAGE WILL BE PAID BY:

SOCIETY OF MANUFACTURING ENGINEERS
ATTN: MARKETING DEPARTMENT
ONE SME DRIVE
P.O. BOX 930
DEARBORN, MICHIGAN 48128

NO POSTAGE
NECESSARY
IF MAILED
IN THE
UNITED STATES

BUSINESS REPLY CARD

FIRST CLASS   PERMIT NO. 1103   DEARBORN, MICH., U.S.A.

POSTAGE WILL BE PAID BY:

SOCIETY OF MANUFACTURING ENGINEERS
ATTN: MARKETING DEPARTMENT
ONE SME DRIVE
P.O. BOX 930
DEARBORN, MICHIGAN 48128

BUSINESS REPLY CARD

FIRST CLASS   PERMIT NO. 1103   DEARBORN, MICH., U.S.A.

POSTAGE WILL BE PAID BY:

SOCIETY OF MANUFACTURING ENGINEERS
ATTN: MARKETING DEPARTMENT
ONE SME DRIVE
P.O. BOX 930
DEARBORN, MICHIGAN 48128